MICROBIAL BEHAVIOUR
'IN VIVO' AND 'IN VITRO'

Other Publications of the
Society for General Microbiology

THE JOURNAL OF GENERAL MICROBIOLOGY
(Cambridge University Press)

SYMBIOTIC ASSOCIATIONS
THIRTEENTH SYMPOSIUM OF THE SOCIETY
(Cambridge University Press)

MICROBIAL CLASSIFICATION
TWELFTH SYMPOSIUM OF THE SOCIETY
(Cambridge University Press)

MICROBIAL REACTION TO ENVIRONMENT
ELEVENTH SYMPOSIUM OF THE SOCIETY
(Cambridge University Press)

MICROBIAL GENETICS
TENTH SYMPOSIUM OF THE SOCIETY
(Cambridge University Press)

VIRUS GROWTH AND VARIATION
NINTH SYMPOSIUM OF THE SOCIETY
(Cambridge University Press)

THE STRATEGY OF CHEMOTHERAPY
EIGHTH SYMPOSIUM OF THE SOCIETY
(Cambridge University Press)

MICROBIAL ECOLOGY
SEVENTH SYMPOSIUM OF THE SOCIETY
(Cambridge University Press)

BACTERIAL ANATOMY
SIXTH SYMPOSIUM OF THE SOCIETY
(Cambridge University Press)

MECHANISMS OF MICROBIAL PATHOGENICITY
FIFTH SYMPOSIUM OF THE SOCIETY
(Cambridge University Press)

AUTOTROPHIC MICRO-ORGANISMS
FOURTH SYMPOSIUM OF THE SOCIETY
(Cambridge University Press)

ADAPTATION IN MICRO-ORGANISMS
THIRD SYMPOSIUM OF THE SOCIETY
(Cambridge University Press)

THE NATURE OF VIRUS MULTIPLICATION
SECOND SYMPOSIUM OF THE SOCIETY
(Cambridge University Press)

THE NATURE OF THE BACTERIAL SURFACE
FIRST SYMPOSIUM OF THE SOCIETY
(Blackwell's Scientific Publications Limited)

MICROBIAL BEHAVIOUR, 'IN VIVO' AND 'IN VITRO'

FOURTEENTH SYMPOSIUM OF THE
SOCIETY FOR GENERAL MICROBIOLOGY
HELD AT THE
ROYAL INSTITUTION, LONDON
APRIL 1964

CAMBRIDGE
Published for the Society for General Microbiology
AT THE UNIVERSITY PRESS
1964

PUBLISHED BY
THE SYNDICS OF THE CAMBRIDGE UNIVERSITY PRESS

Bentley House, 200 Euston Road, London, N.W. 1
American Branch: 32 East 57th Street, New York 22, N.Y.
West African Office: P.O. Box 33, Ibadan, Nigeria

Printed in Great Britain at the University Printing House, Cambridge
(Brooke Crutchley, University Printer)

CONTRIBUTORS

BUXTON, E. W., Rothamsted Experimental Station, Harpenden, Hertfordshire.

DEVERALL, B. J., Department of Botany, Imperial College, London.

FOSTER, M. A., Microbiology Unit, Department of Biochemistry, University of Oxford.

HENDERSON, D. W., Microbiological Research Establishment, Porton, Salisbury, Wiltshire.

HOLLAND, J. J., Department of Microbiology, University of Washington, Seattle, Washington, U.S.A.

INGRAM, P. L., Department of Pathology, Royal Veterinary College, London.

KEPPIE, J., Microbiological Research Establishment, Porton, Salisbury, Wiltshire.

MACKANESS, G. B., Department of Microbiology, University of Adelaide, South Australia.

MARIAT, F., Institut Pasteur, Paris, France.

SMITH, H., Microbiological Research Establishment, Porton, Salisbury, Wiltshire.

STANDFAST, A. F. B., The Lister Institute of Preventive Medicine, Elstree, Hertfordshire.

WALLEN, V. R., Plant Research Institute, Ottawa, Canada.

WEITZ, B., The Lister Institute of Preventive Medicine, Elstree, Hertfordshire.

WOODS, D. D., Microbiology Unit, Department of Biochemistry, University of Oxford.

CONTENTS

EDITORS' PREFACE

When this Symposium was being considered, the title 'Where has test tube microbiology taken us?' was suggested. A more formal title was adopted, but we have a sneaking regard for the colloquial one because it captured so well the aim of the Symposium: to make microbiologists stop and consider how far laboratory experiments were explaining microbial behaviour in nature.

The examination of micro-organisms under defined conditions in pure culture has been fundamental to the advance of microbiology, especially in the elucidation of processes, such as energy production, which are essential to all micro-organisms. In nature, however, we are concerned with the behaviour of micro-organisms in a complex environment. Since environment influences the genotype by selection and the phenotype by variation, some phenomena demonstrated in the laboratory may not occur under more natural conditions, and vice-versa. Also there is a growing awareness that many questions cannot be answered solely by experiments outside the natural environment. For example, how does a micro-organism enter its natural host and spread, or fail to spread, either locally or systemically? What happens to known micro-organisms which apparently disappear, such as influenza virus in summertime? Why are some diseases more contagious than others?

This Symposium is restricted to microbial behaviour in animals and in plants although the theme also applies to micro-organisms in natural environments such as soil, water and sewage. The contributors were asked to consider how much was known about a few examples of microbial behaviour *in vivo* and how far studies *in vitro* have explained this behaviour; also they were asked to examine the possibilities of studying microbial behaviour *in vivo* either directly or by more realistic experiments *in vitro*. This task has been tackled honestly and without prejudice, and taking a broad view of the contributions the answers to the questions posed are as follows. Knowledge of microbial behaviour *in vivo* is superficial. Little is known of the exact mechanisms involved. Although some microbial behaviour *in vitro* reflects that *in vivo*, there can be wide and misleading discrepancies, especially for bacteria and fungi that are grown in the laboratory on defined non-living media. The behaviour of viruses and bacteria in tissue culture, '*in vitro*' in the context of this Symposium, is nearer that *in vivo* but divergencies also occur here owing to the de-differentiated state of the host cells in culture. Finally, although difficult, it is possible to make more than superficial studies of microbial behaviour *in vivo* and realistic experiments *in vitro* can help in this respect; the results of such studies have been encouraging.

Inevitably the meanings of 'natural behaviour', 'behaviour *in vivo*' and 'behaviour *in vitro*' have been discussed, the first two especially in relation to the important group of pathogenic micro-organisms whose natural habitat is man. Unlike other micro-organisms that can be studied experimentally *in vivo* in their natural host, these must be examined mostly in laboratory animals which, although living, are unnatural hosts. However, it seems agreed that if we had more knowledge of host–pathogen relations in these animals it might be helpful in explaining natural microbial behaviour in man. As much caution should, however, be expended in extrapolating results obtained from such 'artificial' hosts to man, as from experiments *in vitro* to laboratory animals.

The Symposium shows that microbiologists are little people working in a large world only tiny areas of which have been explored. Its function will have been fulfilled if they are stimulated to question how far their experiments both *in vitro* and *in vivo* reflect the behaviour of micro-organisms in nature.

Our thanks are due to Dr J. Keppie, Dr J. Postgate and other colleagues for help with the editorial work.

H. SMITH

Microbiological Research Establishment,
Porton, Salisbury, Wiltshire

JOAN TAYLOR

Salmonella Reference Laboratory,
Central Public Health Laboratory,
Colindale, N.W. 9

MICROBIAL BEHAVIOUR IN NATURAL AND ARTIFICIAL ENVIRONMENTS

H. SMITH

Microbiological Research Establishment, Porton, Wiltshire

The isolation of micro-organisms from nature and their study in pure culture under more defined conditions in the laboratory has been essential for the advance of microbiology. What has been accomplished for most bacteria, fungi and some protozoa, is now being attempted for viruses and rickettsia with the aid of improved methods of tissue culture. The causative organisms of many microbial processes have been recognized, enabling their behaviour to be investigated more thoroughly. In addition, the study of a few micro-organisms in pure culture has led to spectacular advances in knowledge of fundamental processes common to most micro-organisms and probably to all living things, for example protein and nucleic acid synthesis. However, it is sobering to turn from unity to diversity and to inquire into the mechanisms responsible for microbial differences in nature. The main metabolic processes of many microbial species are almost certainly similar but why, for example, is one smooth strain of *Brucella abortus* virulent and another avirulent, or why is smallpox fatal and cow-pox relatively innocuous? Here, small idiosyncrasies of microbial properties are involved, which would escape the attention of unitarian investigators and almost certainly depend for their revelation on the appropriate environmental conditions. Many microbial differences can be demonstrated in pure culture in laboratory media. Often these properties are of practical importance for distinguishing between various micro-organisms; but are they characteristic of microbial behaviour in nature where different and more complex conditions occur? This is the hub of our symposium. The success of laboratory cultures in elucidating general processes has been stressed, and undoubtedly some of the peculiarities observed in these cultures occur in nature, for example toxin production by *Clostridium tetani*. But are enough aspects of natural behaviour being detected in the laboratory and is there a need now to supplement our present knowlege by studying micro-organisms in more natural environments? Contributors to *Microbial Genetics* (Hayes & Clowes, 1960) and *Microbial Reaction to Environment* (Meynell & Gooder, 1961), have made clear that changes in environment alter phenotypic expression, effect

selection, and influence genetic transfer and mutation. Hence, differences between microbial behaviour in nature and that in the laboratory
are to be expected, and these differences may have ecological, medical
and economic importance. In other Symposia of the Society over the
past decade, contributors have commented on this and noted our lack
of understanding of micro-organisms in nature. Van Heyningen (1955)
and Brian (1955), dealing with the role of toxins in animal and plant
disease respectively, considered that many of the numerous toxins
demonstrated in laboratory cultures had yet to be shown to be produced in infected hosts and important in disease. In a discussion of the
nutrition of soil micro-organisms, Gibson (1957) noted the disadvan-
tages of enrichment cultures in displaying the zymogenous population
of the soil rather than the autochthonous; she stressed the lack of
knowledge of the more slowly growing organisms, and looked forward
to more sensitive methods for analysis of mixed cultures in their natural
habitat. Woods & Tucker (1958) wrote: 'In chemotherapy, we are concerned in reality with the metabolism of the parasite within the special
environment of the host, and with the metabolism of the infected host
rather than the normal host. Information on both these subjects is woefully
fragmentary.' Last year, Smith (1963) deplored the excessive study of the
isolated symbionts of lichens in pure culture, and advocated a closer examination of algae and fungi as they occurred together in the intact lichen-
thallus. These and similar opinions of other contributors to past Symposia
are echoed elsewhere (Winogradsky, 1949; Smith, Keppie & Stanley, 1953;
Dimond & Waggoner, 1953; Dubos, 1954; Smith, 1958; Bloch, 1960).

While many microbiologists advocate studying microbial behaviour
under natural conditions, few of them do so. This is because their
morale for overcoming the difficulties is constantly sapped by the
attractive ease of working with laboratory cultures. Hence, a symposium
to encourage the study of microbial behaviour in more natural environments appears to be a timely sequel to its predecessors. The difficulties of
such work, the techniques which might be applied, and the advantages
that might accrue from it can be discussed. The worth of any relevant
studies can be examined and the behaviour of some micro-organisms
under natural and artificial conditions can be compared. Perhaps most
of all, the symposium can encourage the design of more realistic
laboratory experiments which simulate microbial behaviour in nature.
The symposium should appeal to microbiologists who are interested in
the actual behaviour of micro-organisms in nature, rather than in their
performance under highly artificial conditions, however laudable and
profitable the interest in that performance might be.

In seeking more information about microbial behaviour in the complex environment of nature, the symposium is in vogue with present trends in other fields of biology and science. Hitherto scientific treatment of complex systems has been by analysis, i.e. isolating simple components which could be studied under defined conditions where disturbing influences had been eliminated. Now, new interest is developing in the functioning of unresolved complex systems, such as the study of cybernetics (Wiener, 1961), the design of computers and of automatic control mechanisms, operational research and system engineering (Ackoff, 1959; Von Bertalanffy, 1962). These studies can build on information gained from analytical techniques by seeing how far this information applies when the component parts function together. Studies of complex systems as a whole without investigating their component parts have also been advocated (Ashby, 1958; Von Bertalanffy, 1962).

While the general principle of this symposium applies to all microbiology, it is not surprising that contributions centre around microbial pathogenicity in plants and animals. There is no finer example of the subtleties of microbial behaviour being influenced by a natural environment than the mechanisms which determine virulence; and knowledge of the general processes of microbial growth has not contributed much to the understanding of these mechanisms. Furthermore, because this is a medically and economically important field, much work has been done with laboratory cultures; hence there is abundant information from such cultures to compare with general observations on infectious disease and with the few experimental studies that have been made with micro-organisms from infected plants or animals.

Since only micro-organisms in plants and animals are discussed, and not those in soil, water and other environments, the title of the Symposium is 'Microbial Behaviour in Vivo and in Vitro'. What do the Latin phrases mean or, more important, in what sense are they used in the literature? In vitro describes an experimental environment even if the apparatus is no longer essentially of glass, and always creates the true impression of artificial conditions. In vivo describes experiments with plants and animals; but sometimes its use creates a false impression of a natural environment, especially in animal experiments with micro-organisms whose customary habitat is man. The study of micro-organisms as they occur naturally in soil, water, sewage, plants and animals may be difficult but it is always practicable. Comparable studies in man are limited, and restricted to diagnosis and the applications of therapeutic or prophylactic measures. Laboratory animals must be used in most experiments which, although they take place *in vivo*, are often highly unnatural. Infections with *Shigella shigae*, *Salmonella typhi* and *Streptococcus pyogenes* in mice and other laboratory animals bear little resemblance to human dysentery, typhoid, and streptococcal infections respectively. Some organisms, e.g. *Haemophilus influenzae*, will not infect laboratory animals unless injected with mucin to inhibit the host defences. Also, in most animal experiments the mode of infection is unlike natural infection in man. Nevertheless, despite the frequent lack of similarity between the disease syndrome in man and laboratory animals, many results from animal experiments appear to be applicable to man, because they have indicated measures which have proved effective for human use. Furthermore, there is still much to be learned from the study of infections in laboratory animals, because in them most mechanisms of microbial pathogenicity remain a mystery; if these mechanisms could be elucidated it is probable that more information applicable to infections of man would become available.

The remainder of this chapter deals with microbial virulence in animals by way of illustrating the theme of the Symposium. General observations which encourage a closer study of organisms grown *in vivo* are followed by a description of the techniques for doing so and examples of the results obtained. The examples selected concern mainly pathogenic bacteria because the time is opportune for their study *in vivo*: much fundamental knowledge has already accumulated from experiments with laboratory cultures as well as an insight into their limitations for explaining some aspects of pathogenicity. Only a few remarks are made about viruses; these may need more study in tissue culture before further work in animals. The nearly synonymous terms 'pathogenicity' and 'virulence' have been used in the manner suggested by Miles (1955), and 'immunogenicity' means ability to produce active immunity against disease.

THE ENVIRONMENT *IN VIVO* AND STUDIES OF PATHOGENICITY

Microbial pathogens possess metabolic processes which enable them, in contrast to non-pathogens, to produce disease in a suitable host. The general nature of these bases for pathogenicity—ability to grow and multiply in the host, toxin and aggressin production—have been reviewed elsewhere (Dubos, 1945; Howie & O'Hea, 1955; Smith, 1958; Braun, 1960). Since virulence is often a fleeting characteristic and detectable only *in vivo*, knowledge of the factors responsible for it is vague. Some microbial factors may be produced by the pathogen only *in vivo*, others may be formed more *in vivo* than *in vitro*. This is indicated by the loss of virulence and change in morphology during artificial culture, and the reversal of this process by animal passage.

It is understandable that some virulence factors are not formed by the pathogen *in vitro*, because at present the decisive nutritional conditions *in vivo*, i.e. those of the host tissues under microbial attack, are not reproducible *in vitro*. These conditions are not even normally physiological, for example the changes during inflammation and within a phagocyte when it becomes infected (Suter & Hulliger, 1960). Virulence factors could be formed as a reaction to these semi-pathological conditions. Some pathogens preferentially attack certain tissues, for example *Brucella abortus* attacks the bovine cotyledons, and *Corynebacterium diphtheriae* the throat and tonsils. Only recently (Smith *et al.* 1962*b*) have we learned something of the peculiar nutritional requirements responsible for one example of tissue specificity; in most cases these requirements are not understood and not reproducible *in vitro*.

The rapid development of mammalian cell tissue culture (Ross & Syverton, 1957; Swim, 1959; Ross, Treadwell & Syverton, 1962) has allowed fundamental studies of viral isolation and multiplication under relatively simple experimental conditions. The convenience of tissue culture prompts its use for problems of microbial pathogenicity and this will be discussed later. However, it should be emphasized that growth of pathogens in tissue culture is a long way from growth *in vivo*; and as a result either some aspects of pathogenicity may be missed or some abnormal products may receive attention in such studies. Most tissue culture cells are different from their parent cells from the whole animal, a difference often accompanied by a change in susceptibility to certain viruses (Kaplan, 1955; Ross & Syverton, 1957; Swim, 1959). The pathogenicity of virus changes in tissue culture (Ross & Syverton, 1957; Edney, 1957; Swim, 1959). Enders (1952) stated: 'Just as repeated transfer *in vitro* may bring about alterations in the virulence or pathogenicity of bacteria, so serial passage of viruses in tissue culture has frequently led to changes in their capacity to induce disease when inoculated into a susceptible animal. Most often under these conditions a decrease in virulence has been noted. In some cases, however, especially when an agent is propagated in tissue derived from its natural host, an increase in pathogenic properties may follow.' The attenuation during tissue culture of vaccinia virus for the rabbit (Rivers & Ward, 1933), yellow fever virus for man and monkeys (Theiler & Smith, 1937) and poliomyelitis strains for monkeys or mice (Ross & Syverton, 1957; Edney, 1957) are examples of changes in the virulence of viruses in

tissue culture. The artificial nature of tissue culture cells has been discussed recently (Ross *et al.* 1962).

Since conditions *in vivo* cannot at present be reproduced *in vitro*, the
study of organisms *in vivo* might elucidate hitherto unknown aspects
of microbial pathogenicity. In advocating this approach, there is no
intention to disparage studies *in vitro* but rather to build on them.
Many virulence factors have been recognized *in vitro*, for example the
capsular polysaccharides of *Streptococcus pneumoniae* (Felton &
Bailey, 1926). However, in a diseased animal virulent organisms are
more likely to produce their whole armoury of factors, some of which
might be recognized *in vivo* and then reproduced *in vitro*. In addition,
some potential virulence factors found *in vitro* may be unimportant *in
vivo* (e.g. lycomarasmin in plant disease; Brian, 1955). In three practical
fields, vaccination, chemotherapy and the study of mixed infections so
often found in clinical disease, there are indications that the examination
of organisms growing *in vivo* might be rewarding.

A qualitative or quantitative difference between the production of
virulence factors *in vivo* and *in vitro* is indicated by the efficacy of living
attenuated vaccines when killed vaccines are relatively ineffective, for
examples S 19 vaccine for contagious abortion and B.C.G. vaccine for
tuberculosis. These attenuated organisms persist in some tissues of the
host for relatively long periods (Pierce, Dubos & Schaefer, 1956;
Taylor & McDiarmid, 1949). Hence, they can produce immunizing
antigens *in vivo* at the most effective site. Killed vaccines may be ineffective
because sterilization destroyed the essential antigens, but an alternative
explanation is that immunizing antigens are not formed to any significant
extent *in vitro*. It is also possible that some immunizing antigens are
produced *in vivo* as a result of the interaction of bacterial products with
components of the host's tissues.

Differences, possibly connected with virulence, between bacteria
grown *in vivo* and *in vitro* may explain some of the contrasting effects of
drugs in the two environments. Some drugs active *in vitro* are inactive
in vivo; often this is due to inactivation of the drug by the host, or to the
pathogen being inaccessible to the drug (Dubos, 1954; Woods & Tucker,
1958; Knox, 1958). However, the organisms themselves can become
drug resistant, usually by selection but sometimes phenotypically
(Meynell, 1961). Phenotypic resistance may be correlated with failure
of growth (Meynell, 1961), but other changes, for example the production *in vivo* of a more resistant wall or capsule, could play a role,
especially as phenotypic resistance can occur before growth ceases
(Meynell, 1958). Some drugs are active *in vivo* but not *in vitro* (Hart &

Rees, 1955; Lacey, 1958). Sometimes this is due to the production of an active form of the drug by the host tissues, for examples Prontosil rubrum and pentavalent arsenicals (Woods & Tucker, 1958). This is not always the case for polyoxyethylene ethers were active against *Mycobacterium tuberculosis in vivo*, but not *in vitro* and Hart & Rees (1955) considered it improbable that these materials were converted to an antibacterial form because of their intrinsic chemical stability. They suggested that the compounds modified the surface of tubercle bacilli *in vivo* so they became more sensitive to the intracellular environment. It is possible that the surface layers produced by the bacteria *in vivo* had a protective function and played some role in virulence (see below; the difference in surface properties of virulent tubercle bacilli grown *in vivo* and *in vitro* and the protective surface material on *Brucella abortus* grown *in vivo*). Similarly, small doses of other drugs which are non-bactericidal *in vitro* may act *in vivo* by interfering with the production or action of virulence factors (Lacey, 1958). Hence screening procedures *in vitro* may well miss potentially effective drugs which might interfere with virulence factors *in vivo*, leaving the host able to destroy pathogenic bacteria as it normally deals with non-pathogenic ones.

Often clinical disease is due to a mixed infection or infection complicated by the growth of commensals. Detailed knowledge of such mixed infections is fragmentary, except that animal experiments have indicated that one organism can have a profound effect *in vivo* on the behaviour of another. Examples are the experiments of Henderson described in this Symposium and the stimulation of growth of *Candida albicans* and various enteric pathogenic bacteria by removing the commensal population with antibiotics (Ainsworth, 1955; Watkins, 1960; Lankford, 1960). Studies on mixed infections *in vivo* suffer from the lack of knowledge of single infections *in vivo* and of mixed cultures *in vitro*. However, even if this knowledge were forthcoming, it would be unlikely to explain all the behaviour of mixed infections, and studies *in vivo* could still be necessary.

TECHNIQUES FOR STUDYING MICROBIAL BEHAVIOUR *IN VIVO*

Information on microbial behaviour *in vivo* can be obtained in several ways. First, pathogenic bacteria and their products can be separated from the diseased host for biological examination and for chemical, metabolic, and serological study *in vitro*. Secondly, the behaviour of organisms growing *in vivo* and their repercussion on the host can be examined either in the whole animal or in restricted areas of its tissues.

Thirdly, as half way between experiments *in vivo* and *in vitro*, observations can be made in tissue or organ culture. Finally, tests *in vitro* can be made more relevant to microbial behaviour *in vivo*. These procedures will be considered in turn.

The isolation and examination of organisms grown
in vivo *and of their products*

Until recently the chemistry and metabolism of bacteria grown *in vivo* had not been studied, although extracellular products present in exudates from infected animals had received some attention. Methods were needed to separate reasonably pure organisms from infected animals.

A method, applicable to infections in which the causal organism grows freely in the blood or in exudates, was developed by Smith *et al.* (1953) using *Bacillus anthracis* in guinea pigs. Bacteria of adequate purity were obtained in sufficient quantity for chemical and metabolic examination. Guinea pigs (800–1000 g.) were infected intraperitoneally and intrapulmonarily; at death the thoracic and peritoneal exudates were collected and by differential centrifugation, body fluids and bacteria were separated from blood cells. *B. anthracis* (1·5–2·0 g. dry wt.) and 1·5–2·5 l. of body fluids were obtained from 100 guinea pigs; the bacteria were almost free from blood cells. Later Smith, Keppie, Cocking & Witt (1960) collected *Pasteurella pestis* (*c.* 1·5 g. dry wt.) from 100 guinea pigs; and in small-scale experiments, the following organisms were similarly obtained: *Streptococcus pyogenes*, *Strep. pneumoniae*, *Listeria monocytogenes* and *Staphylococcus aureus*. Recently, this method was used by Gellenbeck (1962) and by Beining & Kennedy (1963) for studies on *Staph. aureus*, and by Fukui, Delwiche, Mortlock & Surgalla (1962) for work with *P. pestis*. With the development of differential centrifugation in density gradients and the separation of different organisms in two-phase systems (Tiselius, Porath & Albertsson, 1963), the method might be extended to the separation of individual organisms from a mixed infection. If interest centres around an extracellular product, the body fluid in which the maximum multiplication of the organisms occurs should be examined (Smith, Keppie & Stanley, 1955). A toxin liberated by the organisms might be immediately fixed on the host tissues; therefore, if no toxin can be detected, infected fluid should be withdrawn and incubated *in vitro* to see if toxin appears.

Intracellular pathogenic bacteria must be freed from tissue cells before isolation. Hanks (1951) and Gray (1952) isolated *Mycobacterium lepraemurium* from infected testicular tissue. After removal of coarse tissue from homogenates by slow centrifugation, the bacilli were separated from tissue components by centrifugation through sucrose and KCl solutions. Segal & Bloch (1956) separated *M. tuberculosis* from the lungs of moribund mice by a similar method used later by others (Artman & Bekierkunst, 1961 *a*). Bacteria were separated from tissue components in homogenates by differential centrifugation in cold isotonic glucose. The bacteria (4 mg. per infected mouse lung) appeared to be free from much tissue contamination. Smith *et al.* (1961) modified this method to separate *Brucella abortus* from the cotyledons collected from an infected pregnant cow. About 1–2 g. (dry wt.) organisms were obtained, together with foetal fluids (4–5 l.) containing extracellular products. Differential centrifugation did not remove all tissue debris; this was finally eliminated by treatment with trypsin.

Bacteria separated by these methods should be examined for contamination by host material. This contamination is easier to detect in organisms originally extracellular than in those which were intracellular. During the isolation of *Bacillus anthracis* and *Pasteurella pestis* from infected guinea pigs, contaminating blood cells could be easily counted under the microscope. However, considerable contamination of *Brucella abortus* by the fine debris formed during initial homogenization was not detected easily in this way; electron microscopy and comparison of dry weights with corresponding suspensions of organisms grown *in vitro* were needed. Soluble host components may be absorbed by the bacteria, for example, a mouse diphosphopyridine nucleotide splitting enzyme was retained by *Mycobacterium tuberculosis* (Artmann & Bekierkunst, 1961*a*, *b*; Bekierkunst & Artmann, 1960, 1962). In any comparison with organisms grown *in vitro*, and especially in metabolic experiments, the organisms grown *in vitro* should be pretreated with tissue extracts from normal or infected animals.

Bacteria separated by the methods described can be used for metabolic experiments, for antigenic analyses, and for biological and chemical studies. Extracellular bacterial products or easily removed surface components are present in infected body fluids or tissue extracts and it is difficult to separate and identify them. First, there is the general difficulty of isolating small quantities of products from large amounts of host constituents. Secondly, an infected body fluid or tissue extract may be biologically active and apparently contain a bacterial product, but this is not certain unless it is supported by serological evidence which may be difficult to obtain. The third and major difficulty, which does not confront workers with plant pathogens *in vivo*, is antibody production in the donor animals which might result in an important antigen being isolated as its complex with antibody. This is unlikely in acute infections such as those occurring after intraperitoneal injection of *Bacillus anthracis* or *Pasteurella pestis*, when the animal dies in 2–3 days, but in chronic infections these antigen–antibody complexes may create a problem. A solution might be to grow the organism in foetus or in eggs. An important extracellular or easily removed surface component, revealed by studies *in vivo*, might be best identified by reproducing it first *in vitro*, thereby minimizing the difficulties of fractionation, for example, the components of the anthrax toxin or the immunizing material from *Brucella abortus* (see later).

It is important to point out defects of the method described above for preparing organisms grown *in vivo* for studies on pathogenicity. All the methods collect organisms from a late stage of the disease for the simple reason that a maximum yield of bacteria is desired. However, bacteria may vary in their possession of different virulence factors throughout the course of an infection and, for a full picture, organisms should be studied at an early stage of the disease. Some of the techniques described below for the study of organisms in localized areas of host tissues may be of use for such studies. Another defect of these methods for studying organisms grown *in vivo* is that the organisms may change rapidly as soon as they are removed from the animals, despite the fact that they are kept at 0–2°. In this symposium, Woods and Foster develop this point with regard to using such bacteria for metabolic experiments.

Observations on organisms growing in vivo: *the pathology of infection and disease*

To understand the mechanisms of microbial pathogenicity and to design cogent biological tests for microbial products, precise information on the pathology of infectious disease is required. This is often lacking, especially for the first few hours of infection, when important aspects of invasive action might be indicated such as the production of antiphagocytic or antibactericidal compounds. Also, only in a few cases (Smith, 1960) has any precise knowledge supplemented the clinical picture of the disease syndrome. In most diseases, not only are the important toxic products of the pathogen *in vivo* unknown, but so is the nature of the disease syndrome and the cause of death.

Experiments on infection throughout the whole animal

In most animal tests for virulence, toxicity and immunogenicity of micro-organisms, there are no detailed examinations of microbial behaviour in the various tissues as infection proceeds. The organisms are injected, the animal dies or contracts the disease and is examined post mortem. Sometimes, the spread of infection is followed by examinations of blood, lymph nodes, spleen and other tissues. More could be learned about the behaviour of pathogenic bacteria *in vivo* by improving conventional animal tests, and by studying more closely the spread of infection especially in the early and late stages of the disease. Instead of injecting the infecting organisms, the course of disease could be studied after infection by a more natural route; examples are the experiments of Henderson, Lancaster, Packman & Peacock (1956) with various pathogenic bacteria administered by the respiratory route, and those with enteric organisms given orally (Meynell, 1963).

The degree to which a pathogenic organism multiplies in the host is unknown. At any instant, the number of viable bacteria can be estimated, but how many bacteria have died since the infection became established, or have been removed as, for example, in intestinal infections? Viable counts represent the resultant of bacterial division and bacterial destruction or removal. Meynell & Subbaiah (1963) described a method of estimating the true division rates of bacteria *in vivo*. A biochemically detectable genome was introduced into the pathogen by abortive transduction (Stocker, Zinder & Lederberg, 1953). At division, the donated genome passed to only one of the two daughter cells and so on. Hence, the proportion of cells carrying the donated genome was halved at each generation, provided unlabelled and labelled cells were killed or removed to an equal extent. The number of bacterial divisions that had occurred at any instant was indicated by comparing the proportion of viable cells containing the genome with the proportion in the original inoculum. The applicability of this method will depend on how many pathogenic organisms will undergo abortive transduction; but the principle of measuring the division rate *in vivo*, by making use of a characteristic retained by a known proportion of the progeny only, was used by Meynell (1959) previously, and is worthy of development.

Insight into the nature of toxins produced by pathogenic organisms may be gained from chemical and physiological observations on infected animals approaching death, when the main pathological syndrome is becoming apparent (Smith, 1960). Experiments on heavily infected animals are technically difficult because of the necessity to use general safety precautions; difficulties in the removal of a bacteria from fluids without affecting the composition of these fluids; and difficulties in assessing any direct effect of the metabolic activity of pathogenic bacteria on results attributed to the host's condition, for example the infecting organisms use of glucose and the reduction of the host's blood sugar. One solution of these difficulties was indicated by studies on anthrax. The infecting organisms were allowed to grow in the host until it was about to succumb to the disease; then almost all the infecting organisms were removed by antibiotic treatment. This did not save the host; the pathological syndrome was irreversible and was studied without interference from bacterial growth.

Some study of the metabolism of pathogenic organisms is possible in conventional animal experiments. The importance in virulence of favourable nutritional conditions was demonstrated in mice when avirulent, nutritionally deficient, mutants of *Salmonella typhi* (Burrows, 1960) and *Klebsiella pneumoniae* (Garber, 1960) were restored to virulence by injecting their specifically required nutrients. During the terminal bacteraemia of anthrax in guinea pigs, there is a remarkably consistent multiplication of organisms and increase of toxin in the blood (Keppie, Smith & Harris-Smith, 1955; Tempest & Smith, 1957). By determining the change in free amino acid content of infected plasma at successive stages of the bacteraemia, Smith & Tempest (1957) showed that *Bacillus anthracis* used relatively large quantities of glutamine, threonine, tryptophan and glycine. Several precautions were taken to ensure that recognition of any utilization by the bacteria was not obscured by compensatory mechanisms of the tissues. In particular, samples of infected plasma were withdrawn and incubated for several hours *in vitro*; they yielded a similar pattern of amino acid utilization. To investigate the more specific role of metabolites in growth and toxin production, the effects of injecting metabolite analogues were studied; it was impossible to withdraw metabolites as in experiments *in vitro*. A number of metabolite analogues inhibited bacterial multiplication and toxin production *in vivo*, and reversal of their effects by appropriate metabolites suggested hypoxanthine, adenine, methionine, alanine, phenylalanine, and tryptophan were important for growth of *B. anthracis* and that pyrimidines and nicotinamide were needed for toxin synthesis. Not one of the amino acids which appeared important for growth *in vivo* was implicated in a study *in vitro* (Oikawa, 1956).

Germ-free animals have been used in experiments on pathogenicity, especially for continuing the study of intestinal commensals, pathogenic bacteria and of their interaction (Lev, 1963).

The artificial nature of experiments with animals in relation to pathogenic action in man has been noted. It is possible, now that treatment of bacterial disease with drugs is so effective, that more experiments could be allowed in human volunteers under strictly controlled conditions. Thus, by a controlled experiment with volunteers in Minnesota State Prison under the auspices of the World Health Organization, Spink, Hall, Finstad & Mallet (1962) showed that two live vaccines against brucellosis, *Brucella abortus* 19-BA and *B. melitensis* Rev. (1), were unsafe for use in man.

Experimental infection in localized areas of the host

The main use of these methods is to investigate the early invasion of tissues by micro-organisms and the effect of microbial and other products on this invasion.

Conventional methods of skin or peritoneal infection, followed by killing the animals at time intervals to examine sequential skin sections or peritoneal smears, has been used to study the invasive mechanisms of *Bacillus anthracis* (Cromartie, Bloom & Watson, 1947) and *Pasteurella pestis* (Burrows & Bacon, 1954). Miles, Miles & Burke (1957) have shown the potentiality of experimental skin infection for studying the primary lodgement of pathogenic organisms. The rapid development of cytochemical staining procedures and the use of fluorescent antibody and radioactive materials should extend the scope of the method. Also, the techniques for isolating and examining single animal cells developed for studies on immunity (Nossal & Mäkelä, 1962) might prove useful. In these experiments on antibody production by single cells, bacteria were added to the micro-droplets to indicate the type of antibody that had formed. If the single cells were phagocytes obtained from an infected peritoneum or lymph node, the micro-droplets would probably contain also the infecting organisms grown *in vivo*, if not, they could be introduced. Hence, the reaction between the bacteria and cell could be observed together with the effect on this reaction of any materials added to the micro-droplets.

The methods described have the disadvantage that sequential observations are made on different animals. For continuous observation of a developing infection in a single animal, the rabbit ear chamber has been used in studies with *Mycobacterium tuberculosis* (Ebert & Barclay, 1950), *Pseudomonas pseudo-mallei* (Miller & Clinger, 1961*a*) and *Histoplasma capsulatum* (Barclay & Winberg, 1963). Miller & Clinger (1961*b*), described an ingenious arrangement for frequently observing abdominal viscera of infected rabbits. Part of the abdominal wall was replaced by a transparent plastic sheet overlaid with another containing a zip-fastener. Four weeks after the operation, the animals were infected and the zip-fastener allowed frequent observation of the abdominal contents without resort to further surgery. Similar plastic windows might be developed for other animals and used in conjunction with Algire cells (see below).

A technique particularly applicable to studies of the pathogenicity of enteric organisms in view of the artificial nature of experiments with ordinary laboratory animals, is that of infecting ligated rabbit gut loops (De & Chatterje, 1953). Living pathogenic organisms distend the loops by causing oedema and exudate. Bacterial counts can be made and histological sections of the gut lining indicate the degree of inflammation and desquamation of the mucosal epithelium. Taylor, Maltby & Payne (1958) and Jenkin & Rowley (1959) used this technique to study the action of pathogenic *Escherichia coli* and *Vibrio cholerae* respectively. Nikonov, Khokhlova, Bichul & Timofeeva (1959) claimed to have increased the lytic activity of cholera phage *in vivo*, by passaging the phage on *V. cholerae* grown in loops of ligated guinea-pig gut.

All the methods so far described suffer from the disadvantage that the pathogenic organism, although localized for a time, is not enclosed in the area under examination; hence, bacterial losses and the continuing influx of inflammatory responses make quantitative work difficult. How can pathogenic organisms and their reaction with the cellular and humoral defences be studied in an enclosed area within the host under as natural conditions as possible? Enclosures in dialysis tubing are unsatisfactory because many important serum components cannot enter. A chamber, the Algire cell, developed for studying the growth of cancer cells in the peritoneal cavities of mice and other animals (Algire, Borders & Evans, 1958) may provide the answer. The membrane of the chamber is a cellulose acetate filter, which prevents the escape of bacteria and of animal cells but allows free access of serum components. Circular pieces of the filter are sealed to either side of plastic rings of various sizes which contain sealable holes for inoculation and sampling. Chambers are placed in the peritoneum and removed later usually post mortem. The zip-fastener technique, described above, might be modified so that the chambers could be withdrawn and sampled at relatively frequent intervals. The chamber could contain the pathogenic organism either alone or mixed with cells from a 'buffy coat' or an inflammatory exudate. The pathogenic organism might be allowed to invade the peritoneum, so forming a mixture of organisms grown *in vivo* and the constituents of the inflammatory response; then some of this mixture could be transferred to an Algire cell for further incubation and observation in either the same or a different animal.

Tissue culture

In tissue culture used for the growth of viruses, the animal cells divide and lose some of the characteristics of their parents (see above). Perhaps a less artificial system is maintenance culture, in which cells obtained directly from the host do not divide but are maintained in culture a sufficiently long time for an experiment to be completed. Such cultures, especially with monocytes from inflammatory reactions, have been used for growing intracellular organisms (Suter & Hulliger 1960; Elberg, 1960); the ability of pathogenic organisms to grow intra-cellularly in these cells has paralleled the virulence of strains which were indistinguishable by other laboratory tests. Monocytes collected from 2- to 4-day-old inflammatory reactions are used because they are easily maintained *in vitro* for relatively long periods. Polymorphonuclear cells have only been studied in short-term phagocytosis experiments, because in culture they are difficult to keep in a healthy state. Recently, intra-cellular survival and growth of *Brucella abortus* was studied in a system using the mixed cells of bovine 'buffy coat' (10–15 % polymorpho-nuclear cells, 2–5 % monocytes) maintained in culture for 40 hr. (Pearce *et al.* 1962; Smith & FitzGeorge, 1963). This procedure allowed chemical investigations of the compounds responsible for survival and growth of virulent *B. abortus* in cells of a natural host. Such tests might be used for screening drugs against intracellular parasites infecting man, if human 'buffy coat' was used.

Maintenance culture may be of limited use for comparing microbial behaviour in cells from different animals. In our hands, the tests with bovine cells yielded consistent results when the same batch of cells was used for comparing different organisms or products (Smith & Fitz-George, 1963). However, the behaviour of *Brucella abortus* varied in cells from the same animal taken on different days. The variation was such that any comparison between microbial behaviour in different cells, e.g. those from normal and immune animals, on any one day was not practicable (Macrae & Smith, 1963).

A technique designed for radiological work (Trowell, 1959) which maintains fresh tissue slices in good condition for 10–14 days might be applied to studies of pathogenicity especially for work on tissue specificity.

Experiments in vitro *to elucidate microbial behaviour* in vivo

Perhaps the experiments *in vitro* that can shed most light on microbial behaviour *in vivo* are those dealing with nutrition and metabolism. This is particularly true of studies on tissue and host specificity, and for detecting metabolic differences between avirulent and virulent strains, which might explain the inability of the former to grow in the tissues of the host. The effects of extracts from tissues normally invaded by the pathogen *in vivo* on its growth and metabolism *in vitro* should be examined, together with materials known to be present in these extracts. Relatively small quantities of extracts may be added to cultures in laboratory media or attempts may be made to grow the pathogenic organism in the whole extract. Sometimes bactericidal materials may be present, necessitating the use of a diffusate or a heated extract. If a growth stimulant is detected and identified its metabolism by the pathogen can be studied. The scope of this approach is indicated by experiments which led to the recognition of a growth stimulant—erythritol— as the cause of the predilection of *Brucella abortus* for various bovine foetal tissues (Smith *et al.* 1962*b*; Keppie, 1964). Continuous culture could be used in such studies, and the methods used by Hungate (1963) for studying rumen bacteria might be helpful especially if mixed infections were involved.

In designing biological and immunological tests *in vitro* for microbial processes and products responsible for pathogenicity, the multitude of microbial products unconnected with pathogenicity should be excluded as far as possible, Most serological tests such as agglutination, precipitation or complement fixation are designed primarily for diagnosis, and detect many antigens or antibodies which may be unimportant in pathogenicity or immunogenicity. Detectable 'antibody' is often produced in disease. This helps diagnosis, but it does not indicate the degree of immunity unless the 'antibody' is protective. More cogent tests *in vitro* for the understanding of the bases of pathogenicity include interference with the phagocytosis and interference with the bactericidal action of serum or white cell extracts; results from these tests should be interpreted cautiously as regards reflecting microbial behaviour *in vivo*. Wood, Smith & Watson (1946) showed that organisms can resist phagocytosis in tests where both phagocytes and bacteria are in suspension, yet be susceptible *in vivo*, because they are trapped against immovable surfaces. They repeated 'surface phagocytosis' *in vitro* using filter paper as the trapping surface. This system or those involving chemotaxis through a membrane are perhaps the best tests to use *in vitro* for detecting materials interfering with phagocytosis. Again, a substance may interfere with the bactericidal action of serum *in vitro*, but this may have no relevance *in vivo* because the bactericidal material in serum might be an artifact: for example, some of the heat-stable

β-lysins might not be true serum components but derived from blood platelets during clotting (Hirsch, 1960). Finally, if a material produced *in vitro* appears to be important for pathogenicity in these and other tests, proof of its production *in vivo* should be obtained, for example by precipitating it or by neutralizing its biological activity with an anti-serum prepared against live virulent organisms.

EXAMPLES OF THE USE OF ORGANISMS GROWN *IN VIVO* AND OF THEIR DIFFERENCES FROM ORGANISMS GROWN *IN VITRO*

Work on organisms grown *in vivo* is described, which revealed hitherto unknown virulence mechanisms or immunogenic factors, some of which were later reproduced *in vitro*. In addition less complete studies are included, which at present indicate only a difference between organisms grown *in vivo* and *in vitro*, but which point out the need to search *in vivo* for unknown virulence attributes. Older work which was reviewed by Smith (1958, 1960) is summarized briefly as background for the review of more recent material.

Bacillus anthracis

Prior to work with organisms grown *in vivo* almost nothing was known about the killing power of *Bacillus anthracis*. Neither the general nature of the lethal effect on the host nor the products responsible for it were known. In particular no lethal toxin had been found in artificial cultures. The position was clarified by work on anthrax in the guinea pig (Smith, 1958, 1960). The main points were: (1) the massive terminal bacteraemia was not the primary cause of death since removal of the organisms did not prevent death, which was therefore probably caused by toxin; (2) the fatal syndrome was oligaemic secondary shock; (3) a search for an oedema-producing factor, revealed in the plasma of infected guinea pigs the specifically neutralizable anthrax toxin, which killed mice and guinea pigs in secondary shock; (4) the immunogenic toxin contained at least two components which acted synergistically, one of which was an immunizing antigen which had been previously demonstrated in a defined medium; and (5) the toxin originally found *in vivo* was reproduced *in vitro* first in a medium containing serum and later in a semi-synthetic medium.

Recently, the anthrax toxin has been demonstrated in other infected animals including rhesus monkeys and shown to be responsible for death (Klein *et al.* 1962). Fractionation of the toxin produced *in vitro* has increased the number of known components to three (Stanley & Smith, 1961; Beall, Taylor & Thorne, 1962). All have been purified and have intriguing interactions in tests for toxicity and immunogenicity (Stanley & Smith, 1963). Two of these components are proteins and the third a chelating agent which contains protein, some carbohydrate and phosphorus. In addition to its lethal action, the toxin helps primary invasion by inactivating phagocytes and interfering with serum bactericidins (Keppie, Harris-Smith & Smith, 1963).

The fact that three interacting factors comprise this toxin makes it difficult to know exactly what occurs *in vivo*. What proportions of the three factors are produced by any particular strain and what are the optimal proportions needed for full toxicity or immunogenicity? It is almost impossible to investigate the latter because of the large numbers of animals needed for comprehensive titrations (Stanley & Smith, 1963). A vaccine based on one factor may not confer resistance against a strain in whose toxin the other two factors predominate.

Pasteurella pestis

Virulent and avirulent strains of *Pasteurella pestis*, indistinguishable by any test *in vitro*, behaved differently in the peritoneal cavities of mice (Burrows & Bacon, 1954). Initially both strains were phagocytozed, but the avirulent strain was progressively destroyed, whereas the virulent strain became resistant to phagocytosis within $\frac{1}{2}$ hr., multiplied and killed the host. Similar studies were made on their behaviour in the lungs of infected guinea pigs (Fukui, Lawton, Janssen & Surgalla, 1957). After investigating their results by experiments *in vivo* and *in vitro*, Burrows & Bacon (1956a, b) concluded that phagocytosis resistance was not solely due to capsular fraction I, because some strains became resistant to phagocytosis *in vitro* and *in vivo* in the absence of visible capsulation. This resistance was associated with two new antigens V and W, although solutions containing them had no antiphagocytic action. The association of V and W antigens with virulence has been confirmed (Fukui *et al.* 1960).

A puzzling feature of studies with *Pasteurella pestis* was that live virulent organisms killed both guinea pigs and mice, whereas only mice were killed by a product, murine toxin, obtained from cultures *in vitro*. Similarly, live avirulent vaccines immunized both experimental animals but a product, fraction I, obtained from cultures *in vitro*, only immunized mice if injected without adjuvant. To resolve these anomalies Smith *et al.* (1960), Cocking, Keppie, Witt & Smith (1960), and Keppie, Cocking, Witt & Smith (1960) worked with organisms and products from infected guinea pigs. No extracellular toxin comparable to the anthrax toxin was demonstrable, but an extract of the organisms killed guinea pigs as well as mice. The guinea pig toxin was later produced *in vitro* and had two components which acted synergistically. Cell-wall material from the organisms grown *in vivo* immunized guinea pigs and mice, although it contained little capsular fraction I. Possibly *in vivo*, the cell-wall material acts as an adjuvant to fraction I, because artificial adjuvants render small amounts of fraction I immunogenic for guinea pigs (Spivack *et al.* 1958).

Pasteurella multocida

This organism causes haemorrhagic septicaemia in cattle and buffaloes. Bain (1960), in phagocytosis and serum bactericidal tests, showed that test organisms isolated from infected animals reacted differently to mixtures of phagocytes and sera from immune and normal animals. No differences could be detected if the test organisms were grown *in vitro*.

Coccidioides immitis

This pathogenic fungus is dimorphic. Parasitic in man or animals, it forms a spherule or sporangium in which endospores develop and later grow into spherules. Normally on laboratory media it forms a mycelium bearing arthrospores, but Converse (1955, 1957) produced *in vitro* the more parasitic spherule form. The possibility arose that vaccines prepared with the spherule form would be more immunogenic than vaccines made from the arthrospore or mycelial form of the fungus. Levine and his colleagues (Levine, Cobb & Smith, 1960, 1961; Levine, Miller & Smith, 1962) showed this to be so in experiments making use of nasal infection of mice and respiratory infection of monkeys.

Brucella abortus

Several workers have studied *Brucella abortus* in monocyte maintenance culture (Holland & Pickett, 1958; Elberg, 1960; Stinebring, Braun & Pomales-Lebrón, 1960; Stinebring, 1962). Strains of differing virulence, which were indistinguishable by most laboratory tests, behaved differently in these cultures; the more virulent strains grew more rapidly in these cells than did the less virulent strains. Also *B. abortus* grown within guinea pig monocytes had an increased ability compared with organisms from artificial cultures to survive and multiply intracellularly *in vitro* and in animals, and to resist the bactericidin of normal bovine serum (Stinebring *et al.* 1960; Stinebring, 1962).

Recently *Brucella abortus* and its products were isolated from foetal fluids and tissues of infected pregnant cows for studies on pathogenicity and immunity (Smith *et al.* 1961, 1962*b*; Pearce *et al.* 1962; Williams, Keppie & Smith, 1962; Smith, Keppie, Pearce & Witt, 1962*a*; Smith & FitzGeorge, 1963; Macrae & Smith, 1963). Three problems were investigated: the cause of the predilection of *B. abortus* for some bovine foetal tissues, the nature of the immunogens and the basis for the ability of virulent cells to survive and multiply within bovine phagocytes. Investigations of the first problem are reviewed by Dr J. Keppie in this Symposium.

Crude cell-wall preparations from *Brucella abortus* grown *in vivo* were immunogenic but the cytoplasmic constituents were not. Similar results were obtained with organisms grown *in vitro*, but the following new fact emerged from the studies *in vivo*. The finding that some filtered fluids from infected animals were immunogenic led to the production *in vitro* of an immunizing culture filtrate, first from a complex medium and then from a defined medium. The use of a defined medium aided fractionation and a purified product was obtained, which immunized guinea pigs and mice in quantities of less than 1 μg., and which interfered with the bactericidal action of bovine serum on *B. abortus*. It is not known whether this material is an extracellular or an easily removed surface component. The remainder of the immunogenic material in the cell walls seemed strongly bound. Investigations are being made to determine whether or not the culture filtrate material and the cell-wall material are identical.

With regard to intracellular growth, virulent *Brucella abortus* from infected foetal fluids resisted the bactericidal action of bovine phagocytes better than did the same strain grown *in vitro*. The lethal effect of bovine white blood cells occurred first extracellularly before all organisms were phagocytozed and later intracellularly. Bacterial cell-wall material interfered with the bactericidal action. Although cell-wall material from *B. abortus* grown *in vitro* interfered with the extracellular bactericidal activity as well as cell-wall material from *B. abortus* grown *in vivo*, the latter appeared to be superior in combating the subsequent intracellular killing. The inhibitory activity of cell-wall material was specifically neutralized by an antiserum prepared against live virulent organisms; it appeared to be connected with immunogenic activity because the purified immunogen also interfered with the extracellular bactericidal action of bovine phagocytes.

Intestinal pathogens

Recently various experiments *in vivo* have demonstrated factors involved in pathogenicity of these organisms other than their endotoxins. This had been indicated by Hill, Hatswell & Topley (1940) who showed that mice selectively bred for resistance to the relevant endotoxin were not more resistant to oral infection by *Salmonella typhimurium*.

Olitzki & Godinger (1963) and Olitzki & Kaplan (1963) found that *Salmonella typhi* (Ty2) isolated from infected mice was more virulent than the same strain grown *in vitro*. The organisms grown *in vivo* were resistant to the bactericidal action of various sera when examined directly *in vitro*, but were as susceptible as strains grown *in vitro* if they were stimulated by broth or tissue extracts to grow and divide *in vitro*. Extracts from infected tissues were not toxic, but promoted infections with *S. typhi* in normal mice, interfered with the bactericidal action of sera and contained soluble bacterial antigens; extracts from normal tissues had a weaker activity in promoting infections. Although no antigen additional to the well-known H, O and Vi antigens could be detected in the extracts, the Vi antigen alone was not responsible for their virulence-enhancing effect, because extracts of tissues from mice infected with organisms other than *S. typhi* but which produced the Vi antigen were not as effective as those from mice infected with *S. typhi*.

Growth of enteric pathogens such as *Salmonella* spp., *Shigella* spp. or *Vibrio cholerae* is suppressed by the normal gut flora, and if the latter are removed by streptomycin, subsequent infections with pathogenic organisms flourish (Watkins, 1960; Lankford, 1960; Lev, 1963; Meynell & Subbaiah, 1963; Meynell, 1963). The mechanisms of this phenomenon are becoming clearer from results of experiments *in vivo* and *in vitro*. Formal *et al.* (1961), working with germ-free animals, showed that growth of *Shigella flexneri* was suppressed by *Escherichia coli*, which *in vitro* did not produce a colicine or any other inhibitory factor. Meynell & Subbaiah (1963), using their new method for estimating bacterial division *in vivo*, showed that the contents of the colon and caecum of a normal mouse had a bacteriostatic and weakly bactericidal action towards orally administered *Salmonella typhimurium*. Meynell (1963) showed that this activity was dependent on an E_h of -0.2 V. and the production of much volatile fatty acid by the normal flora, which are mainly obligate anaerobes. A similar environment *in vitro* had an identical inhibitory and weak bactericidal action on *S. typhimurium*. Removal of the normal flora with streptomycin raised the E_h to $+0.2$ V.

and reduced the concentration of volatile fatty acid, thereby producing conditions favourable for the multiplication of salmonellae.

Pathogenic streptococci

Over the past decade, Watson and his colleagues (review by Smith, 1958; Barkulis, 1960) have worked with toxic materials produced in rabbit skin lesions by group A streptococci; as a result, Watson (1960) ascribed a new role to the erythrogenic toxins found in skin lesions, which to some extent explained the pathogenesis and sequelae seen in this infection. In small doses, these toxins were relatively innocuous but they increased the susceptibility of a host to subsequent doses of themselves, streptolysin 0 and to a number of Gram-negative endotoxins; in rabbits, the second injection caused a syndrome like the generalized Shwartzman reaction, producing myocardial necrosis, cortical necrosis of the kidney and death. In large doses, the erythrogenic toxins were lethal when injected alone. Three immunologically distinct toxins were recognized, one of which had not been demonstrated before. All group A streptococci formed one or more toxins *in vivo*, but only some of them, the so-called toxigenic strains, produced toxin in culture media. No group B or group C streptococci examined formed erythrogenic toxin *in vivo*.

Von Răska & Rotta (1956) and Barkulis (1960) showed that extracts from infected skin lesions or other organs could enhance the persistence of streptococci *in vivo*; whether or not this property is due to erythrogenic toxins is unknown.

Mycobacterium tuberculosis

Segal & Bloch (1956, 1957) showed that *Mycobacterium tuberculosis* from infected mouse lungs differed from the same strain grown *in vitro* by being more virulent, more easily dispersed in saline or water (indicating a difference in surface structure), and less easily stimulated to respire under the influence of various substrates. Recently Bekierkunst & Artman (1960, 1962) and Artman & Bekierkunst (1961*a*, *b*), working with cell-free preparations from *M. tuberculosis* grown *in vivo*, showed that at least part of this inability to react to various substrates was because a diphosphopyridine nucleotide splitting enzyme had been absorbed from the host. The diphosphopyridine nucleotide splitting enzyme was present in normal mice, but increased enormously in infected mice, so that it not only inhibited the nucleotide linked dehydrogenases of the pathogen but also those of the infected host. The relationship of these results to the pathogenicity of *M. tuberculosis* is still obscure.

Pathogenic staphylococci

The need to design new measures against the increasing numbers of antibiotic-resistant staphylococci should encourage research into the basis of their virulence. None of the numerous factors which are produced in greater amount by virulent than by avirulent strains *in vitro* appears to be of major importance (Burns & Holtman, 1960; Smith, 1962). The possibility that virulence factors are produced *in vivo* and not *in vitro* has been broached (Smith, 1962; Koenig, Melly & Rogers, 1962). Investigations of this possibility should be encouraged by the recent appearance of two papers indicating clearly the differences between staphylococci grown *in vivo* and *in vitro*.

Gellenbeck (1962) showed that the respiratory responses of staphylococci harvested from infected guinea pigs to various substrates was higher than those of the same strain grown *in vitro*; the latter however had a higher endogenous respiration. The various substrates contained known and unknown factors to provide a wide basis for comparison: glucose, a synthetic medium, a broth, guinea-pig serum, and filtered exudate from the cavities of infected guinea pigs.

Beining & Kennedy (1963) not only compared the organisms from guinea pigs and broth culture but also their extracellular products. The organisms from guinea pigs were more virulent for mice and rabbits than those grown *in vitro*, and their respiration was stimulated to a greater degree by addition of glucose. The two types differed markedly in the degree and nature of their agglutinability by the same antiserum. In contrast to the strain grown *in vitro*, that obtained from guinea pigs failed to grow in tellurite glycine agar which is used to isolate potentially pathogenic staphylococci; this is a matter of concern in diagnosis, since some pathogenic staphylococci growing *in vivo* may never be detected. Growth of the organisms from guinea pigs was markedly inhibited by human γ-globulin; more so than that of the organisms grown *in vitro*. Extracellular products from the two types of organisms were different when examined serologically by the Oudin and Ouchterlony gel diffusion techniques. The product from the organisms grown *in vivo* killed mice, and contained far more haemolysin, leucocidin, hyaluronidase, and deoxyribonuclease than did the non-lethal product from artificial cultures. The differences between these solutions of extracellular products could be quantitative rather than qualitative. Despite these marked differences between the two types of organisms, they were practically identical in tests for bound and soluble coagulases, bacteriophage type, antibiotic sensitivity pattern, the usual fermentation reactions, and various enzyme reactions.

This survey indicates that pathogenic bacteria grown *in vivo* can be studied, that differences exist between organisms grown *in vivo* and *in vitro*, and that microbial properties important in virulence and immunogenicity which are revealed by studies *in vivo* can be reproduced *in vitro*. The survey therefore illustrates the general theme of this Symposium and I hope it will encourage studies of micro-organisms in more natural environments.

REFERENCES

ACKOFF, R. L. (1959). Games, decisions, and organizations. *General Systems*, 4, 145.

AINSWORTH, G. C. (1955). Pathogenicity of fungi in man and animals. In *Mechanisms of Microbial Pathogenicity. Symp. Soc. gen. Microbiol.* 5, 242.

ALGIRE, G. H., BORDERS, M. L. & EVANS, V. J. (1958). Studies of heterografts in diffusion chambers in mice. *J. nat. Cancer Inst.* 20, 1187.

ARTMAN, M. & BEKIERKUNST, A. (1961 a). Studies on *Mycobacterium tuberculosis* H37RV grown *in vivo. Amer. Rev. resp. Dis.* 83, 100.

ARTMAN, M. & BEKIERKUNST, A. (1961 b). *Mycobacterium tuberculosis* H37RV grown *in vivo*: nature of the inhibitor of lactic dehydrogenase of *Mycobacterium phlei. Proc. Soc. exp. Biol., N.Y.* 106, 610.

ASHBY, W. R. (1958). General systems theory as a new discipline. *General Systems*, 3, 1.

BAIN, R. V. S. (1960). Mechanism of immunity in haemorrhagic septicaemia. *Nature, Lond.* 186, 734.

BARCLAY, W. R. & WINBERG, E. (1963). Histoplasmosis: *in vivo* observations on immunity, hypersensitivity, and the effects of silica and amphotericin B. *Amer. Rev. resp. Dis.* 87, 331.

BARKULIS, S. S. (1960). Biochemical properties of virulent and avirulent strains of group A haemolytic streptococcus. *Ann. N.Y. Acad. Sci.* 88, 1034.

BEALL, F. A., TAYLOR, M. J. & THORNE, C. B. (1962). Rapid lethal effect in rats of a third component found upon fractionating the toxin of *Bacillus anthracis. J. Bact.* 83, 1274.

BEINING, P. R. & KENNEDY, E. R. (1963). Characteristics of a strain of *Staphylococcus aureus* grown *in vivo* and *in vitro. J. Bact.* 85, 732.

BEKIERKUNST, A. & ARTMAN, M. (1960). Studies of *Mycobacterium tuberculosis* H37RV grown *in vivo*: inhibitor of lactic acid dehydrogenase in normal and infected mice. *Proc. Soc. exp. Biol., N.Y.* 105, 605.

BEKIERKUNST, A. & ARTMAN, M. (1962). Tissue metabolism in infection: DPN-ase activity, DPN levels, and DPN-linked dehydrogenases in tissues from normal and tuberculous mice. *Amer. Rev. resp. Dis.* 86, 832.

BLOCH, H. (1960). Biochemical properties of virulent and avirulent strains of *Mycobacterium tuberculosis. Ann. N.Y. Acad. Sci.* 88, 1075.

BRAUN, W. (editor) (1960). Biochemical aspects of microbial pathogenicity. *Ann. N.Y. Acad. Sci.* 88, 1021.

BRIAN, P. W. (1955). The role of toxins in the etiology of plant diseases caused by fungi and bacteria. In *Mechanisms of Microbial Pathogenicity. Symp. Soc. gen. Microbiol.* 5, 294.

BURNS, J. & HOLTMAN, D. F. (1960). Biochemical properties of virulent and avirulent staphylococci. *Ann. N.Y. Acad. Sci.* 88, 1115.

Burrows, T. W. (1960). Biochemical properties of virulent and avirulent strains of bacteria: *Salmonella typhosa* and *Pasteurella pestis*. *Ann. N.Y. Acad. Sci.* **88**, 1125.

Burrows, T. W. & Bacon, G. A. (1954). The basis of virulence in *Pasteurella pestis*: comparative behaviour of virulent and avirulent strains *in vivo*. *Brit. J. exp. Path.* **35**, 134.

Burrows, T. W. & Bacon, G. A. (1956a). The basis of virulence in *Pasteurella pestis*: the development of resistance to phagocytosis *in vitro*. *Brit. J. exp. Path.* **37**, 286.

Burrows, T. W. & Bacon, G. A. (1956b). The basis of virulence in *Pasteurella pestis*: an antigen determining virulence. *Brit. J. exp. Path.* **37**, 481.

Cocking, E. C., Keppie, J., Witt, K. & Smith, H. (1960). The chemical basis of the virulence of *Pasteurella pestis*. II. The toxicity for guinea pigs and mice of products of *Past. pestis*. *Brit. J. exp. Path.* **41**, 460.

Converse, J. L. (1955). Growth of spherules of *Coccidioides immitis* in a chemically defined liquid medium. *Proc. Soc. exp. Biol., N.Y.* **90**, 709.

Converse, J. L. (1957). Effect of surface active agents on endosporulation of *Coccidioides immitis* in a chemically defined medium. *J. Bact.* **74**, 106.

Cromartie, W. J., Bloom, W. L. & Watson, D. W. (1947). Studies on infection with *Bacillus anthracis*. I. A histopathological study of skin lesions produced by *B. anthracis* in susceptible and resistant animal species. *J. infect. Dis.* **80**, 1.

De, S. N. & Chatterje, D. N. (1953). An experimental study of the mechanism of action of *Vibrio cholerae* on the intestinal mucous membrane. *J. Path. Bact.* **66**, 559.

Dimond, A. E. & Waggoner, P. E. (1953). Effect of lycomarasmin decomposition upon estimates of its production. *Phytopathology*, **43**, 319.

Dubos, R. J. (1945). *The Bacterial Cell*, p. 188. Cambridge, Mass.: Harvard University Press.

Dubos, R. J. (1954). *Biochemical Determinants of Microbial Disease*, p. 12. Cambridge, Mass.: Harvard University Press.

Ebert, R. H. & Barclay, W. R. (1950). Effect of chemotherapy on tissue response to tuberculous infection as observed *in vivo*. *J. clin. Invest.* **29**, 810.

Edney, M. (1957). Variation in animal viruses. *Annu. Rev. Microbiol.* **11**, 23.

Elberg, S. S. (1960). Cellular immunity. *Bact. Rev.* **24**, 67.

Enders, J. F. (1952). In *Viral and Rickettsial Infections of Man*, p. 126. Ed. by T. M. Rivers & F. L. Horsfall. Philadelphia: Lippincott.

Felton, L. D. & Bailey, G. H. (1926). Biologic significance of the soluble specific substances of pneumococci. *J. infect. Dis.* **38**, 131.

Formal, S. B., Dammin, G., Sprinz, H., Kundel, D., Schneider, H., Horowitz, R. E. & Forbes, M. (1961). Experimental *Shigella* infections. V. Studies in germ-free guinea pigs. *J. Bact.* **82**, 284.

Fukui, G. M., Delwiche, E. A., Mortlock, R. P. & Surgalla, M. J. (1962). Oxidative metabolism of *Pasteurella pestis* grown *in vitro* and *in vivo*. *J. infect. Dis.* **110**, 143.

Fukui, G. M., Lawton, W. D., Ham, D. A., Janssen, W. A. & Surgalla, M. J. (1960). The effect of temperature on the synthesis of virulence factors by *Pasteurella pestis*. *Ann. N.Y. Acad. Sci.* **88**, 1146.

Fukui, G. M., Lawton, W. D., Janssen, W. A. & Surgalla, M. J. (1957). Response of guinea-pig lungs to *in vivo* and *in vitro* cultures of *Pasteurella pestis*. *J. infect. Dis.* **100**, 103.

Garber, E. D. (1960). The host as a growth medium. *Ann. N.Y. Acad. Sci.* **88**, 1187.

Gellenbeck, S. M. (1962). Aerobic respiratory metabolism of *Staphylococcus aureus* from an infected animal. *J. Bact.* **83**, 450.

Gibson, J. (1957). Nutritional aspects of microbial ecology. In *Microbial Ecology. Symp. Soc. gen. Microbiol.* **7**, 22.

GRAY, C. T. (1952). The respiratory metabolism of murine leprosy bacilli. *J. Bact.* **64**, 305.

HANKS, J. H. (1951). Measurement of the hydrogen transfer capacity of myco-bacteria. *J. Bact.* **62**, 521.

HART, P. D'A. & REES, R. J. W. (1955). In *Experimental Tuberculosis*, p. 299. Ed. by G. E. W. Wolstenholme & M. P. Cameron. London: Churchill.

HAYES, W. & CLOWES, R. C. (editors) (1960). *Microbial Genetics. Symp. Soc. gen. Microbiol.* **10**. Cambridge University Press.

HENDERSON, D. W., LANCASTER, M. C., PACKMAN, L. & PEACOCK, S. (1956). The influence of a pre-existing respiratory infection on the course of another super-imposed by the respiratory route. *Brit. J. exp. Path.* **37**, 597.

HILL, A. B., HATSWELL, J. M. & TOPLEY, W. W. C. (1940). The inheritance of resistance, demonstrated by the development of a strain of mice resistant to experimental inoculation with a bacterial endotoxin. *J. Hyg., Camb.* **40**, 538.

HIRSCH, J. G. (1960). Comparative bactericidal activities of blood serum and plasma serum. *J. exp. Med.* **112**, 15.

HOLLAND, J. J. & PICKETT, M. J. (1958). A cellular basis of immunity in experi-mental Brucella infection. *J. exp. Med.* **108**, 343.

HOWIE, J. W. & O'HEA, A. J. (editors) (1955). *Mechanisms of Microbial Patho-genicity. Symp. Soc. gen. Microbiol.* **5**. Cambridge University Press.

HUNGATE, R. E. (1963). Symbiotic associations: the rumen bacteria. In *Symbiotic Associations. Symp. Soc. gen. Microbiol.* **13**, 266.

JENKIN, C. R. & ROWLEY, D. (1959). Possible factors in the pathogenesis of cholera. *Brit. J. exp. Path.* **40**, 474.

KAPLAN, A. S. (1955). The susceptibility of monkey kidney cells to polio virus *in vivo* and *in vitro*. *Virology*, **1**, 377.

KEPPIE, J. (1964). Host and tissue specificity. In *Microbial Behaviour, in Vivo and in Vitro. Symp. Soc. gen. Microbiol.* **14**, 44.

KEPPIE, J., COCKING, E. C., WITT, K. & SMITH, H. (1960). The chemical basis of the virulence of *Pasteurella pestis*. III. An immunogenic product obtained from *Past. pestis* which protects both guinea pigs and mice. *Brit. J. exp. Path.* **41**, 577.

KEPPIE, J., HARRIS-SMITH, P. W. & SMITH, H. (1963). The chemical basis of the virulence of *Bacillus anthracis*. IX. Its aggressins and their mode of action. *Brit. J. exp. Path.* **44**, 446.

KEPPIE, J., SMITH, H. & HARRIS-SMITH, P. W. (1955). The chemical basis of the virulence of *Bacillus anthracis*. III. The role of the terminal bacteraemia in death of guinea pigs from anthrax. *Brit. J. exp. Path.* **36**, 315.

KLEIN, F., HODGES, D. R., MAHLANDT, B. G., JONES, W. I., HAINES, B. W. & LINCOLN, R. E. (1962). Anthrax toxin: Causative agent in the death of rhesus monkeys. *Science*, **138**, 1331.

KNOX, R. (1958). The chemotherapy of bacterial infections. In *The Strategy of Chemotherapy. Symp. Soc. gen. Microbiol.* **8**, 288.

KOENIG, M. G., MELLY, M. A. & ROGERS, D. E. (1962). Factors relating to the virulence of staphylococci. III. Antibacterial versus antitoxic immunity. *J. exp. Med.* **116**, 601.

LACEY, B. W. (1958). Mechanisms of chemotherapeutic synergy. In *The Strategy of Chemotherapy. Symp. Soc. gen. Microbiol.* **8**, 247.

LANKFORD, C. E. (1960). Factors of virulence of *Vibrio cholerae*. *Ann. N.Y. Acad. Sci.* **88**, 1203.

LEV, M. (1963). Studies on bacterial association in germ-free animals and animals with defined floras. In *Symbiotic Associations. Symp. Soc. gen. Microbiol.* **13**, 325.

LEVINE, H. B., COBB, J. M. & SMITH, C. E. (1960). Immunity to coccidioidomycosis induced in mice by purified spherule, arthrospore, and mycelial vaccines. *Trans. N.Y. Acad. Sci.* **22**, 436.

LEVINE, H. B., COBB, J. M. & SMITH, C. E. (1961). Immunogenicity of spherule-endospore vaccines of *Coccidioides immitis* for mice. *J. Immunol.* **87**, 218.

LEVINE, H. B., MILLER, R. L. & SMITH, C. E. (1962). Influence of vaccination on respiratory coccidioidal disease in cynomolgous monkeys. *J. Immunol.* **89**, 242.

MACRAE, R. M. & SMITH, H. (1963). The chemical basis of the virulence of *Brucella abortus*. VI. Studies on immunity and intracellular growth. *Brit. J. exp. Path.* (in the Press).

MEYNELL, G. G. (1958). The effect of sudden chilling on *Escherichia coli*. *J. gen. Microbiol.* **19**, 380.

MEYNELL, G. G. (1959). Use of superinfecting phage for estimating the division rate of lysogenic bacteria in infected animals. *J. gen. Microbiol.* **21**, 421.

MEYNELL, G. G. (1961). Phenotypic variation and bacterial infection. In *Microbial Reaction to Environment*. *Symp. Soc. gen. Microbiol.* **11**, 174.

MEYNELL, G. G. (1963). Antibacterial mechanisms of the mouse gut. II. The role of E_h and volatile fatty acids in the normal gut. *Brit. J. exp. Path.* **44**, 209.

MEYNELL, G. G. & GOODER, H. (editors) (1961). *Microbial Reaction to Environment.* *Symp. Soc. gen. Microbiol.* **11**. Cambridge University Press.

MEYNELL, G. G. & SUBBAIAH, T. V. (1963). Antibacterial mechanisms of the mouse gut. I. Kinetics of infection by *Salmonella typhimurium* in normal and streptomycin-treated mice studied with abortive transductants. *Brit. J. exp. Path.* **44**, 197.

MILES, A. A. (1955). The meaning of pathogenicity. In *Mechanisms of Microbial Pathogenicity*. *Symp. Soc. gen. Microbiol.* **5**, 1.

MILES, A. A., MILES, E. M. & BURKE, J. (1957). The value and duration of defence reactions of the skin to the primary lodgement of bacteria. *Brit. J. exp. Path.* **38**, 79.

MILLER, R. & CLINGER, D. I. (1961 a). Melioidosis pathogenesis in rabbits. I. *In vivo* studies in the rabbit ear chamber. *Arch. Path.* (*Lab. Med.*), **71**, 629.

MILLER, R. & CLINGER, D. I. (1961 b). Melioidosis pathogenesis in rabbits. II. A simplified surgical technique for *in vivo* observations of pathologic changes in abdominal viscera. *Arch. Path.* (*Lab. Med.*), **71**, 635.

NIKONOV, A. G., KHOKHLOVA, A. M., BICHUL, K. C. & TIMOFEEVA, R. I. (1959). Cholera phage. *Zh. mikrobiol. epidemiol. immunobiol.* **30**, 90.

NOSSAL, G. J. V. & MÄKELÄ, O. (1962). Elaboration of antibodies by single cells. *Annu. Rev. Microbiol.* **16**, 53.

OIKAWA, T. (1956). The amino-acid requirement of *Bacillus anthracis*. I. Simple amino-acid media for the growth of *Bacillus anthracis*. *Nippon Saikengaku Zasshi.* **11**, 1099.

OLITZKI, A. L. & GODINGER, D. (1963). Comparative studies of *Salmonella typhi* grown *in vivo* and *in vitro*. I. Virulence, toxicity, production of infection-promoting substances and DPN-ase activity. *J. Hyg., Camb.* **61**, 1.

OLITZKI, A. L. & KAPLAN, O. (1963). Comparative studies on *Salmonella typhi* grown *in vivo* and *in vitro*. II. The effect of extracts from normal and infected organs on the bactericidal serum action on strains grown *in vivo* and *in vitro*. *J. Hyg., Camb.* **61**, 21.

PEARCE, J. H., WILLIAMS, A. E., HARRIS-SMITH, P. W., FITZGEORGE, R. B. & SMITH, H. (1962). The chemical basis of the virulence of *Brucella abortus*. II. Erythritol, a constituent of bovine foetal fluids which stimulates the growth of *Br. abortus* in bovine phagocytes. *Brit. J. exp. Path.* **43**, 31.

PIERCE, C. H., DUBOS, R. J. & SCHAEFER, W. B. (1956). Differential characteristics *in vitro* and *in vivo* of several strains of BCG. III. Multiplication and survival *in vivo*. *Amer. Rev. Tuberc.* **74**, 683.

RIVERS, T. M. & WARD, S. M. (1933). Further observations on the cultivation of vaccine virus for Jennerian prophylaxis in man. *J. exp. Med.* **58**, 635.

ROSS, J. D. & SYVERTON, J. T. (1957). Use of tissue cultures in virus research. *Annu. Rev. Microbiol.* **11**, 459.

ROSS, J. D., TREADWELL, P. E. & SYVERTON, J. T. (1962). Cultural characterization of animal cells. *Annu. Rev. Microbiol.* **16**, 141.

SEGAL, W. & BLOCH, H. (1956). Biochemical differentiation of *Mycobacterium tuberculosis* grown *in vivo* and *in vitro*. *J. Bact.* **72**, 132.

SEGAL, W. & BLOCH, H. (1957). Pathogenic and immunogenic differentiation of *Mycobacterium tuberculosis* grown *in vitro* and *in vivo*. *Amer. Rev. Tuberc.* **75**, 495.

SMITH, D. C. (1963). Experimental studies of lichen physiology. In *Symbiotic Associations. Symp. Soc. gen. Microbiol.* **13**, 31.

SMITH, D. D. (1962). Experimental staphyloccal infection in mice. *J. Path. Bact.* **84**, 359.

SMITH, H. (1958). The use of bacteria grown *in vivo* for studies on the basis of their pathogenicity. *Annu. Rev. Microbiol.* **12**, 77.

SMITH, H. (1960). Studies on organisms grown *in vivo* to reveal the bases of microbial pathogenicity. *Ann. N.Y. Acad. Sci.* **88**, 1213.

SMITH, H. & FITZGEORGE, R. B. (1963). The chemical basis of the virulence of *Brucella abortus*. V. The basis of intracellular survival and growth in bovine phagocytes. *Brit. J. exp. Path.* (in the Press).

SMITH, H., KEPPIE, J., COCKING, E. C. & WITT, K. (1960). The chemical basis of the virulence of *Pasteurella pestis*. I. The isolation and the aggressive properties of *Past. pestis* and its products from infected guinea pigs. *Brit. J. exp. Path.* **41**, 452.

SMITH, H., KEPPIE, J., PEARCE, J. H., FULLER, R. & WILLIAMS, A. E. (1961). The chemical basis of the virulence of *Brucella abortus*. I. Isolation of *Br. abortus* from bovine foetal tissue. *Brit. J. exp. Path.* **42**, 631.

SMITH, H., KEPPIE, J., PEARCE, J. H. & WITT, K. (1962a). The chemical basis of the virulence of *Brucella abortus*. IV. Immunogenic products from *Brucella abortus* grown *in vivo* and *in vitro*. *Brit. J. exp. Path.* **43**, 538.

SMITH, H., KEPPIE, J. & STANLEY, J. L. (1953). A method for collecting bacteria and their products from infections in experimental animals, with special reference to *Bacillus anthracis*. *Brit. J. exp. Path.* **34**, 471.

SMITH, H., KEPPIE, J. & STANLEY, J. L. (1955). The chemical basis of the virulence of *Bacillus anthracis*. V. The specific toxin produced by *B. anthracis in vivo*. *Brit. J. exp. Path.* **36**, 460.

SMITH, H. & TEMPEST, D. W. (1957). The uptake of amino acids during the terminal bacteraemia in guinea pigs infected with *Bacillus anthracis*. *J. gen. Microbiol.* **17**, 731.

SMITH, H., WILLIAMS, A. E., PEARCE, J. H., KEPPIE, J., HARRIS-SMITH, P. W., FITZGEORGE, R. B. & WITT, K. (1962b). Foetal erythritol: A cause of the localisation of *Brucella abortus* in bovine contagious abortion. *Nature, Lond.* **193**, 47.

SPINK, W. W., HALL, J. W., FINSTAD, J. & MALLET, E. (1962). Immunisation with viable brucella organisms: Results of a safety test in humans. *Bull. World Hlth Org.* **26**, 409.

SPIVACK, M. L., FOSTER, L., LARSEN, A., CHEN, T. H., BAKER, E. E. & MEYER, K. F. (1958). The immune response of the guinea pig to the antigens of *Pasteurella pestis*. *J. Immunol.* **80**, 132.

STANLEY, J. L. & SMITH, H. (1961). Purification of factor I and recognition of a third factor of the anthrax toxin. *J. gen. Microbiol.* **26**, 49.

STANLEY, J. L. & SMITH, H. (1963). The three factors of anthrax toxin: Their immunogenicity and lack of demonstrable enzymic activity. *J. gen. Microbiol.* **31**, 329.

STINEBRING, W. R. (1962). Characteristics of intracellularly grown *Brucella abortus*. *J. infect. Dis.* **111**, 17.

STINEBRING, W. R., BRAUN, W. & POMALES-LEBRÓN, A. (1960). Modified serum resistance of bacteria following intracellular residence. *Ann. N.Y. Acad. Sci.* **88**, 1230.

STOCKER, B. A. D., ZINDER, N. D. & LEDERBERG, J. (1953). Transduction of flagellar characters in salmonella. *J. gen. Microbiol.* **9**, 410.

SUTER, E. & HULLIGER, L. (1960). Non-specific and specific cellular reactions to infections. *Ann. N.Y. Acad. Sci.* **88**, 1237.

SWIM, H. E. (1959). Microbiological aspects of tissue culture. *Annu. Rev. Microbiol.* **13**, 141.

TAYLOR, A. W. & MCDIARMID, A. (1949). The stability of the avirulent character of *Brucella abortus*, strain 19 and strain 45/20 in lactating and pregnant cows. *Vet. Rec.* **61**, 317.

TAYLOR, J., MALTBY, M. P. & PAYNE, J. M. (1958). Factors influencing the response of ligated rabbit-gut segments to injected *Escherichia coli*. *J. Path. Bact.* **76**, 491.

TEMPEST, D. W. & SMITH, H. (1957). The effect of metabolite analogues on growth of *Bacillus anthracis* in the guinea pig and on the formation of virulence-determining factors. *J. gen. Microbiol.* **17**, 739.

THEILER, M. & SMITH, H. H. (1937). The effect of prolonged cultivation *in vitro* upon the pathogenicity of yellow fever virus. *J. exp. Med.* **65**, 767.

TISELIUS, A., PORATH, J. & ALBERTSSON, P. (1963). Separation and fractionation of macromolecules and particles. *Science*, **141**, 13.

TROWELL, O. A. (1959). The culture of mature organs in a synthetic medium. *Exp. Cell Res.* **16**, 118.

VAN HEYNINGEN, W. E. (1955). The role of toxins in pathology. In *Mechanisms of Microbial Pathogenicity. Symp. Soc. gen. Microbiol.* **5**, 17.

VON BERTALANFFY, L. (1962). General system theory—a critical review. *General Systems*, **7**, 1.

VON RÁSKA, K. & ROTTA, J. (1956). Die persistenz von streptokokken der gruppe A nach intranasaler infektion. *Schweiz. Z. Path.* **19**, 356.

WATKINS, H. M. S. (1960). Some attributes of virulence in Shigella. *Ann. N.Y. Acad. Sci.* **88**, 1167.

WATSON, D. W. (1960). Host–parasite factors in group A streptococcal infections: pyrogenic and other effects of immunologic distinct exotoxins related to scarlet fever toxins. *J. exp. Med.* **111**, 255.

WIENER, N. (1961). *Cybernetics*, 2nd edn. New York: Wiley and Sons Inc.

WILLIAMS, A. E., KEPPIE, J. & SMITH, H. (1962). The chemical basis of the virulence of *Brucella abortus*. III. Foetal erythritol, a cause of the localization of *Brucella abortus* in pregnant cows. *Brit. J. exp. Path.* **43**, 530.

WINOGRADSKY, S. (1949). *Microbiologie du Sol*. Paris: Masson et Cie.

WOOD, W. B., SMITH, M. R. & WATSON, B. (1946). Studies on the mechanism of recovery in pneumococcal pneumonia. IV. The mechanism of phagocytosis in the absence of antibody. *J. exp. Med.* **84**, 387.

WOODS, D. D. & TUCKER, R. G. (1958). The relation of strategy to tactics: Some general biochemical principles. In *The Strategy of Chemotherapy. Symp. Soc. gen. Microbiol.* **8**, 1.

METABOLIC CONSIDERATIONS RELATING TO THE LIFE OF BACTERIA *IN VIVO*

D. D. WOODS AND M. A. FOSTER

Microbiology Unit, Department of Biochemistry, University of Oxford

The purpose of this review is to attempt to assess the possible significance of present knowledge of bacterial metabolism, almost entirely derived from studies *in vitro*, to an understanding of the life of bacteria in association with an animal or plant. It is not the intention, nor is it considered to be profitable at this stage, to examine in detail such disconnected and fragmentary information as is available concerning the metabolism of bacteria actually within the host tissues or of bacteria abstracted in bulk from host tissues. Indeed it will be shown that in the latter case the information might well be misleading and irrelevant, and this further reduces the field which it might otherwise be worth while to survey. We do not wish to be in the position of the White Knight, an inventive gentleman in other respects:

> I'll tell thee everything I can;
> There's little to relate.*

However, increasing knowledge of the detailed mechanism of bacterial metabolism, and particularly how it may adjust itself to experimental changes in the environment, make it possible to predict some of the factors which may influence for better or worse the life *in vivo* of a bacterium; it is mainly these matters that will be discussed. Unfortunately it is only possible in almost every case to say 'may influence'; a more positive statement demands a knowledge in fine and quantitative detail of the chemical nature, especially with regard to organic components, of the environment in which the bacterium would find itself *in vivo*. Such information is in most cases not available nor can it readily be obtained until suitable micro-methods are devised. The gross composition of certain body fluids (e.g. blood, urine, cerebrospinal fluid) is known in some animals, but not in terms of the steady-state concentrations of free amino acids, nucleic acid derivatives and energy sources that may influence the metabolic activities of the bacterium. And even less is known in these respects of intracellular fluids, interstitial fluid between cells, the fluid phase of the mucosa of the throat and lungs and so on—all places where infection may be initiated.

* *Alice Through the Looking Glass*, by Lewis Carrol.

At least two phases of life *in vivo* ought to be borne in mind, in which the biochemical problems may differ substantially. The first is the initiation and establishment of infection, perhaps by a single organism; this is the most important phase and, unfortunately, the aspect which can be least predicted from studies *in vitro*, since most metabolic work is done with comparatively large numbers of organisms where individual variations are averaged out. It may well be the (temporarily) atypical organism which succeeds initially *in vivo*. The second phase is the life of the bacterium once the infection is established. Here the position is more hopeful, although it must be remembered that bacteria which become 'locked away' (e.g. in a tubercle or gall) may to a considerable extent then be creating their own environment.

Any suggestions emanating from work *in vitro* about biochemical factors which may influence life *in vivo* must be largely speculative and cannot be critically assessed at the present time. Nevertheless, the exercise should be made since it will at least pin-point some of the matters concerning which it would now be worth while to seek further factual information. It will not be possible to give chapter and verse to substantiate every statement since it will be necessary to draw on general knowledge ranging over the whole field of the biochemistry and nutrition of micro-organisms; in general, only references to reviews covering certain fields will be quoted.

GENERAL CONDITIONS OF LIFE *IN VIVO*

Continuous or batch culture?

When the host is complex, and has differentiated tissues, it will have some form of circulatory or translocatory system. The bacterium will therefore be in an environment which is constantly renewed in the sense that there will be, due to the activities of the host, both a continual replenishment of growth metabolites and a continual removal of waste products of metabolism. The situation is essentially similar for organisms living in an intestinal tract. The system is therefore analogous in many ways to continuous culture and less like the batch cultures that are the source of so much of the bacterial material used for metabolic studies *in vitro*. When infection is confined to a particular tissue, a proportion of the bacteria themselves are not simultaneously removed, as in continuous culture, and the position is half-way between batch and continuous cultures; it is more comparable to a culture of bacteria within a semi-permeable membrane in a continuous flow of medium. Again the two phases of infection mentioned above may have to be distinguished, since the conditions during the initiation phase in a restricted locality may more closely resemble batch culture.

Although relatively little work has been done on the metabolism of bacteria growing under conditions of continuous or quasi-continuous culture, it is clear that low concentrations of metabolites are used efficiently and need not be present at the higher initial concentrations usually needed for the batch culture of a similar quantity of new bacteria. In this sense such conditions are an advantage to the bacterium. But in any attempt to assess the magnitude of such an effect in terms of the composition of tissue fluid it must be borne in mind that once the infection is established the steady-state environment can no longer be equated exactly to that of the uninfected host since the high metabolic activity of the microbe will itself modify it. Continuous culture technique is reviewed by Řičica (1958).

Aerobiosis

The host, be it animal or plant, is overall a strictly aerobic organism and has efficient mechanisms for the transport of O_2 to all tissues. Oxygen will therefore normally be available, with certain exceptions mentioned later, to bacteria living *in vivo*, though it does not follow that all of them can, or do, use it. The steady-state O_2 tension in an actively metabolizing tissue (e.g. liver, muscle) or in certain body fluids (e.g. cerebrospinal fluid) may be quite low; nevertheless, O_2 will be there and the bacteria, with their high metabolic rate, may compete successfully for it. Many bacteria are facultative aerobes; the aerobic processes produce far more utilizable energy per unit of metabolizable substrate than do the anaerobic processes; this is reflected in the higher growth yield with increasing degree of aerobiosis. The capacity to use available O_2 to advantage is therefore present both in facultative and strict aerobes; with the facultative organisms the infection can be maintained or slowly increased until more O_2 becomes available as the result, for example, of the spread of the organism to a different and more aerobic location.

The growth mechanism of some bacteria (e.g. streptococci) is essentially anaerobic but they can tolerate O_2 even though they do not effectively use it. The strict anaerobes which can tolerate little, if any, O_2 present a different problem. Here the difficulty for life *in vivo* is the initiation of infection; once established these organisms themselves excrete metabolic products which are highly reducing and create the necessary anaerobic environment. It is perhaps noteworthy that infections by strict anaerobes (e.g. gangrene) commonly originate in tissues where there is failure in the circulatory system, and therefore in O_2 supply, or in wounds where again tissue damage has interrupted the circulation. An interesting example of a change in the micro-environment permitting the establishment of *Clostridium tetani* was described by Fildes (1929). He concluded that 'A slight inflammatory process is brought about by the combined action of bacteria of lower virulence and foreign particles or bodies: this inflammation results in a vicious circle, i.e. some degree of capillary thrombosis and hindrance in oxygen diffusion, and increased local demand for oxygen by leucocytes and bacteria. These conditions induce local asphyxia....'

The massive bacterial populations which exist normally in healthy hosts in, for example, the lower intestinal tract and the rumen of animals, are also living under essentially anaerobic conditions. The particular problems involved in such borderline cases of bacteria living *in vivo* and their contribution to the host's economy are outside the scope of the present article; they were dealt with in last year's Symposium of this Society on *Symbiotic Associations* (Nutman & Mosse, 1963).

Carbon dioxide

Associated with the aerobic life of the plant or animal host is the production of CO_2 as the main end product of energy metabolism. The bacterium will always therefore find itself in an environment in which CO_2 is available. This may be of importance especially from the point of view of the initiation of an infection. Some bacteria require CO_2 as an essential growth factor; it is often a routine in studies of bacterial nutrition to incubate in a gas phase containing 5 % CO_2. With other bacteria CO_2 is not essential, but either promotes earlier growth (i.e. shortens the lag phase) or is needed for growth from small inocula. It appears to act as a 'starter' and to be required for the initial formation of intermediates in the tricarboxylic acid cycle, i.e. the main, and catalytic, pathway in aerobic energy yielding reactions; once instituted this pathway is itself the principal immediate source of the CO_2 of aerobic respiration.

NUTRITIONAL CONSIDERATIONS

Bacteria vary widely in their competence to synthesize for themselves various organic compounds required as structural units or catalysts in the formation of new cell material. These include amino acids, nucleic acid derivatives (e.g. purines and pyrimidines), B-group vitamins, i.e. components of the structure of coenzymes or prosthetic groups of enzymes, and several other substances whose cellular function is not yet clear. There is a recent review by Guirard & Snell (1962). These substances belong to the class usually called essential metabolites; when an organism is unable to achieve the synthesis of one or more of them they become required growth factors. Individual species differ greatly in the number and type of growth factors required, from none at all (e.g. *Escherichia coli*) to every amino acid plus most B-group vitamins and nucleic acid derivatives (e.g. *Streptococcus equinus*). In a broad consideration of all infecting organisms in relation to every infected animal or plant tissue it is not possible to make any generalization that the kinds of organism found most *in vivo* are those with the most exacting nutritional requirements.

There is a basic similarity in fundamental metabolic reactions of all types of living organisms and it therefore follows, and has been demonstrated in practice, that the essential metabolites of bacteria are in general the same as those of the infected host. It is thus commonly assumed that the bacterium will find available in the host tissues any growth factor which it may require. This is not necessarily the case. It was pointed out earlier that the steady-state concentration in tissues of free amino acids, purines and pyrimidines, etc., is usually not known. They may, with some organisms, be liberated from host protein or nucleic acid by extracellular enzymes of the bacterium (see next section). In other cases the metabolite may be present in host tissue in a form not utilizable by the bacterium due to impermeability or other causes. For example, pantothenic acid, a B-group vitamin essential for many bacteria, is present in animal tissues mainly in its more elaborate functional coenzyme form (coenzyme A) and this supports the growth of very few pantothenate-requiring bacteria. Similarly, the form in which folic acid exists in host tissues will not support the growth of many bacteria requiring its metabolic precursor, *p*-aminobenzoic acid; indeed the chemotherapeutic success of the sulphonamides depends on this very fact (Woods, 1962).

The presence specifically in a given tissue of a growth factor required or used preferentially by a particular organism could in theory account for the high tissue specificity found in many infections. An interesting example where this has been proved to be the case will be described in another contribution to this Symposium.

One further point requires emphasis and is of importance particularly in the context of initiation of infection. Work on the nutritional requirements of bacteria *in vitro* has been concentrated for the most part on growth factors for which the need is absolute, and growth in many cases does not reach the density of that obtained on a complex, chemically undefined medium. There has been less attention to factors which, though not essential, increase either the rate or extent of growth. When they have been studied (e.g. the peptide-like strepogenin in the case of *Streptococcus haemolyticus*) they have proved difficult to define. This field would repay further study as would also the possibility, for which indications have arisen from time to time, that growth *in vitro* from really small inocula (e.g. of the order of 1–10 organisms) may require additional or special factors; a possible role of CO_2 has been mentioned earlier.

ENZYMIC CONSIDERATIONS
Exocellular enzymes

The presence of high molecular weight compounds in the host tissue is a characteristic which is shared by all living organisms and as was mentioned above these materials might be expected to provide an organism *in vivo* with a rich source of energy and raw materials for growth. However, bacteria do not appear to possess any mechanism for the ingestion of such substances comparable with phagocytosis or pinocytosis in cells of higher organisms. Active uptake by bacteria is normally restricted to relatively small molecules although it is known that, usually under highly specific conditions, native deoxyribonucleic acid can gain access. The production of either extracellular or surface-bound enzymes for the preliminary degradation of macromolecules would be of considerable advantage to an organism in a host tissue. A wide variety of bacterial exoenzymes degrading proteins, polysaccharides, nucleic acids and lipids has been described from studies *in vitro*. The subject is reviewed by Rogers (1961) and Pollock (1962). The frequency of production of such enzymes by pathogens in particular strongly suggests that they play an important part in the establishment and maintenance of infection.

In many cases there are difficulties in the formal demonstration of the true extracellular nature of such enzymes, particularly in distinguishing them from enzymes released on autolysis. However, in the present context, this distinction is mainly an academic one since the autolytic production of enzyme from a small proportion of the bacterial population would confer an advantage comparable with true extracellular production by the entire population. Since the translocation systems of the host will tend to remove an extracellular enzyme from the site of infection (a situation not encountered in studies *in vitro*) it might appear more advantageous for the bacterium to produce surface-bound enzymes for the handling of macromolecules; on the other hand, this very dispersion may well be important in establishment of infection, as in the production of extracellular enzymes (in this case toxic ones) by *Corynebacterium diphtheriae* and *Clostridium welchii*.

Also of importance in considering the effect of exoenzymes *in vivo* as opposed to *in vitro* is the possibility of synergism of the bacterial exo-enzyme with host enzymes released on autolysis; the production of a single exoenzyme by the microbe might, for example, cause lysis of host cells with release of enzymes degrading several types of macromolecule. Studies *in vitro* show that the production of exoenzymes varies with environment, but since the available information is fragmentary and often incidental to other work, it is difficult to generalize about the factors likely to control exoenzyme formation, release or activity *in vitro* and *in vivo*. However, there is strong evidence that the formation of some exoenzymes is induced by the substrate or related substances; this in itself raises the interesting question, not yet answered even in studies *in vitro*, of the mechanism of induction by substances which do not penetrate the bacterium.

Inhibition and repression of enzymes

One of the most striking and important advances during the past decade in our understanding of the biochemical basis of bacterial growth has been the discovery and elucidation of at least some of the mechanisms whereby bacteria automatically control the rate of their own metabolic processes, especially those associated with biosynthesis. If a micro-organism is to make the most efficient use of raw materials present in the environment, the individual metabolic pathways must be so regulated that the right amount of a given product is available at the right time for reaction with similarly regulated products of other path-ways. Overproduction of any product would be wasteful both of energy-yielding substrates and of any specific raw material. Thus for maximum efficiency a bacterium competent to synthesize all amino acids needs to produce them at different rates reflecting the amino acid composition of its own protein. Furthermore, when a bacterium can synthesize a given substance it would benefit the overall economy of growth if the synthesis of this substance ceased if it were already avail-able in the environment and could permeate the organism.

Two mechanisms whereby the end-product of a synthetic pathway controls the rate of its own synthesis are now known and have been demonstrated for a variety of products, especially amino acids and nucleic acid derivatives, although they have been studied with relatively few bacteria. They are illustrated schematically in Fig. 1 and have been reviewed by Vogel (1957) and Pardee (1959). Briefly, in one mechanism, usually referred to as enzymic repression, the end-product prevents the *formation* of an enzyme or enzymes catalysing earlier steps in the pathway. The second mechanism, often called feed-back inhibition, depends on the inhibition by the end-product of the activity of an enzyme catalysing an early step. Both types of control may operate in the same pathway and indeed in a number of cases in relation to the same enzyme. In both types also the enzyme affected includes the one catalysing the first step of the independent pathway leading to the product, i.e. the control is not normally exerted, though there are exceptions, on the formation of intermediates which are also required for the synthesis of other products (e.g. R in Fig. 1). Later enzymes may also be affected especi-

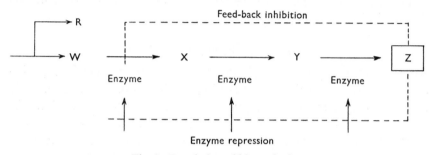

Fig. 1. Regulation of biosynthesis.

ally in enzymic repression. The maximum conservation of resources is thus achieved, and when a product is either already available in the environment in sufficient concentration, or is being formed by the bacterium in excess of requirement, further synthesis will be diminished or cease.

The effect of enzymic repression will not be immediate since, although new enzyme will not be formed, existing enzyme will continue to function; it has been suggested that this type of control serves to adjust the enzymic constitution of the bacterium to the general composition of any new environment in which it may find itself. Inhibition of enzyme activity will however be immediate and would reflect moment to moment variation in the concentration of the product due to changes either in its rate of utilization or formation. There have recently been some cases in which enzymes are actively and selectively removed in response to changes in the environment. When this occurs with a repressible enzyme, this method of control will also be rapidly effective.

Control of certain catabolic energy-producing reactions is achieved in a slightly different and less specific manner. The presence, for example, of glucose or other utilizable energy source leads to suppression of the formation of bacterial enzymes catabolizing amino acids (as energy sources) and thus conserves them for use in protein synthesis.

In general the newly discovered control mechanisms would appear to favour life of bacteria *in vivo*, since metabolic activities no longer required because of the presence of their products in the infected tissue can be temporarily suppressed and the raw materials and energy so saved diverted to fortify other pathways whose product is not available. The presence of particular metabolites in higher concentration in specific tissues might thus contribute to the tissue specificity shown in many infections by leading indirectly to more efficient growth of a given species, which normally makes these metabolites for itself, in that tissue. Implications with regard to bacteria extracted in bulk from infected tissue will be considered in a later section.

Induced enzyme formation

Another much-studied and widespread character of bacteria, especially certain organisms, which may be of importance in the present discussion is the ability of some species to produce new enzymes (i.e. inducible enzymes) in response to the presence of the substrate in the environment. The phenomenon, like enzyme repression, is a purely phenotypic one and the new enzyme ceases to be formed immediately the inducing substrate is withdrawn; it is to be clearly distinguished from the selection, by the presence of the substrate, of spontaneous mutants having a new enzymic ability. There is a good review by Pardee (1962). A bacterium *in vivo* may therefore be able to form enzymes to utilize substrates which were not present in its environment before it entered the host. It is possible that induced enzyme formation may help to explain tissue specificity of infection. Clearly, if a tissue contains a novel component, or is rich in a general tissue component, and a particular bacterium can respond to its presence by forming an enzyme which can metabolize it usefully, then that bacterium will have a unique advantage in such a tissue. Unusual compounds may well be found also in traumatic or already diseased tissues due to autolysis or degradation.

CONSIDERATIONS OF BIOCHEMICAL GENETICS

The striking modern advances in this field of microbiology, especially in relation to the mechanism whereby genetic material (deoxyribonucleic acid) directs the production of specific enzyme proteins and thus ultimately controls every phase of metabolism, does not unfortunately at the present state of knowledge in the other relevant fields appear to have any direct implications for the life of bacteria *in vivo*, as opposed to their life *in vitro*. It is clear that fortuitous mutants which have a changed enzymic constitution more suited for growth in the new environment of the host tissues would tend to be selected. The conception that one gene or region of the DNA strand directs the production of a single enzyme makes it clear that any overall changes to suit the general nature of the environment would require a succession of mutations, each of the right type; the known probability of this is very slight and adjustment of this kind is more likely to occur by the phenotypic mechanisms outlined in the last main section. However, mutation at a single locus could be advantageous if it resulted in the production, for example, of an exoenzyme attacking a host macromolecule or in the deletion of a permease permitting the entry into the bacterium of a host metabolite toxic to it. Furthermore, as already discussed for phenotypic enzyme induction, the acquisition of an enzyme, in this case as a result of a genotypic change, able to attack some substance qualitatively or quantitatively unique to a particular tissue may well contribute to tissue specificity.

Assuming that a mutant with a metabolic advantage for growth *in vivo* arises within a population of infecting organisms, modern work on microbial genetics suggests the possibility that it may become predominant more rapidly than selection pressure alone would indicate. Not only will the progeny of the mutant bacterium itself carry the new property, but it may also be transferred to other bacteria of the original population, which will then in turn breed true for the new character and again transfer it. Such transference is known to take place with various bacteria under specific conditions by at least three mechanisms: transformation, conjugation and transduction (Hotchkiss, 1955; Jacob & Wollman, 1961).

In transformation the DNA of the donor bacterium is released by death and lysis; a small segment controlling a particular character permeates the recipient organism and presumably recombines with the genome. The phenomenon was first discovered as a consequence of experiments *in vivo* (Griffith, 1928); the simultaneous injection into mice of heat-killed virulent pneumococci type III and living avirulent pneumococci type II led to the death of the animal and the isolation from it of living, virulent type III organisms. The material transferred in this case is the fragment of DNA controlling the formation of an enzyme catalysing the synthesis of a specific polysaccharide which forms the capsule of the virulent type III organism. Many different characters, in some cases known to be dependent on the formation of a single enzyme, are now known to be transmissible by this mechanism in a variety of bacteria.

Conjugation has been described so far in very few bacteria, i.e. some strains of *Escherichia coli*, *Salmonella typhimurium*, *Serratia marcescens* and *Pseudomonas aeruginosa*. In this type of genetic transfer a portion (or even the whole) of the genome of the donor organism passes into the recipient. In transduction, which is dependent on the concurrent presence of a bacteriophage, a small segment only of the DNA of the donor DNA becomes associated with the phage DNA and passes into the recipient bacterium with the phage at the time of infection. In both conjugation and transduction the transferred material is integrated with the recipient DNA and a wide variety of characters, in many cases known to be due to the activity of a single enzyme, have been interchanged between bacteria of the same species and even between different species (e.g. *E. coli* K_{12} and *Shigella* spp.). Transduction has been mainly studied in relatively few bacteria (*E. coli*, *S. typhimurium* and *Bacillus subtilis*) although transducing phages of other species are known (e.g. *Staphylococcus aureus*, *P. aeruginosa* and *Proteus mirabilis*).

It is as yet difficult to assess what role direct transfer of genetic material may play in bacterial life *in vivo*. More investigations *in vitro* need to be made, particularly of the possibility of conjugation and transduction, with a wide variety of microbes known to be able to live in plants or animals. Simultaneous infection of a host with mutants of the same species bearing different genetic markers and the analysis of re-isolated organisms for recombinants is clearly also a valuable approach. Since transduction has been described in *Salmonella*, and since sewage and thus possibly faeces are a rich source of bacteriophages, it is conceivable that this phenomenon might be important for the more rapid selection of mutants of this organism suited to life in the intestinal tract and thus contribute to its pathogenicity.

BACTERIA ABSTRACTED FROM THE HOST

It would appear at first sight that valuable information about the metabolism *in vivo* of bacteria would be obtained by direct studies of their biochemical properties when abstracted in bulk from host tissues and not further cultivated before test. Dr H. Smith will no doubt comment in his contribution on whether the deliberate massive infection of a particular tissue for experimental purposes will yield organisms truly comparable with those normally found *in vivo*. Furthermore, as has been pointed out by Smith (1958), the methods used to free the bacteria from host material are sometimes drastic and may affect especially the bacterial surface. Serious objections also arise from the matters considered in earlier sections of the present review, particularly those bearing on phenotypic changes induced by the environment. The moment the bacterium is removed from the host its environment will change and its enzymic constitution may be modified. Enzymes whose formation has been repressed by their end-product will at once start to be synthesized again and enzymes whose formation has been induced by the presence of their substrate will cease to be formed. If it were technically possible, and this seems doubtful, valuable information could be obtained from a study of the time course of appearance or disappearance of selected enzymic activities immediately after the collection of the bacteria; this should indicate whether the enzyme has been induced or held repressed *in vivo* and thus provide evidence as to whether the substrate or end-product has been present at an effective concentration in the host tissue. Such studies would demand that possible changes in the bacteria should be stopped (e.g. by low temperature) while the bacteria were still in the host and that subsequent manipulations up to the time of the enzymic test should all be at low temperature. Alternatively, the use during the isolation procedures of a specific inhibitor of protein synthesis subsequently removed might give clear-cut results.

The situation would be different, and possibly more hopeful, in the case where the host environment had selected a mutant. The changed potential enzymic constitution reflecting the genetic change would not alter when the special environment was removed; a careful comparison of the metabolism and nutrition of the abstracted organism with that of the parent organism used for infection might, in purely experimental infections, give useful results. However, in most natural infections no such parent strain will be available for comparison since the bacterium passes directly from one host to another. Furthermore, any differences might be masked by the phenotypic changes noted above; it must be borne in mind that the mutation may well be not to the straightforward production of a new enzyme, but to the ability to produce it if a repressor is absent or an inducer present.

GENERAL COMMENT

It will be realized that in a wide survey of this type it has been necessary to over-simplify to some extent and at times to make dogmatic statements for the sake of clarity. Certainly it would be possible to criticize everything that has been said in the light of some particular observation or other in a specific case. Indeed, this is the crux and inherent difficulty of the whole matter, for every bacterial infection of, or association with, a host tissue represents a different biological situation which ought to be considered as a separate biochemical entity in which the metabolic activities of host and bacterium are interwoven if not integrated.

It is difficult to draw any significant conclusions from the present survey. On the whole, modern knowledge of the metabolism of bacteria and its control tends to suggest that, at any rate once they are established, bacteria ought to derive advantage from the living environment. From considerations of metabolism alone it is perhaps even tempting to inquire why more bacteria do not take advantage of such facilities; no doubt contributors viewing the matter from other aspects will tell us why. This conclusion does not however pin-point the critical importance of the first stage of life *in vivo*, the establishment of the infection. There is a strong case for more biochemical investigations *in vitro* of factors which influence growth from very small inocula.

This article began, or almost began, with 'Alice'; a final reading of the manuscript suggests that it is more in tune with the last four lines of Hilaire Belloc on 'The Microbe'

> But Scientists, who ought to know,
> Assure us that they must be so...
> Oh! let us never, never doubt
> What nobody is sure about!

REFERENCES

FILDES, P. (1929). Bacillus tetani. In *A System of Bacteriology*, p. 298. London: Medical Research Council.

GRIFFITH, F. (1928). The significance of pneumococcal types. *J. Hyg., Camb.* **27**, 113.

GUIRARD, B. M. & SNELL, E. E. (1962). Nutritional requirements of micro-organisms. In *The Bacteria*, **4**, 33. Ed. I. C. Gunsalus & R. Y. Stanier. New York: Academic Press.

HOTCHKISS, R. D. (1955). Bacterial transformation. *J. cell. comp. Physiol.* **45**, Suppl. 2, 1.

JACOB, F. & WOLLMAN, F. L. (1961). *Sexuality and the Genetics of Bacteria.* London: Academic Press.

NUTMAN, P. S. & MOSSE, B. (editors) (1963). *Symbiotic Association. Symp. Soc. gen. Microbiol.* **13**. Cambridge University Press.

PARDEE, A. B. (1959). Mechanisms for control of enzyme synthesis and enzyme activity in bacteria. In *Regulation of Cell Metabolism*, p. 295. Ed. G. E. W. Wolstenholme & C. M. O'Connor. London: Churchill.

PARDEE, A. B. (1962). The synthesis of enzymes. In *The Bacteria*, **3**, 577. Ed. I. C. Gunsalus & R. Y. Stanier. New York: Academic Press.

POLLOCK, M. R. (1962). Exoenzymes. In *The Bacteria*, **4**, 121. Ed. I. C. Gunsalus & R. Y. Stanier. New York: Academic Press.

ŘIČICA, J. (1958). Continuous culture techniques. In *Continuous Cultivation of Micro-organisms: a Symposium*, p. 75. Ed. I. Malek. Prague: Czechoslovak Academy of Sciences.

ROGERS, H. J. (1961). The dissimilation of high molecular weight substances. In *The Bacteria*, **2**, 257. Ed. I. C. Gunsalus & R. Y. Stanier. New York: Academic Press.

SMITH, H. (1958). The use of bacteria grown *in vivo* for studies on the basis of their pathogenicity. *Annu. Rev. Microbiol.* **12**, 77.

VOGEL, H. J. (1957). Repression and induction as control mechanisms of enzyme biogenesis: the 'adaptive' formation of acetylornithinase. In *The Chemical Basis of Heredity*, p. 276. Ed. W. D. McElroy & B. Glass. Baltimore: The Johns Hopkins Press.

WOODS, D. D. (1962). The biochemical mode of action of the sulphonamide drugs. *J. gen. Microbiol.* **29**, 687.

HOST AND TISSUE SPECIFICITY

J. KEPPIE

Microbiological Research Establishment, Porton, Wiltshire

Many diseases of microbial origin are host-specific and this is particularly true of those diseases affecting man. Another feature of infectious disease is the striking tendency of micro-organisms to invade specific cells or to localize in particular organs of the host. Both phenomena are so little understood that they constitute a major challenge to those investigating the pathogenesis of infectious diseases.

This Symposium seeks to throw light on microbial behaviour *in vivo*, and my task is to examine the reasons for host and tissue specificity and to assess how far experiments *in vitro* have helped to explain them. The twin problems in disease specificity have only been investigated in a few instances, and the earlier part of this review consists of little more than a general discussion of the occurrence of the phenomena and of speculation on what the key factors might be. Most of the material has been selected from studies of bacterial infections together with a few examples from virology. Recently, sufficient experimental data to enable us to reach some definite conclusions have accumulated on two topics, namely: the role of urease in the localization of bacteria in the kidney, and the basis of host and tissue specificity in brucellosis. These investigations have been described separately in the two final sections because they illustrate how realistically designed experiments *in vitro*, based on and complementary to experiments *in vivo*, can yield information of value in explaining host and tissue specificity in microbial disease.

HOST SPECIFICITY

In infectious disease 'host specificity' implies that individual micro-organisms only produce disease in a single species of host. This concept may have become too widely accepted merely because some pathogens such as pneumococci, meningococci, shigellae, typhoid, diphtheria and leprosy bacilli, and poliomyelitis virus appear to be restricted to man, whereas others, such as *Mycobacterium johnei*, *Vibrio foetus*, or the virus of foot and mouth disease seem only to infect cattle and related species. However, the problem is not as simple as this. Apart from the fact that comprehensive attempts to infect other animals with many of the above-mentioned pathogens have never been made, many pathogens show no specificity, but cause disease in a wide range of host-species. Among the many organisms in the latter group are those causing tularaemia, leptospirosis, tuberculosis, listeriosis, anthrax, actino-mycosis, plague, salmonellosis, and psittacosis.

This variation from the apparent host specificity of some organisms to the complete non-specificity of others, is almost to be expected since the host-parasite relationships recognizable today are the outcome of a complex evolutionary process. They have been influenced by the degree of contact between man, domestic animals, and wild life (and thus between their respective microbiological flora). This contact has been diminishing since man instituted the earliest rules of hygiene. However, as Meyer (1948) has indicated in his article entitled 'The Animal Kingdom, a Reservoir of Human Disease' a number of 'infec-tion chains' between the various divisions of the animal kingdom re-main today. At least seventy-five diseases of domestic and wild animals are of potential public health significance. The infective agents may pass to man from the lower animals by direct or indirect contact, by the handling and ingestion of infected animal food, or through the inter-vention of insect vectors. These interactions have produced pathogenic organisms some of which only occur in man, others which only occur in a single animal species, and also those which infect and cause disease in a wide range of host-species. An extreme type of host specificity has been demonstrated in laboratory animals where the hereditary con-stitution of strains within a host-species determines whether or not the animals are resistant to a pathogen. Thus, Lurie, Abramson & Hepples-ton (1952) exposed two strains of rabbits to a respiratory challenge of human tubercle bacilli and in order to produce a comparable degree of lung infection, the challenge given to resistant rabbits had to be 16 times greater than that given to the susceptible ones. Chandler (1961, 1962) found that C 57 black mice were more susceptible to oral or intravenous *Mycobacterium johnei* than were Swiss white or CBA brindle strains. Strains of mice also vary in susceptibility to infection with *Salmonella typhimurium* and some of the factors influencing this variation in resistance were described by Gowen (1960). These results, and others reviewed by Wilson & Miles (1955), indicate clearly that susceptibility to bacterial and viral infections may be genetically determined within a species, thus rendering the term host specificity meaningless unless details of the heredity of the host and the properties of the infective agent are fully considered.

In summary, although the host-range of many micro-organisms is wide, the range of others has become narrowed down to a single species; it is with the latter micro-organisms that this section is concerned. The mechanisms which determine host specificity of particular pathogens are not known, but some speculation about them is possible. To establish itself successfully, a pathogen requires: (1) to overcome the host's defences, and (2) to find an adequate nutritional environment in the host. Different host species could show differential susceptibility to a pathogen by virtue of differences in one or both of these factors.

The host defences in different species

The many defence systems of the body, such as mucous secretions, serum bactericidin, free and fixed phagocytes and fever, have been studied in several species both *in vivo* and *in vitro* in an unsuccessful search for differences between the species which would account for their individual susceptibilities to micro-organisms (see Housewright, 1960). Differences related to species have been found in some mechanisms, such as surface phagocytosis (Wood, 1960), non-specific opsonins (Rowley, 1960) and blood clearance mechanisms (Rogers, 1960). However, such differences, either singly or in concert, do not adequately explain the differing susceptibilities of hosts to a particular pathogen. If different genetic strains of the same host were used for such studies, this might yield information which would explain differences in the relative susceptibility of each strain.

Dubos (1954) reviewed work in which he and others had sought to mimic *in vitro* the conditions facing bacteria during inflammation in the body. Some of the physical and chemical changes which occurred in the host were identified, for example the production of lactic acid and ketone bodies; pathogens such as staphylococci and tubercle bacilli were shown to be sensitive to these compounds. However, we cannot be certain that these effects have any bearing on host specificity because there is still a dearth of comparative data from the different host species.

The occurrence in susceptible hosts of adequate nutrition for the pathogen

Studies of pathogens in artificial culture have shown that in addition to sources of nitrogen, carbon, etc., many organisms require factors, both inorganic and organic, for maximal growth and for the production of virulence factors. It is conceivable that the availability of such specific nutrients might determine host-parasite relations. Until recently there was no evidence linking a nutrient in a host's tissues with its susceptibility to a pathogen under natural conditions, but there were indications from some experimental work that such a relationship was possible. Pathogenic bacteria were induced to become demanding for particular nutrients by mutation. These artificially prepared organisms were used to challenge laboratory animals, and were found to be virulent or avirulent depending on whether or not the required meta-bolite was available in a host. In mice and guinea pigs, purines were not available, and purine-requiring strains of *Salmonella typhi* were aviru-lent in those two species, until parenteral injection of the missing purine enabled the mutant to multiply sufficiently to prove fatal (Bacon, Burrows & Yates, 1951). Similar results were obtained in mice with purine-requiring mutants of other bacteria (Dubos, 1954; Burrows, 1963) and there is one report of the isolation of a naturally occurring, purine-requiring mutant of *S. typhi* which was avirulent for mice (Formal, Baron & Spilman, 1954). Obviously, this work with artificially prepared mutants indicates that the nutritional conditions of the host can determine the host specificity of a particular pathogen. Studies on brucellosis to be described later have explained, on a nutritional basis, why the brucellae only cause acute placentitis in certain species belong-ing to the order *Artiodactyla* (i.e. cattle, sheep, goats, swine, and deer). The important nutrient for brucellae was erythritol, and a comparative survey of the placentae of susceptible and resistant species showed that the occurrence of erythritol paralleled the host's susceptibility to gross placental infection. This example of how a nutrient can play a decisive role in an infectious disease should encourage similar studies designed to relate the occurrence of a nutrient in a host to its susceptibility to a particular pathogen.

TISSUE SPECIFICITY

In a susceptible host, pathogenic micro-organisms often select a par-ticular tissue for their maximum growth and disease production. After initial penetration into the host, pathogens may establish themselves in the adjacent lymphoid tissue where multiplication is less than maximal and where tissue destruction is minimal. Later, a spread of the pathogen may occur in the lymph and blood, prior to the characteristic tissue localization. Some examples of localization in man are *Corynebacterium diphtheriae* in the throat, *Proteus* spp. in the kidney medulla, gonococci in the epithelium of the urinary tract, shigellae in the mucosa of the large intestine and poliomyelitis virus in the anterior horn cells of the spinal cord. Examples of localization in hosts other than man are: in sheep, louping-ill virus in the Purkinje cells, and *Fusiformis nodosum* in hoof tissue; in cattle, *Mycobacterium johnei* in the intestinal mucosa,

Corynebacterium renale in the kidney medulla and *Brucella abortus* in the placenta.

Some of the factors which might determine tissue specificity of a micro-organism would seem to be: (*a*) the route of entry to the body, (*b*) differences in the inhibitory activity of the tissues and (*c*) differences in the suitability of the tissues for the growth of the pathogen.

The route of entry to the body

Some diseases are localized near to the natural point of entry of the pathogen to the host. Examples are cholera, typhoid, and dysentery in the intestine, leprosy, ringworm, and staphylococcal infection in the skin, and influenza and pneumococcal pneumonia in the lungs.

However, Wilson & Miles (1955) emphasized that most bacteria tend to lodge and multiply at a site remote from their point of entry and that bacterial species differ sharply in their choice of tissues in which to localize. An example they chose from human medicine was the fate of several common organisms such as pneumococci, meningococci, diphtheria bacilli, haemolytic streptococci and influenza bacilli. All are transmitted by droplet infection but cause lesions as diverse in location and type as primary pneumococcal pneumonia, meningitis, diphtheria, streptococcal tonsillitis, and secondary pneumonia. Wilson (1957) described several further examples of tissue localization in which the sites had not been influenced by the route of entry of the pathogen: *Streptococcus viridans* in the endocardium, *Leptospira icterohaemorrhagiae* in the epithelial cells of kidney and liver, *S. pyogenes* type 12 in the kidney, and *Staphylococcus aureus* phage type 71 in impetigo of the skin. Some organisms selected epithelial tissues, others the interstitial cells; also some viruses favoured the cytoplasm and others the cell nucleus.

It would appear that the route of entry does not play a major role in tissue specificity.

Differences in the inhibitory activity of the tissues

Once an organism had gained a footing, and had acquired its armoury of invasins by growth *in vivo*, one might expect that defence mechanisms would no longer play a part in determining tissue localization. However, certain tissues may be more inhibitory than others to individual pathogens, and consideration must be given to the part such activities may play in the localization.

The subject of antimicrobial substances in the tissues was reviewed by Dubos (1954), Wilson & Miles (1955), Skarnes & Watson (1957), and Hirsch (1960). Bactericidal basic proteins or polypeptides have been isolated from extracts of spleen, brain, thymus, pancreas, and other mammalian organs. They have been found to be active against organisms as diverse as *Staphylococcus aureus, Bacillus anthracis, Escherichia coli* and *Streptococcus pyogenes* both *in vitro* and in some case *in vivo*; e.g. a synthetic polylysine reduced certain virus infections in the chick embryo (Green & Stahmann, 1953; Green, Stahmann & Rasmussen, 1953). Dubos, Hirsch and colleagues (see Hirsch, 1960) investigated the resistance of the kidneys of the guinea pig to tuberculosis by means of appropriate experiments *in vitro*. Spermine isolated from bovine kidney was not tuberculostatic *in vitro* unless activated by a specific enzyme—spermine oxidase—which occurred in guinea-pig kidney, but not in guinea-pig serum or rabbit kidney. The isolation and identification of the antimycobacterial product of the enzymic reaction was not possible because it was unstable in concentrated solutions, and spermine was too toxic in experimental animals for trials *in vivo*. Hirsch (1960) thought that the role of the spermine–spermine oxidase system in tissue resistance was a matter for speculation.

When seeking a reason for the vulnerability of the kidney to coliform infection, Beeson & Rowley (1959) found that the bactericidal power of normal human or rabbit serum for *Escherichia coli* was diminished by suspensions of kidney from several species of laboratory animals but not by their other tissues. The loss of activity by the serum was due to an anti-complementary effect, and this was 5–15 fold stronger in rabbit kidney than in other rabbit tissues. Similarly, kidney tissue of man, calf, ox and pig were all anti-bactericidal. The active material was associated with the tissue particles, and inactivated the fourth component of complement; it was involved in the production of ammonia, which is known to be anti-complementary, and it may have been renal glutaminase. Beeson & Rowley postulated that the local inactivation of complement in kidney tissue could allow *E. coli* to localize there in contrast to other tissues in which ammonia production does not occur. However, other pathogens equally susceptible to host defence mechanisms involving complement do not localize in the kidney, and this leaves the issue in doubt.

Differences in the suitability of the tissues for the growth of the pathogen

The nutrient and general physio-chemical conditions in individual tissues may be more suitable for the growth of some pathogens than others. Localized occurrence of one or more important metabolites could be the basis of tissue specificity in disease. The two investigations in the next section have this as their theme. For the moment, experimental work will be described which indicates that while tissue specificity might be studied in the chick embryo and in organ culture, studies *in vitro* in tissue culture may yield misleading results.

The chick embryo

The behaviour of several bacterial species in developing chick embryos has been described by Goodpasture & Anderson (1937), Gallavan & Goodpasture (1937), Buddingh & Polk (1939) and Buddingh & Womack (1941). A variety of bacteria including *Pasteurella tularensis*, *P. pestis*, the brucellae and the meningococcus showed differences both as regards the tissues in which they multiplied and whether they grew extracellularly or intracellularly. It was surprising that *Brucella melitensis* grew in ectodermal epithelium of the chick, whereas the closely related *B. abortus* and *B. suis* grew in endothelium of mesodermal origin. One result which deserves further investigation was that a strain of *Streptobacillus moniliformis* localized in the synovial membranes of the embryonic joints as it does in natural infection (Buddingh, 1944). In many instances, viruses manifested in the chick embryo the same specificity for liver, lung, nerve, or brain tissue as they do in natural infection, and Goodpasture (1959) and Buddingh (1959) claimed that such studies would help our understanding of tissue specificity. This technique seems worthy of development, but the findings in the chick embryo must not be extrapolated without care to provide information on the more complex parasitism by bacteria *in vivo*.

Organ culture

In virology, fragments of embryonic tissues in organ culture have been used for analysing tissue specificity. Bang & Niven (1958) showed that the following combinations of virus and embryonic tissues resulted in lesions characteristic for each virus: influenza virus in ferret trachea, herpes and vaccinia in human skin. The use of organ culture here, and on the other occasions reviewed by Bang & Luttrell (1961), confirmed that both characteristic lesions and metabolic changes could be demonstrated in specific embryonic tissues. However, more extensive comparative studies would be required to furnish complete information on specific affinities between viruses and embryonic tissues.

Tissue culture

Richardson (1959) investigated the growth of *Brucella abortus* in primary cultures of adult and foetal bovine tissues. The parasite grew well in the cells from several tissues, for example: foetal skin and kidney, and adult testis, uterine mucosa, spleen and lung. Unfortunately foetal placenta was not used and differences in the extent of the intracellular growth of *B. abortus* in the various tissues were not demonstrable. Hence, this work provided no explanation for the tissue localization of *B. abortus in vivo*.

Prior to the work of Chaproniere & Andrewes (1957) there were few examples of the successful cultivation of a virus in tissue culture, if the cells used were those from a species which was resistant to natural infection by the virus. Although myxomatosis is restricted to the rabbit *in vivo*, these workers showed that *in vitro* myxoma virus would multiply in tissue cultures of cells from man, guinea pigs, rat and hamsters. The subsequent discovery of further examples of this lack of correlation between the susceptibility of an animal and of its cells *in vitro* have robbed tissue culture of much of its value for studying the problem under discussion.

This review of the possible explanations of microbial tissue specificity has offered little helpful fact but merely speculation. The two investigations discussed next contain more factual information and indicate experimental approaches to some of these problems.

THE ROLE OF UREASE IN BACTERIAL LOCALIZATION IN THE KIDNEY MEDULLA

Infections in the tubules, papillae and pelvis of the medulla of the kidney are usually due to *Proteus* spp., *Pseudomonas* spp., or *Escherichia coli*. The enterococcus in man, and *Corynebacterium renale* in cattle, may also infect the kidney medulla but the incidence of Gram-positive infections is relatively low. There is a gradation in the degree of persistence of infection and the severity of kidney damage caused by the above organisms. Thus, *C. renale* and *Proteus* spp. persist and cause severe pyelonephritis, whereas *E. coli* infection is usually self limiting and is followed by healing (Katz & Buordo, 1962). Separate investigations with *C. renale* and *P. mirabilis* led to a similar conclusion on why these organisms successfully persist and cause damage in the kidney medulla of their respective hosts.

Corynebacterium renale

Corynebacterium renale infection is found only in cattle, where it causes pyelonephritis characterized by necrosis, and progressive destruction of the medulla of the kidney. Lovell (1946) showed that the intravenous injection of several strains of *C. renale* into mice resulted in 70 % having infected kidneys, whereas 10 strains of miscellaneous diphtheroids infected the kidneys of only 14 %. In further experiments in mice, counts of the organisms in the tissues indicated *C. renale* grew actively in the kidneys but was rapidly eliminated from the spleen and lungs (Lovell & Cotchin, 1946). In biochemical studies *in vitro*, *C. renale* utilized urea when growing in bovine urine or in urea-enriched peptone water. In human urine also, cell-free extracts of *C. renale* attacked urea and uric acid but extracts of *C. equi* and *Escherichia coli* did not. A comparative study was then made of the reactions between these extracts and a large number of defined nitrogenous substrates. The attack on urea by *C. renale* was the only reaction producing large amounts of ammonia (Lovell & Harvey, 1950).

Hence the main findings were, first, *Corynebacterium renale* when injected intravenously into mice grew and persisted in the kidneys more than in the spleen and lungs, and more than did other diphtheroids; and secondly *C. renale* had high urease activity. It would seem therefore that the possession of urease by *C. renale* might be a factor involved in its persistence in the kidney in cattle. However, proof that this is the main reason for its tissue specificity would depend on whether (*a*) organisms such as *C. ovis* having a high urease activity cause medulla infection and whether (*b*) all organisms causing this type of infection have urease.

Proteus mirabilis

Shapiro, Braude & Siemienski (1959) used organisms commonly associated with pyelonephritis in man to produce this condition experimentally in rats. Localization of the organisms in the kidneys following intracardial injection was encouraged by means of light renal massage. *Proteus mirabilis* caused more renal damage than either *Escherichia coli* or enterococci. It induced the formation of renal calculi and grew intracellularly in the tubule cells (Braude & Siemienski, 1960). These two features were thought to be related to urease activity of *Proteus* in view of the following experiments.

Tissue cultures of kidney epithelium were used to study the response of *Proteus mirabilis* and *Escherichia coli* to added urea. Growth of intracellular *Proteus* organisms was stimulated by urea, whereas that of *E. coli* was not. The optimum concentration of urea (0·2 %) was the same as that found in kidney homogenate. In addition to the urease being directly cytotoxic, it was postulated that renal damage might also result from alkalinity due to the production of ammonia. This view was investigated *in vitro* in tissue cultures, and *in vivo* by pH measurements of the urine of rats suffering from pyelonephritis. Marked alkalinity was indicated by both techniques. Further evidence of the importance of urease in the pathogenicity of this localized infection was obtained from the intravenous injection of killed *Proteus* organisms which had retained their urease activity; they caused a typical lesion in the kidney medulla. This work clearly showed that the urease of *P. mirabilis* enabled it to grow better than *E. coli* in kidney cells in tissue culture, and also to produce a more severe lesion experimentally in infected rats.

It would appear from this work that the urease activity of *Proteus mirabilis* is an important factor for its localization in, and toxicity to, the kidney. *Escherichia coli* with no urease activity causes kidney infections which are more amenable to treatment than is *Proteus* infection. Sandford, Hunter & Donaldson (1962) confirmed experimentally in rats that *E. coli* produced a less severe infection than *Proteus*. Fluorescent microscopy showed that the organisms did not grow intracellularly nor were the tubules invaded. The infection did not persist and despite the accumulation of bacterial debris as the infection was overcome no gross changes in the kidney pelvis occurred.

AN EXPERIMENTAL INVESTIGATION OF TISSUE AND HOST SPECIFICITY IN BRUCELLOSIS

Brucellosis is primarily a disease of the pregnant female ungulate, for example, the cow, sheep, goat, and pig among the domestic animals. The commonly occurring host–parasite relationships are *Brucella abortus* in cattle, *B. melitensis* in the sheep and goat, and *B. suis* in the sow. Whereas an acute placentitis with subsequent abortion occurs in the susceptible pregnant females, the disease in non-pregnant ungulates and in other hosts (including man) is different and takes the form of a chronic infection which is confined mainly to the lymphatic system and causes periodic fever and allergic disturbances. In the bull, ram, he-goat, and boar, the infection is to some extent localized in the genitalia, particularly in the seminal vesicles. Recent investigations on brucellosis, first in the bovine and then in other species, have provided an explanation for tissue and host specificity in this disease (Smith *et al.* 1962; Pearce *et al.* 1962; Williams, Keppie & Smith, 1962, 1963; Williams, 1963).

Bovine brucellosis

General pathology of *Brucella abortus* infection in the cow indicated that the disease process was mainly confined to the lumen of the pregnant uterus. We determined the distribution of *B. abortus* organisms among the various adult and foetal tissues in experimentally infected cows (Smith *et al.* 1961).

The pregnant cows were infected intravenously with a virulent strain of *Brucella abortus* and the distribution of the organisms was examined 3–4 weeks later, the animals being killed as they were about to abort. The results are shown in Table 1.

Table 1. *Distribution of* Brucella abortus *in the tissues of experimentally infected pregnant cows when abortion was imminent*

	Approx. total no. ($\times 10^{-12}$) of *B. abortus* per tissue in cows				
Tissue	I	II	III	IV	V
Foetal tissue					
Cotyledon (placenta)	63	90	140	23	27
Allantoic and amniotic fluid	0·4	0·9	3·7	7·9	1·2
Chorion	5·6	2·1	9·9	0·6	3·3
Spleen, kidney, lung, brain, liver, stomach contents	In no cow did the number of *B. abortus* in each of these foetal tissues exceed $0·1 \times 10^{12}$				
Adult tissue					
Caruncle (placenta)*	4·4	11	13	2·4	5·1
Uterine mucosa	—	—	3·0	2·4	0·4
Spleen, lung, iliac gland, kidney, udder, thoracic gland, liver, muscle	In no cow did the number of *B. abortus* in each of these maternal tissues exceed $0·1 \times 10^{12}$				

* It was impossible to separate completely all the foetal tissue from the maternal caruncles.

The total yield of organisms in each pregnant cow ranged from 0·3 to $1·5 \times 10^{14}$, and in all cases over 90 % of the organisms occurred on the foetal side of the placental barrier. Apart from an apparent infection of the caruncles due to difficulty in separating the infected foetal placentae from their maternal attachments, the gross infection was entirely within foetal tissues. Of the latter, only three tissues significantly contributed to the total yield of *Brucella abortus*: the placenta (cotyledons), the foetal fluids including the purulent uterine débris, and the chorion; they contained respectively *c.* 64, 27 and 9 % of the yield.

These results were consistent with the classical pathology, which had suggested that the disease was a placentitis confined within the uterus. The main reason for *Brucella abortus* concentrating in foetal tissue cannot be solely the absence of immunity mechanisms in this tissue. If it were so, other maternal pathogens might reasonably be expected to localize similarly, but this does not happen. A more specific reason for the preferential growth of the pathogen seemed likely, and one possibility was the occurrence of a metabolite in foetal tissues which, if not essential, was at least preferentially selective for the growth of *B. abortus*. This reasoning received support from the following experiments.

A comparison of saline extracts of bovine tissues as growth media for
  Brucella abortus

Various foetal and adult tissues from several pregnant cows were macerated with isotonic phosphate-saline buffer at pH 7·4. The extracts were centrifuged, adjusted to pH 7·0 and passed through cellulose acetate filters. These extracts were used as growth media for *Brucella abortus* (strain 544) under standard conditions (shake cultures, 37°, 5 % CO_2/air). The growth of *B. abortus* in the extracts was compared with that in a 'standard' medium of digest broth plus Fildes's peptic-digest of sheep blood. The extract of foetal cotyledon tissue was the best medium; the lag phase was short and the exponential growth rate was greater than in any other extract. This was also true when the carbohydrate contents of all extracts had been adjusted with glucose to equiv. 5 mg. glucose/ml. The active nutrients were small molecules since the diffusate prepared by dialysis of the foetal cotyledon extract was equally effective as a medium for the growth of *B. abortus*.

The isolation of erythritol—a growth stimulant for Brucella abortus—*first from infected and then from normal bovine foetal fluids*

At this stage, the results from a different study of the pathogenesis of *Brucella abortus* became relevant to the topic. A material which promoted the growth of *B. abortus* within bovine phagocytes had been fractionated from a mixture of infected allantoic and amniotic fluids collected from a cow which was about to abort after infection with *B. abortus*. The biological test which detected this material was described by Pearce *et al.* (1962). It was based on the ability of virulent brucellae to grow intracellularly. The mixed white blood cells from the buffy layer of bovine blood were allowed to phagocytose agar-grown *B. abortus* (strain 544) under defined conditions. Any extracellular organisms were then killed by the addition of fresh bovine serum plus streptomycin (2 μg./ml.). Samples of materials being tested were added after the organisms were intracellular. The cells were incubated in rotating tubes for 2 days in a continuous flow of 5 % CO_2/air, and the intracellular growth of the test organism in the controls and in the treated cells was determined by plate counts. Materials enhancing intracellular growth produced higher counts (2–5-fold) than control samples. The material in infected bovine foetal fluids which promoted intracellular growth was fractionated and crystallized; 0·3 μg. of the crystals were active in the biological test detecting intracellular growth. The active material also stimulated extracellular growth of *B. abortus* (strain 544, shake cultures, 37°, 5 % CO_2/air) in three other media (see below). At first, the active material was believed to be a product of *B. abortus*, but when further infected and normal foetal fluids were examined, it became apparent that the material was probably a constituent of normal foetal fluids. This was confirmed by fractionating normal allantoic fluid by a similar method to that used for infected foetal fluid. The fractionation is summarized in Table 2.

Table 2. *The purification of the growth factor for* Brucella abortus *in bovine allantoic and amniotic fluids*

Fractionation	% original growth enhancement activity	Amount of material
Allantoic and amniotic fluids	100	540 ml.
Diffusate	85	16·5 g.
Eluate from mixed bed resin (IR 45, ZK 225)	85	5·13 g.
Eluate from basic resin IRA 400 (OH)	50	0·34 g.
Paper chromatography *n*-butanol:pyridine:water (6:4:3)		
Crystallization from acetone	25	0·085 g.

The crystalline growth stimulant was shown to be erythritol by the usual criteria.

Erythritol

The optical isomer D-threitol proved inactive in the biological test when tested at concentrations tenfold greater than erythritol.

Immediately the possibility arose that growth stimulation by erythritol could be the cause of the preferential growth of *Brucella abortus* in foetal tissue. This idea soon received support. In experiments summarized below, growth-stimulating ability and erythritol were found only in foetal tissue and they were concentrated in the tissues most prone to heavy infection; in addition, injections of erythritol considerably enhanced infections of *B. abortus* in 1- to 7-day-old calves.

A survey of cow and calf tissue extracts for ability to enhance growth of Brucella abortus *and for the presence of erythritol*

To survey tissue extracts for their ability to stimulate growth of *Brucella abortus*, the intracellular growth test was too complicated; and the system used previously in which the whole tissue extract was used as the culture medium had the disadvantage that the effect of a single nutrient might be masked by the variable nature of the whole extract. The extracts of the tissues were therefore added in small amounts to a basal medium and the growth of B. *abortus* in these mixtures was determined. The media used were: (1) digest broth with added blood-digest, (2) 50 % heated bovine serum in saline, and (3) the defined medium of Rode, Oglesby & Schuhardt (1950). A similar pattern of results was obtained whichever medium was used and was independent of whether the tissue extract was added in the crude form, in equivalent amounts of its diffusate, or as equivalent amounts of the eluate from the ion exchange columns (see Table 2). The results in Table 3, which are

Table 3. *The growth-promoting activity, and erythritol content of tissue extracts from pregnant cows*

	Highest dilution having growth-enhancing activity	Erythritol content (μg./ml.)
Foetal tissues		
Allantoic fluid	1/300	600
Amniotic fluid	1/100	200
Placenta	1/100	60
Chorion	1/100	60
Kidney and liver	1/35	35
Spleen, lung, serum	1/10	30
Cow tissues		
Placenta*	1/50	30
Kidney, liver, spleen, lymph gland, udder, serum	Inactive 1/10	< 10

* These samples could only be incompletely separated from foetal placenta at mid-gestation: two samples which separated cleanly at birth contained < 5μg. erythritol/ml. extract.

representative of analyses of tissue extracts from several pregnant cows, showed that this technique detected a growth stimulant which was confined to certain foetal tissues. The various extracts were also analysed for their erythritol contents (Williams *et al.* 1962; Williams, 1963) which have been included in Table 3. Growth-promoting activity and erythritol were concentrated in certain foetal tissues, namely the placenta, foetal fluids and chorion, which in brucellosis are most prone to heavy infection. This discovery that naturally occurring erythritol enhanced the growth of B. *abortus* at its predilection site in the bovine pregnant uterus, gave added significance to an earlier study, carried out *in vitro*, of the metabolites used by the brucellae. McCullough & Beal

(1951) showed that of the nine carbohydrates used as the sole carbon and energy source in a simple basal medium, erythritol was the best for the three main brucella species.

The enhancement of infection of Brucella abortus in 1- to 7-day-old calves by injections of erythritol

Concurrently with the above experiments a study was also made of its influence on the organism growing in newly born calves. Parenteral injection of erythritol for a week enhanced the spleen counts of the treated calves 3- to 30,000-fold compared with the spleen counts of control animals. Thus, the stimulation of growth of *Brucella abortus* strain 544, by erythritol which had been found initially *in vitro* in the intracellular growth test and then in the laboratory media, was confirmed in experiments with calves—the natural host—*in vivo*.

The presence of erythritol in the genital tissues of the bull

The connexion between erythritol and the localization of *Brucella abortus* in the natural infection received further support when erythritol was found in extracts of bovine seminal vesicle and testis, especially the former (see Table 4).

The connexion between erythritol and the nature of brucellosis in other host species

Once the picture had become clear in cattle, investigations were made: (1) to see if the predilection sites of brucella organisms in other host species contained erythritol; and (2) to see if the growth of *Brucella melitensis* and *B. suis* was likewise stimulated by erythritol.

A survey for erythritol in other species of ungulates and in non-ungulates

Severe localized brucella infection occurs naturally in both males and females of other ungulates such as the sheep, goat and pig. In contrast, acute localized brucellosis does not occur in certain other species such as man, rabbit, guinea pig and rat, although they can harbour a chronic infection. Appropriate tissue samples from all these species were therefore analysed for their erythritol content.

The results from this survey shown in Table 4 emphasized that erythritol occurred in the placenta of only four of the eight species. Those four species, namely cattle, sheep, goats, and pigs are closely related taxonomically, and are all highly susceptible to acute localized brucellosis. Erythritol is therefore almost certainly the cause of host specificity as well as tissue specificity in *Brucella abortus* infection, inasmuch as its presence is necessary for the acute placentitis which is the complete expression of the disease produced by this organism in a host.

Table 4. *The erythritol content (μg./ml. extract) of tissues from susceptible and non-susceptible species*

	Susceptible				Non-susceptible			
	Ox	Sheep	Goat	Pig	Man	Rabbit	Guinea pig	Rat
Maternal serum	< 2	< 2	< 2	< 2	—	—	—	—
Foetal placenta	60	45	25	20	< 2	< 2	< 2	< 2
Seminal vesicle	35	15	60	8	—	—	—	—
Testis	3	5	8	3	—	—	—	—

The influence of erythritol on the growth of Brucella melitensis *and* Brucella suis

Heretofore the single virulent strain 544 of *Brucella abortus* had been used as the test organism. However the two allied microbial species— *B. melitensis* and *B. suis*—also cause characteristic lesions in the placenta and seminal vesicles of susceptible hosts. Since erythritol had been located in these tissues, it was necessary, in order to round off this study, to show that erythritol stimulated the growth of these organisms. Erythritol (0·4–4 μg./ml.) produced twofold enhancement of the growth of *B. melitensis* (strain 6015) in digest broth and heated serum-saline. Enhancement of *B. suis* (KG 25) was more difficult to demonstrate, but this was possible in the defined medium of McCullough & Dick (1943) (erythritol 4 μg./ml.) and on occasions in heated swine serum-saline (40 μg./ml.).

Table 5. *Increased infection rate in guinea pigs, one week after challenge with either* Brucella melitensis *or* Brucella suis, *caused by parenteral injection of erythritol*

Challenge	Dose of erythritol (g.)		No. infected No. in group	No. infected (%)
	1st day	Daily		
B. melitensis (6015), 500 viable organisms	1	0·2	20/24	83
	Nil	Nil	1/20	5
B. suis (KG 25), 100 viable organisms	0·04	0·008	18/30	60
	Nil	Nil	7/90	8

These results were supported by experiments *in vivo*: injections of erythritol enhanced the degree of infection in guinea pigs challenged with *Brucella melitensis* (6015) and with *B. suis* (KG 25)—see Table 5.

Erythritol is a growth stimulant for three important members of the brucella species, and this, together with its presence in appropriate foetal tissues of susceptible hosts, explains why such animals suffer the characteristic severe localized placentitis.

The relationship of erythritol to the virulence and metabolism of Brucella abortus

The identification of erythritol as a growth stimulant of *Brucella abortus* and the probable cause of its localization in placental tissue, immediately posed two questions. First, is the growth of virulent strains stimulated by erythritol more than that of avirulent strains? Secondly, what is the function of erythritol in the metabolism of *B. abortus*?

The relationship between the response to erythritol by various strains of Brucella abortus and their virulence

The evidence that erythritol stimulated the growth of the brucella group had so far been obtained from a study of fully virulent strains. The response of a number of strains, of differing degrees of virulence for the guinea pig, was studied both in dilute digest-broth and in a synthetic medium. The growth of all fully virulent strains such as 544, 2308, A5, and 1503 responded well to the addition of erythritol, whereas the response to similar concentrations of erythritol of attenuated strains such as S19, 45/20, 205/19 or 58/20, and avirulent strains such as 45/0, 11 or 99 was negligible.

This interesting finding must have some bearing on the relative pathogenicity of strains of *Brucella abortus*, because it means that virulent strains will be equipped to utilize the plentiful supply of erythritol within the pregnant uterus of susceptible hosts, subsequent to their having resisted the body defences.

The utilization of erythritol by Brucella abortus

Anderson & Smith (1963) showed that erythritol was used as an energy source by a virulent strain of *Brucella abortus* growing in a synthetic medium. The yield of the organism was enhanced twofold and it used one and a half times its own weight of erythritol despite the presence of a high concentration of glucose. Radiotracer studies showed that 25 % of the ^{14}C was incorporated into the organism, 45 % appeared as CO_2, and 30 % was excreted into the medium.

A brief recapitulation of the progressive stages of this investigation will show how the pattern of fact finding *in vivo*, followed by related experiments *in vitro*, served to advance the work in the manner suggested by the theme of this symposium. Thus, confirmation of the view that *Brucella abortus* localized in the uterus was obtained by counting the number of organisms in the infected tissues *in vivo*. The first clue to the existence of a host-specific and tissue-specific growth stimulant for *B. abortus* came through utilizing the stimulation of intracellular growth of the organism in bovine phagocytes as the routine biological assay for foetal fluids. The testing of saline extracts of the host's tissues as culture media had a strong link with conditions *in vivo*, and the encouraging results led to the extensive use of growth enhancement experiments *in vitro* for comparing tissue extracts, for detecting activity in the fractionation samples, and for measuring the enhancement of the growth of all three brucella species by erythritol. Finally the ability of erythritol to enhance the growth of all three brucella species was confirmed by experiments *in vivo*.

CONCLUSION

In this paper, the lack of information which would explain host and tissue specificity has been emphasized, but experimental approaches to these problems have been indicated by work on brucellosis and kidney infections. The future search along these lines for the mechanisms which lead to specific relationships between micro-organisms and either a host or a tissue should offer plenty of scope for realistic experiments both *in vivo* and *in vitro*. Research on the defence mechanisms of the body which may differ from tissue to tissue or from host to host can be carried out to a certain extent *in vitro*, albeit usually with the help of freshly collected blood, serum, lymph, or tissue; but a frequent return should be made to experiments *in vivo*, preferably in the natural host, for confirmatory purposes. If, however, the basis of the host or tissue specificity seems to be related to a simple nutrient such as erythritol a great deal of the investigation can be pursued on tissue extracts by the usual culture techniques *in vitro* and the results finally confirmed by experiments *in vivo*.

REFERENCES

ANDERSON, J. D. & SMITH, H. (1963). The metabolism of erythritol by a virulent strain of *Brucella abortus*. *J. gen. Microbiol.* **31**, xxiii.

BACON, G. A., BURROWS, T. W. & YATES, M. (1951). The effect of biochemical mutation on the virulence of *Bacterium typhosum*: the loss of virulence of certain mutants. *Brit. J. exp. Path.* **32**, 85.

BANG, F. B. & LUTTRELL, C. N. (1961). Factors in the pathogenesis of virus diseases. *Advanc. Virus Res.* **8**, 199.

Bang, F. B. & Niven, S. F. J. (1958). A study of infection in organized tissue cultures. *Brit. J. exp. Path.* **39**, 317.

Beeson, P. B. & Rowley, D. (1959). The anticomplementary effect of kidney tissue: its association with ammonia production. *J. exp. Med.* **110**, 685.

Braude, A. I. & Siemienski, J. (1960). Role of bacterial urease in experimental pyelonephritis. *J. Bact.* **80**, 171.

Buddingh, G. J. (1944). Experimental *Streptobacillus moniliformis* arthritis in the chick embryo. *J. exp. Med.* **80**, 59.

Buddingh, G. H. (1959). In *Viral and Rickettsial Infections of Man*, 3rd edition, p. 199. Ed. T. M. Rivers & F. L. Horsfall. Philadelphia: Lippincott.

Buddingh, G. J. & Polk, A. D. (1939). The pathogensis of *Meningococcus meninigitidis* in the chick embryo. *J. exp. Med.* **70**, 499.

Buddingh, G. J. & Womack, F. C. (1941). Observations on the infection of chick embryos with *Bacterium tularense, Brucella,* and *Pasteurella pestis. J. exp. Med.* **74**, 213.

Burrows, T. (1963). Virulence of *Pasteurella pestis* and immunity to plague. *Ergebn. Mikrobiol.* **37** (in the Press).

Chandler, R. L. (1961). Infection of laboratory animals with *Mycobacterium johnei.* IV. Comparative susceptibility to infection of C 57, C.B.A., and Swiss White Mice. *J. comp. Path.* **71**, 233.

Chandler, R. L. (1962). Infection of laboratory animals with *Mycobacterium johnei.* V. Further studies of the comparative susceptibility of C 57 black mice. *J. comp. Path.* **72**, 198.

Chaproniere, D. M. & Andrewes, C. H. (1957). Cultivation of rabbit myxoma and fibroma viruses in tissues of non-susceptible hosts. *Virology,* **4**, 351.

Dubos, R. J. (1954). *Biochemical Determinants of Microbial Diseases.* Cambridge, Mass.: Harvard University Press.

Formal, S. B., Baron, L. S. & Spilman, W. (1954). Studies on the virulence of a naturally occurring mutant of *Salmonella typhosa. J. Bact.* **68**, 117.

Gallavan, M. & Goodpasture, E. W. (1937). Infection of chick embryos with *Haemophilus pertussis* reproducing lesions of whooping cough. *Amer. J. Path.* **13**, 927.

Goodpasture, E. W. (1959). Cytoplasmic inclusions resembling Guarnieri bodies and other phenomena induced by mutants of the virus of fowlpox. *Amer. J. Path.* **35**, 213.

Goodpasture, E. W. & Anderson, K. (1937). The problem of infection as presented by bacterial invasion of the chorio-allantoic membrane of chick embryos. *Amer. J. Path.* **13**, 149.

Gowen, J. W. (1960). Genetic effects in non specific resistance to infectious disease. *Bact. Rev.* **24**, 192.

Green, M. & Stahmann, M. A. (1953). Inhibition of mumps virus multiplication by a synthetic polypeptide. *Proc. Soc. exp. Biol., N.Y.* **83**, 852.

Green, M., Stahmann, M. A. & Rasmussen, A. F. (1953). Protection of embryonated eggs infected with infectious bronchitis or Newcastle disease virus by polypeptides. *Proc. Soc. exp. Biol., N.Y.* **83**, 641.

Hirsch, J. G. (1960). Antimicrobial factors in tissues and phagocytic cells. *Bact. Rev.* **24**, 133.

Housewright, R. D. (1960). The 1959 Fort Detrick symposium on non-specific resistance to infection. *Bact. Rev.* **24**, 1.

Katz, Y. J. & Buordo, S. R. (1962). Chronic pyelonephritis. *Annu. Rev. Med.* **13**, 481.

Lovell, R. (1946). Studies on *Corynebacterium renale.* I. A systematic study of a number of strains. *J. comp. Path.* **56**, 196.

LOVELL, R. & COTCHIN, E. (1946). Studies on *Corynebacterium renale*. II. The experimental pathogenicity for mice. *J. comp. Path.* **56**, 205.

LOVELL, R. & HARVEY, D. G. (1950). A preliminary study of ammonia production by *Corynebacterium renale* and some other pathogenic bacteria. *J. gen. Microbiol.* **4**, 493.

LURIE, M. B., ABRAMSON, S. & HEPPLESTON, A. G. (1952). On the response of genetically resistant and susceptible rabbits to the quantitative inhalation of human type tubercle bacilli and the nature of the resistance to tuberculosis. *J. exp. Med.* **95**, 119.

McCULLOUGH, N. B. & BEAL, G. A. (1951). Growth and manometric studies on carbohydrate utilization of *Brucella*. *J. infect. Dis.* **89**, 266.

McCULLOUGH, N. B. & DICK, L. A. (1943). Growth of *Brucella* in a simple chemically defined medium. *Proc. Soc. exp. Biol., N.Y.* **52**, 310.

MEYER, K. F. (1948). The animal kingdom, a reservoir of human disease. *Ann. intern. Med.* **29**, 326.

PEARCE, J. H., WILLIAMS, A. E., HARRIS-SMITH, P. W., FITZGEORGE, R. B. & SMITH, H. (1962). Erythritol, a constituent of bovine foetal fluids which stimulates the growth of *Brucella abortus* in bovine phagocytes. *Brit. J. exp. Path.* **43**, 31.

RICHARDSON, M. (1959). Parasitization *in vitro* of bovine cells by *Brucella abortus*. *J. Bact.* **78**, 769.

RODE, L. J., OGLESBY, G. & SCHUHARDT, V. T. (1950). The cultivation of brucellae on chemically defined media. *J. Bact.* **60**, 661.

ROGERS, D. E. (1960). Host mechanisms which act to remove bacteria from the blood stream. *Bact. Rev.* **24**, 50.

ROWLEY, D. (1960). Antibacterial systems of serum in relation to non specific immunity to infection. *Bact. Rev.* **24**, 106.

SANDFORD, J. P., HUNTER, B. W. & DONALDSON, P. (1962). Localization and fate of *Escherichia coli* in haematogenous pyelonephritis. *J. exp. Med.* **116**, 285.

SHAPIRO, A. P., BRAUDE, A. I. & SIEMIENSKI, J. (1959). Haematogenous pyelonephritis in rats. IV. Relationship of bacterial species to the pathogenesis and sequelae of chronic pyelonephritis. *J. clin. Invest.* **38**, 1228.

SKARNES, R. C. & WATSON, D. W. (1957). Antimicrobial factors of normal tissues and fluids. *Bact. Rev.* **21**, 273.

SMITH, H., KEPPIE, J., PEARCE, J. H., FULLER, R. & WILLIAMS, A. E. (1961). I. Isolation of *Brucella abortus* from bovine foetal tissue. *Brit. J. exp. Path.* **42**, 631.

SMITH, H., WILLIAMS, A. E., PEARCE, J. H., KEPPIE, J., HARRIS-SMITH, P. W., FITZGEORGE, R. B. & WITT, K. (1962). Foetal erythritol: A cause of the localisation of *Brucella abortus* in bovine contagious abortion. *Nature, Lond.* **193**, 47.

WILLIAMS, A. E. (1963). The chemical basis for the preferential growth of *Brucella abortus* in bovine foetal tissue. Thesis for Ph.D. degree, University of London.

WILLIAMS, A. E., KEPPIE, J. & SMITH, H. (1962). Foetal erythritol, a cause of the localization of *Brucella abortus* in pregnant cows. *Brit. J. exp. Path.* **43**, 530.

WILLIAMS, A. E., KEPPIE, J. & SMITH, H. (1963). Erythritol, its effect on the growth of various strains of *Brucella abortus* and its possible significance in infections with *Brucella melitensis* and *Brucella suis*. *J. gen. Microbiol.* **31**, xxiii.

WILSON, G. S. (1957). The selective action on bacteria of various factors inside and outside the animal body with particular reference to their effect on virulence. In *Microbial Ecology. Symp. Soc. gen. Microbiol.* **7**, 338.

WILSON, G. S. & MILES, A. A. (1955). In *Topley and Wilson's Principles of Bacteriology and Immunity*, 4th edn. chap. 4. London: Edward Arnold.

WOOD, W. B. (1960). Phagocytosis with particular reference to encapsulated bacteria. *Bact. Rev.* **24**, 41.

THE CORRELATION OF PROPERTIES
IN VITRO WITH HOST–PARASITE
RELATIONS

A. F. B. STANDFAST

The Lister Institute of Preventive Medicine, Elstree, Hertfordshire

INTRODUCTION

The elucidation of the relationship between host and parasite and the analysis of the factors responsible for this relationship is complicated and difficult. Four factors have added to the difficulties of the analysis *in vitro* of this phenomenon *in vivo*. First, both host and parasite are living organisms independently capable of variation and modification, which may be genetically controlled, but each responding also to a greater or lesser degree to the presence of the other. In bacteria the genetic control of certain factors pertaining to virulence is well known (Stocker, 1959); in animals, variations in susceptibility in different strains of mice to the same preparation of pertussis toxin (Piersma *et al.* 1962) and to infection with *Salmonella typhi* (Landy, Gaines & Sprinz, 1957) and in guinea-pigs variation in response to diphtheria toxoids (Schiebel, 1943) have been noted. Secondly, the bacterial cell contains a large number of substances which are antigenic, substances as common in free living as in parasitic bacteria. Few of these antigens play any role in pathogenicity or immunity, that is, affect the obvious features of the host–parasite relationship. Thirdly and somewhat paradoxically, the ease with which we can grow on artificial media in the laboratory practically all the parasitic bacteria, *Mycobacterium leprae* is perhaps the only common exception, has in many ways added to the complication. In the laboratory on artificial culture media bacteria remain living organisms and they may, and do, exhibit a range of phenomena which have nothing to do with their parasitic life. Finally, the ease with which we can infect some laboratory animals with practically any of the pathogenic bacteria means that we must always keep in mind in any analysis of the host–parasite relationship the possible differences between infection in the usual host and infection in the laboratory animal, which may or may not be naturally susceptible.

We know something about the behaviour of parasitic or pathogenic bacteria in the laboratory, and something about their behaviour under natural conditions in their usual animal hosts. Between these two, so clearly *in vitro* and *in vivo*, we also know a good deal about the behaviour of many pathogens in 'laboratory animals'—and here I use the term for any animal which can be infected in the laboratory but which is not usually infected in the field. Is this *in vivo*? Already we are in difficulties with definitions.

Definitions

The *Oxford English Dictionary* which admits '*in vacuo*' ignores '*in vivo*' and '*in vitro*'. There will be little divergence in ideas about the term *in vitro*—in the test tube, in the laboratory, in artificial culture, probably in pure culture and almost certainly under controlled conditions. *In vivo* seems to have at least two meanings: (1) in the living animal, and (2) in the wild state. Certain bacteria are said to be '*in vitro*' when in culture in the laboratory and '*in vivo*' when living free, for example in the soil. Even when the term means 'in the body' it has been used to mean 'in the bacterial body' as well as 'in the host's body'. In dealing with the parasitic bacteria the term *in vivo* is used solely to mean 'in the host's body', and I use the term in this sense in this paper.

The expression 'host–parasite' must be interpreted in its widest sense. In bacterial infections of man and animals the carrier state may well be the best level of relationship; the best, that is, for host and for parasite. For at this level, the host is least inconvenienced by the presence of the parasite and makes the least effort to rid itself of the parasite. On the other hand, the parasite is catered for and its means of dispersal is adequate; a possible control of overall numbers is the only restriction. Any reference to host–parasite relation must, therefore, include the carrier state as well as the more usually studied levels in which one or other partner is gaining or has gained the upper hand.

Other words which must be used in any study of the host–parasite relationship—resistance and immunity as aspects of the host's response and virulence and pathogenicity as attributes of the parasite—will be used as defined in the Fifth Symposium of this Society. In this Symposium, Miles (1955) wrote 'Virulence, on the other hand, is conveniently reserved for the pathogenicity of a given stable homogeneous strain of a microbe as determined by observation of its action *on the host in relation to which* the statement about virulence is made.' The words which I have italicized are so frequently forgotten.

The fact that because a strain of bacteria is mouse-virulent it will therefore be virulent for man is often implied, but it is not necessarily true, though in certain cases it may be. Thus, Findlay (1951) found that 5 strains of *Salmonella typhi* isolated from the clinically severe Truro outbreak were more mouse-virulent than any of 5 strains from the clinically mild Richmond outbreak; but Miles (1951) found that the mouse-virulence of 21 strains of *Proteus vulgaris* was not associated with the severity or the site of infection in the human source.

Experiments in test tubes

Antigens are just as common amongst free-living as amongst pathogenic bacteria and, since most bacterial antigens are produced *in vitro*, given the correct conditions, their production is clearly not the result of a stimulus *in vivo*; and since most bacterial antigens are easily produced they have been extensively studied in the laboratory. Many antigen-antibody reactions may be regarded as a series of phenomena *in vitro* which do not occur *in vivo* and have no direct counterpart *in vivo*. This does not mean that facts derived by serological methods are not valid, it means they do not assist in the elucidation of the host–parasite relation without very careful correlation with phenomena *in vivo*. Toxin neutralization by antitoxin is an example where the reactions *in vitro* and *in vivo* are probably the same, though one wonders if flocculation occurs *in vivo*.

A group of bacteria in which reactions *in vitro* play little or no part *in vivo* is the genus *Salmonella*. This genus consists of 300–400 species each with a distinct antigenic make-up (Kauffman, 1951) which makes it a simple matter to distinguish the species by tests *in vitro*. Yet *in vivo* many Salmonellas are indistiguishable and it is impossible to tell from the symptoms or the severity of an outbreak of food poisoning which species is responsible.

The genus *Salmonella* also supplies the best example of an exception to the above. *S. typhi* contains an agglutinogen, called the Vi-antigen by Felix & Pitt (1934), that is closely associated with virulence in mice. The protective action of typhoid vaccines and of typhoid antisera for mice is in proportion to the Vi-antigen content of the former and the quantity of Vi-antibody in the latter (Landy & Webster, 1952). By the intracerebral route of infection Vi-positive strains of *S. typhi* were about 40 times more lethal than Vi-negative strains, and if purified Vi-antigen was added to the Vi-negative variant its virulence for mice was raised by a factor of 15 (Gaines, Tulley & Tiggert, 1961). This enhancement only took place with Vi-antigen; it did not depend on the incorporation of the Vi-antigen into the bacterial cell, though the Vi-antigen had to be injected into the mice at the same time as the bacteria. The mode of action of the Vi-antigen is not understood; Gaines *et al.* (1961) showed that it appeared to enhance the growth rate of the Vi negative strain in mice; perhaps by inhibiting phagocytosis.

This is an example of an agglutinogen (activity *in vitro*) determining virulence (activity *in vivo*). But from the point of view of natural typhoid fever, phenomena as elicited in the mouse are phenomena *in vitro*, the mouse being a suitable test tube. Infection of the mouse by *Salmonella typhi* is a laboratory not a natural infection. The Vi-antigen has proved to be of doubtful importance in man, therefore we can explain the activity of the Vi-antigen in man only in terms of man; we cannot deduce the behaviour of the Vi-antigen in man from its behaviour in mice.

It is interesting to note that though the same Vi-antigen is common to *Salmonella typhi*, *S. paratyphi C*, some strains of *Escherichia coli*, *Citrobacter ballerupensis*, and some 'paracolon' organisms it is only in *S. typhi* that the Vi-antigen is connected with mouse-virulence; though with *S. typhi* the source of the Vi-antigen is immaterial. Vi-antigen derived from paracolon organisms enhanced the virulence of *S. typhi* just as effectively as that derived from *S. typhi* itself.

Experiments in laboratory animals

The ease with which most, if not all, pathogenic bacteria can be made to infect some species of laboratory animals has resulted in a great amount of experimental work, much of which has had no connexion with or even basis on natural infection. The introduction of the laboratory animal in cases where experimental work on the natural host is difficult or impossible was inevitable, but it has added to the confusion.

In a few cases we can be reasonably certain that the infection and the mechanism of infection are the same in the laboratory animal and the natural host. The classic bacterial toxins such as tetanus toxin almost certainly kill mice for the same reason and in the same way as they kill man or any other animal; but aside from the toxin diseases such cases are rare.

In most cases we are dealing with two independent phenomena, natural virulence and laboratory virulence, which may well be dependent on different antigens. As we have seen, the Vi-antigen of typhoid is of supreme importance in the laboratory infection of mice but of doubtful importance in man.

Another example is the type specific M antigen of streptococci which is closely associated with virulence in mice, for streptococcal strains without M protein do not kill mice. But only about half of the streptococcal strains isolated from lesions in man have any M protein, which indicates that it does not determine virulence in man. Yet how often do we read in the literature of 'virulent and non-virulent streptococci' with no indication as to whether mouse-virulent (laboratory animal) or man-virulent (natural host) is meant. A good case can be made out for regarding infections of laboratory animals as processes *in vitro* in which laboratory animals are convenient test tubes. This view has at least one advantage, since we are trying to analyse phenomena *in vivo* in terms of experiments *in vitro* we shall at least try to interpret the results of tests in laboratory animals in terms of the natural host and not as an end in themselves. The artificiality of many infections-in-laboratory-animals often escapes the notice of enthusiastic research workers.

The role of bacterial toxins

It would be very convenient to be able to consider bacterial toxins as the basis of all pathological effects but the present state of our knowledge does not allow us to make such a wide generalization.

Bacteria fall into two groups: (1) known toxin producers and (2) those bacteria from which no toxin has yet been isolated. This distinction really depends on the ability of some organisms to produce their main toxin *in vitro* and on the ease with which most of these toxins can be assayed *in vitro* by various methods.

The first group—the known toxin producers—can be divided into two divisions and these can be conveniently, but not very accurately, called the Gram-positive toxin producers and the Gram-negative toxin producers. The former group consists of the clostridia, corynebacteria, streptococci and staphylococci, but to these must be added *Bordetella pertussis* and *Shigella shigae* (for its neurotoxin). The toxins of this group are lethal, often dermonecrotic, antigenic, neutralized by specific antitoxin in multiple proportions, they can be modified by formaldehyde to form toxoid, and their effects are generally specific *in vitro* and *in vivo*. The part played by these toxins in disease varies in importance from *Clostridium tetani* at one end of the range to *B. pertussis* at the other. In tetanus, toxin seems to be the only factor involved in producing disease, once the organism has been deposited mechanically in the tissues of the host with sufficient extraneous matter or sufficient tissue damage to supply the conditions necessary for growth. These conditions must be supplied for *C. tetani* has no invasive powers. In tetanus the neutralization of the toxin with antitoxin, from passive administration or produced as a result of active immunization, is the sole factor in protection. At the other end of the range is *B. pertussis*; this organism produces a potent toxin which is lethal, dermonecrotic, antigenic, neutralized by antitoxin in multiple proportions and can be toxoided (Evans & Maitland, 1937; Evans, 1940). However, what part this toxin plays, if any, in human infection is not known, for human convalescent serum contains no demonstrable antitoxin though it contains all the other pertussis antibodies (Evans, 1947; Winter, 1953). Vaccines used in human prophylaxis do not contain toxoid and do not elicit antitoxin but they are effective in protection against natural infection. In laboratory animals neither toxin nor toxoid protect actively nor does antitoxin protect passively against infection with *B. pertussis*.

The second division—the Gram-negative toxin producers—consists of members of many genera *Salmonella* and *Shigella* (endotoxin), *Escherichia*, *Pseudomonas* and *Vibrio*, most of them Gram-negative. The toxins elaborated by these organisms have much in common, and differ in many respects from the so-called Gram-positive toxins. The Gram-negative toxins are components of the bacterial cell and so are usually called endotoxins. Few of these toxins can be toxoided and though most of them are antigenic, they cannot be neutralized in multiple doses. The reactions *in vitro* of many of the so-called Gram-negative toxins are confusing and I think one is entitled to wonder if many of them are single antigens. However, the symptoms produced in laboratory animals by many of these toxins are more or less the same irrespective of the organism from which the toxin was derived (see also van Heyningen, 1950, 1955).

The Gram-negative toxin producers pass almost imperceptibly into the 'non-toxin' producers. Indeed, until recently we might have said categorically that there was a large group of bacteria that did not produce toxins. In this group would have been *Bacillus anthracis*, until Smith & Keppie (1955) demonstrated a toxin in the plasma of guinea pigs dying of anthrax, which was lethal for mice and guinea pigs and could be neutralized by antiserum. This work must bring about a re-organization of our ideas about the production of toxins for it seems possible that some organisms produce toxins *in vivo* that are not produced *in vitro* or for the production of which *in vitro* we do not know the correct conditions. A similar approach is being used to investigate the production of toxins or toxic substances by *Vibrio cholerae* and *Escherichia coli* following the work of De (De & Chatterje, 1953; Taylor, Maltby & Payne, 1958; Nikonov, Yevseyeva, Bibikova & Bichul, 1958). Of course we may consider that the guinea pig or the ligated loop of rabbit intestine used for toxin production is not strictly '*in vivo*' but is the use of the guinea pig as a convenient test tube and should therefore be regarded as '*in vitro*'. The classical toxins can all be produced in artificial media in the laboratory, more or less easily, though in certain cases special media or special conditions are necessary for maximum production. These conditions do not always correspond to conditions *in vivo*, diphtheria toxin, for example, is only produced in the test tube at much lower iron concentrations than are found in the tissues of the throat. The conditions of production of a toxin *in vitro* and *in vivo* may not be the same, but we are lucky in most cases in being able to demonstrate that the end products are identical. The development of the latest methods may enable us to isolate toxins in many diseases, such as salmonella food poisoning, for which one suspects a toxin is responsible.

For our convenience we can pursue our investigation of the host–parasite relationship under two headings: (1) a survey of the characters of pathogens, which enable them to infect hosts and so cause disease, which we will call 'virulence', and (2) a study of the responses of hosts to the presence of the parasites which we will call 'immunity'. In our quest for the correlation, if any, of properties *in vitro* with host–parasite relations, we shall find two questions that require to be answered:

has the analysis *in vitro* satisfactorily explained the phenomena *in vivo*,

or has the approach *in vitro* led us astray, either by oversimplification or by stressing the wrong correlations?

VIRULENCE

Virulence should be regarded as the resultant of a complex of interacting factors contributed by host and by parasite. There are, in the sense in which I am using the term virulence, at least three stages at which it can be assessed: (*a*) the capacity to make the primary lodgement of the parasite in the host; (*b*) the capacity to grow in the host, which together are expressions of the capacity to overcome the non-specific defence mechanisms of the host; and (*c*) the capacity to overcome at least one of the specific defence mechanisms of the host which arise in response to the infection. Although the parasite usually overcomes these specific defence mechanisms to produce disease, this is not essential.

It seems unlikely, therefore, that a single antigen will wholly determine 'virulence' but an antigen measurable *in vitro* may be a major determinant for one stage in the 'virulence complex'. There is here not enough space to go through the whole range of bacteria and their various animal hosts; we can only examine a few examples of different stages. In doing this I cannot stress too strongly that what holds for one host–parasite relationship will not necessarily hold for another. Each case must be considered separately; because *Salmonella typhimurium* behaves in a certain way in a mouse, that is no reason to believe that it should behave in the same way in the rat.

The primary lodgement in the host

Very little is known about the factors, measurable *in vitro*, which are involved in the primary lodgement of a parasite, and the importance of this stage will vary with the route in natural infection.

One example of the varying effect of different bacterial factors on the primary lodgement is the infection of mice with *Bordetella pertussis*. *B. pertussis* is a natural pathogen of man only, so that the infection of mice is an artificial laboratory process. Norton & Dingle (1935) found that mice could be infected by intracerebral injection and Burnet & Timmins (1937) showed that they could be infected intranasally; infection by either route results in the death of the mouse in about 7 days.

In infections by the intranasal route with *Bordetella pertussis*, toxin plays a part in the primary lodgement of the organism for which a critical toxin:organism ratio may determine infection or no infection. With strains of *B. pertussis* of high or low virulence there is no enhancement of virulence with added toxin, cell-free toxic fluid, but with strains of medium virulence there is a marked increase. The increase in mortality with a strain of high virulence was 61–63 %, with a strain of low virulence 23–24 % and with a strain of medium virulence from 49 to 65 %. It is obvious that toxin is not the only factor in virulence as the addition of more toxin does not automatically increase virulence. Toxin presumably assists in the primary lodgement in the lung, following which there must be growth of the organism if it is to maintain itself in the lung. We must assume that strains of high virulence produce an optimal amount of toxin and an optimal number of organisms able to grow in the lung. Strains of low virulence show little increase of virulence with added toxin because there are not enough organisms in a culture of this strain capable of growing in the lung even if their primary lodgement is assisted by additional toxin.

Pertussis toxin, either in living organisms or as cell-free toxin, can be shown *in vitro* to inhibit the cilia of ciliated epithelium and this action can be neutralized by antitoxin. The organism in droplet infections in man may be able to establish itself on the ciliated epithelium of the respiratory tract by inhibiting ciliary action, thus avoiding one of the host's non-specific defence mechanisms operative against primary lodgement in the lung (Standfast, 1951, 1958).

The ability to grow in the host

There are several cases where the virulence of a parasite can be correlated with a particular substrate necessary for its growth.

Smith and his colleagues (Smith *et al.* 1962; Williams, Keppie & Smith, 1963) showed that erythritol is a preferred substrate for the growth of *Brucella abortus* in the cow, so that the brucellas are to be found in the largest numbers in tissues rich in erythritol. Another example in which virulence depends on the ability of the organism to grow *in vivo* is the laboratory infection of mice with *Salmonella typhi* (Bacon, Burrow & Yates, 1951). These workers found that the growth of *S. typhi* in the mouse depended on the presence in the peritoneal fluid of certain growth factors. Three biochemical mutants of *S. typhi* which required respectively purines, *p*-aminobenzoic acid, and aspartic acid, were of low virulence for mice; but when the essential growth factor was injected intraperitoneally into the mice, the virulence of the mutant was enhanced since the factor-requiring mutant could then grow from a sublethal inoculum to a degree which was fatal to the host.

The ability to overcome specific defence mechanisms

Much work has been done on this aspect of virulence *in vitro* and *in vivo* but almost all the work has been in laboratory animals. An essential property of virulent bacteria is their ability to grow in the host; this, as long as any special growth requirement has been met, means that the organism must overcome, sooner or later, one or more of the host's specific defence mechanisms, which will develop as a result of the presence of the parasite.

Phagocytosis is generally regarded as one of the main defence mechanisms of the host and a vast amount of work *in vitro* has been done to compare phagocytosis of 'virulent' and 'non-virulent' organisms. The ability of 'virulent' organisms to avoid phagocytosis has been demonstrated *in vitro* with pneumococci, streptococci, staphylococci, coli, typhoid and plague organisms amongst others. But phagocytes do not necessarily behave *in vivo* as they behave *in vitro* and experiments done *in vivo* in laboratory animals have given rather different results from tests *in vitro*; the results have varied from case to case. For example, Enders, Shaffer & Wu (1936) found that the reticuloendothelial cells of mice were not able to ingest virulent capsulated pneumococci, but Rowley (1954) found that there was no difference in the rate at which 'virulent' and 'non-virulent' coli organisms were phagocytozed in the peritoneum of the mouse.

Unfortunately many phagocytosis experiments *in vivo* have an air of artificiality about them and it is doubtful whether they bear any resemblance to natural infection. The role of the phagocyte as a defence mechanism rather than a garbage-clearing scheme is not proven.

The pneumococci, and probably other organisms which produce soluble antigens, have another way of overcoming the defence mechanism of the host which can be demonstrated *in vitro* and *in vivo* in laboratory animals, and which ties up with certain findings in the natural host (for the pneumococcus—man). Pneumococcal capsular polysaccharide in a soluble form has an anti-phagocytic action *in vitro* because of its ability to combine with and so neutralize specific antiserum. *In vivo* the polysaccharide can abolish immunity in mice, when it is injected with a dose of pneumococci to which the mice are otherwise immune (Downie, 1937). Furthermore, with some pneumococci it is the amount of capsular material synthesized rather than the presence of the intact capsule which determines the degree of virulence for mice (McLeod & Krauss, 1950). This suggests that in mice the pneumococci overcome the defence mechanism of the host by the production of large quantities of specific polysaccharide which neutralizes the antibodies of the host at least as far as their opsonic effect is concerned, and so inhibits phagocytosis. This explanation, derived from laboratory experiments, *in vitro* and *in vivo*, agrees with findings from fatal cases of pneumonia in man. As long ago as 1917, Cole observed that empyema fluid from cases of pneumonia contained soluble substances which neutralized the protective antibodies in pneumococcal antiserum. Nye & Harris (1937) calculated that the specific polysaccharide present in the lungs of some patients who died of pneumococcal pneumonia would have required 60 l. of potent antiserum to have neutralized it. Frisch, Tupp, Barrett & Pidgeon (1942) examined *post-mortem* lungs from deaths from types I, II, III, VII and VIII pneumococci and isolated specific polysaccharide to a maximum value of 68 g. per lung. The production of soluble polysaccharide which neutralized the antiserum and so interfered with phagocytosis will explain results of Enders *et al.* (1936); and, although there are differences between the various types of pneumococci, virulence for mice correlates mainly with capsule production and hence with type specific polysaccharide.

On the other hand, one may deduce from the results of Rowley (1954) that the virulence of *Escherichia coli* for mice is not due to a soluble antigen. Studies *in vivo* by intraperitoneal injection of coli organisms into mice showed that certain coli strains were 10,000 times more virulent than were the average coli strains. This virulence is a reflexion of the capacity of virulent strains to grow in the mouse's peritoneum, and this capacity is correlated *in vitro* with greater resistance towards the bactericidal powers of complement + antiserum; the cause of this resistance is not known.

Many other characters demonstrable *in vitro* have been correlated with virulence. Beumer (1938) found that certain strains of staphylococci grew in the presence of staphylococcal antitoxin by producing a proteolytic enzyme which destroyed the antitoxic serum component. Rowley & Jenkins (1962) suggested that a strain of *Salmonella typhimurium* was virulent for mice because certain antigens in the parasite were common to the mouse, which could not therefore produce antibodies against them.

These examples are enough to show that few generalizations can be made and that each case must be treated separately. Not only is this true of each organism, but in many cases for each host, and for each route of infection in any one host.

IMMUNITY

If virulence is one of the important facets of the host–parasite relationship from the point of view of the parasite, immunity is the specific response of the host to the presence of the parasite.

The field of specific immunity is so wide that we cannot deal with the whole subject here. I propose therefore to examine one area, namely the protection of man against some common bacterial infections and the enhancement of protection by bacterial vaccines. This field covers the whole range of host–parasite relations and shows problems in, and illustrations of, serology and virulence, as well as of immunity. Artificially induced immunity is based on the same principles as naturally acquired immunity, and the importance of the use of proper vaccines in man has led to much study of the antigens involved and of the assay of vaccines *in vitro* and *in vivo*. The testing of these vaccines in laboratory animals has lead to a study of the virulence and the causes of the virulence of the challenge strains of bacteria used in laboratory animals; some of these studies have been correlated with field trials in man.

For some therapeutic substances, such as the arsenicals and some antibiotics, it is possible to correlate the behaviour *in vivo* with a chemical analysis, a true *in vitro*:*in vivo* correlation; this is not at present possible with any vaccine or antigen. The best we can do is to attempt to demonstrate and estimate particular antigens *in vitro* and then compare their behaviour in laboratory animals with that in man. It is when the assumption is made that because an antigen has behaved in mice in a certain way it will behave in the same way in man that difficulties and discrepancies arise.

We are dealing here with a complex situation. It is customary to talk of 'protective' antigens and hence of 'protective' antibodies, but by this we mean an antigen of the parasite which elicits protection in the host against that parasite; again we usually mean, and should mean, protection of man or the natural host and not protection of laboratory animals.

These 'protective' antigens may have little connexion with antigens determining virulence in pathogenic bacteria, and, at least *in vitro*, seem to play a minor role in the general metabolism of the particular bacteria. Except in the cases where the 'protective' antigen is associated with the capsule, it is impossible to tell from the cultural or morphological appearance of the bacteria whether the culture contains much or little of the antigen. Maximal production of the 'protective' antigen is often shown by old laboratory strains or even by rough strains. Indeed, the most virulent organism is seldom the best producer of 'protective' antigen and this has led one school of thought to refer to such antigens as 'lethal' antigens on the analogy, I suppose, of the lethal gene. Lethal, of course, to the producing organism. For, so the argument runs, the more 'protective' antigen the parasite produces, the greater the antibody response of the host will be and the quicker the host will rid himself of the parasite.

Now let us take four examples of infectious diseases of bacterial origin in man and see how we can correlate our knowledge obtained *in vitro* with findings *in vivo*.

Diphtheria

In man *Corynebacterium diphtheriae* causes a characteristic disease by means of a toxin which diffuses from the bacteria which are really outside the tissues of the body in the throat in the region of the tonsils. The disease diphtheria in man is essentially a toxaemia in which resistance or susceptibility depends on the presence of antitoxin in the host, produced naturally as the result of an infection or artificially as the result of immunization with diphtheria toxoid. So successful has this acquired, purely antitoxic, immunity been that the disease diphtheria has to all purposes been eliminated in many countries. The vaccine used is a modified toxin, i.e. toxoid obtained from culture filtrates treated with formaldehyde to abolish toxicity. Millions and millions of human doses have been prepared from the Park 8 strain of *C. diphtheriae*, a strain which has been cultivated in the laboratories of the world for more than 50 years and which is still the strain of choice.

With diphtheria toxoid we come near to a real test *in vitro* for potency in man—the flocculation test. Unfortunately the test does not always reflect absolutely the behaviour *in vivo*, and the final test for potency must therefore be done in guinea pigs. However, except for this final essential test it is customary to use the measure *in vitro*, the Lf or flocculation equivalent dose, for the measurement of toxoid.

In diphtheria, therefore, we can produce the essential 'protective' antigen for man, concentrate and purify it, assay it in laboratory animals, and be certain that we can produce an antigen which, when used on a large enough scale, will eliminate the disease from a region. Furthermore we can, with diphtheria, by using reagents carefully standardized in the laboratory, test the susceptibility or immunity of children before and after immunization, either by using a skin test (the Schick test) or by taking a sample of a child's blood and assaying the antitoxin content. We can prepare potent diphtheria antitoxin in horses by the administration of suitable doses of diphtheria toxin or toxoid; we can assay the potency of this antitoxin and we know that we shall be able to cure most children who contract the disease by giving them adequate doses of this antitoxin.

There are still unanswered questions about diphtheria; we do not know how or why antitoxin clears diphtheria bacilli from the throats of patients or indeed whether antitoxin plays any part in the disappearance of the organism, but there is no doubt that it neutralizes the toxin. Amies (1954) suggests that diphtheria toxin can inhibit the local inflammatory reaction and so counteract a defence mechanism of the host. The neutralization of toxin by antitoxin followed by a normal inflammatory response might be responsible for the eventual elimination of the organisms. When the disease was common it was well known that it varied in severity, but we know little about the virulence of *Corynebacterium diphtheriae* in man; though it is reflected by its virulence in guinea pigs. McLeod (1943) and his colleagues found that diphtheria strains could be divided into three types by their appearance on certain laboratory media. These types differed culturally and in their virulence for guinea pigs; they were called *gravis*, *intermedius* and *mitis* strains, because of the frequency with which they were isolated in severe or mild cases of clinical disease (*intermedius* is distinguishable in the laboratory as a third type usually associated with severe disease). However, when these three types were tested in the laboratory it was found that there was no difference in their powers of toxin production, nor in the type of toxin produced. Toxin production *in vitro* does not explain the difference in virulence, but toxin production *in vivo* may be different. Either *gravis* strains grow faster *in vivo* or produce more toxin *in vivo*, and this may only apply *in vivo* in man under natural conditions and not *in vivo* in guinea pigs under artificial ones. Two other suggestions have been put forward to explain the enhanced virulence of *gravis* strains. Mueller (1941) pointed out that iron which is known to depress the production of toxin in culture media is present in large quantities in the tissues of the throat. *Gravis* strains are less affected by iron in their production of toxin than *mitis* strains and so should produce more toxin in the throat. Hoyle (1942) found a difference between the lipid antigens of (*a*) *mitis* strains, and (*b*) *gravis* and *intermedius* strains; his 'specific' antigen was the major antigen in (*a*) and minor in (*b*) and his 'group' antigen major in (*b*) and minor in (*a*). The 'group' antigen was also found in other corynebacteria, notably *C. hofmannii*.

Whooping cough

Let us leave bacterial toxins and turn to other bacterial antigens; whooping cough vaccine and its ability to protect man and animals against infection with *Bordetella pertussi* will make an excellent example of a case where our state of knowledge is 'betwixt and between'.

Our knowledge of whooping cough differs in many ways from our knowledge of diphtheria, though here again we have a vaccine which has been used to immunize children with great success and so to eliminate the disease. We have seen that the causative organism *Bordetella pertussis*, although it does not normally infect mice can be made to do so, by intracerebral injection or by instillation into the lung. Virulence for mice by these two routes does not go hand in hand nor does it correspond in either case to virulence in man though in whooping cough epidemics in man there are not the obvious differences in severity which are seen with diphtheria.

Mice can be protected against infection by both routes either actively or passively (Kendrick, Eldening, Dixon & Misner, 1947; North, Anderson & Graydon, 1941; Dolby & Standfast, 1958). When these routes of infection were used as challenges in mice to test whooping cough vaccines it was found that the orders of potency in which the vaccines were arranged were not the same; a vaccine which was potent in protecting mice against an intranasal challenge was not necessarily as potent against an intracerebral challenge and vice versa (Standfast, 1958). In passive protection tests, while most pertussis antisera (human and rabbit) protected against both challenges, some antisera protected

only against an intranasal challenge and some only against an intracerebral challenge. From these experiments it was obvious that there
were at least two antigens, one of which protected mice against an
intranasal challenge and the other against an intracerebral challenge
(Dolby & Standfast, 1958). Some vaccines contained both antigens,
some only one. Whether either of these antigens could protect man
against a natural infection, or whether both antigens were necessary, or
whether some different antigen was necessary for the protection of man
was not known, and no amount of experimentation in the laboratory
in vitro or in laboratory animals could answer this question. Only a
field trial in which vaccinated children, with suitable controls, were
exposed to natural infection under natural conditions, could answer this
question. Such trials were carried out in England by a Medical Research
Council committee. From the results of these trials it was discovered
that the laboratory assay in mice using the intracerebral route for
challenge ran hand in hand with the results found in natural infection
in man (M.R.C., 1951, 1956, 1959). The antigen which protects mice
against challenge by the intranasal route plays no significant part in
human prophylaxis.

With whooping cough we have a reasonably accurate mouse assay
for the essential protective antigen for man based on the correlation of
field trials and laboratory tests. On the other hand, we do not know
with whooping cough, unlike diphtheria, which is the essential antigen;
but the existence of a laboratory test means that we can set about the
isolation and purification of this antigen with some confidence. The
connexion between the laboratory animal and the test tube is not so
clear, but there is no doubt that this will come when we have a pure
'protective' antigen—that is antigenically pure not chemically pure.
We now know, at least, that some antigens of *Bordetella pertussis*
which can be demonstrated *in vitro* in the laboratory and in laboratory
animals, are not the 'protective' antigens for man. Some of these
antigens can be demonstrated in man by their ability to elicit antibodies
in the sera of human convalescent patients or of vaccinated children.
The protective antigen in pertussis has no connexion with the virulence
of the strain. One vaccine known as Pillemer's P.A. which has been
shown to be effective in protecting children (Pillemer, Blum & Lepow,
1954; M.R.C., 1959) was prepared from a strain of *B. pertussis* which
most people would regard as 'rough'.

The protective antigen is not the only substance that reacts and is
possibly antigenic in man. We have already noted that there are other
antigens in pertussis vaccines; more important is the toxicity associated

with some vaccines (Gerwe, 1961). Certain batches of vaccine cause more reactions in children than other batches but this has nothing to do with their immunizing potency. All children will not react to a 'bad' batch of vaccine, the difference may be between 2 % reactors to a 'good' vaccine and 20 % reactors to a 'bad' one. On the other hand, we have specifications for many toxicity tests in laboratory animals ranging from skin reactions in day-old mice to pyrogen tests in year-old rabbits; the need is for correlation between reactions in children and one laboratory test, but the collection of this kind of evidence is slow and difficult.

To sum up, we can say that in whooping cough there is a test in laboratory animals for the human immunizing antigen and, apart from diphtheria and tetanus vaccines, this is the only bacterial vaccine commonly used in man about which this can be said. We do not know what the 'protective' antigen is and we do not know how it works; we can measure the virulence of pertussis strains in laboratory animals but we do not really know on what this virulence depends and we do not know anything about virulence in man, except that it is not reflected in laboratory animals.

Typhoid

In typhoid we have much the same pattern, but the correlations are quite different. With *Salmonella typhi* strains there is close correlation between measurements *in vitro*, and *in vivo*, of virulence and immunity in mice and other laboratory animals. However, this correlation only holds in laboratory animals none of which suffer naturally from infection with *S. typhi*. The correlation is closely associated with the Vi-antigen; here we should note that many Vi-negative strains of *S. typhi* can still kill mice (Standfast, 1960b; Gaines *et al.* 1961) so that the Vi-antigen is not the only factor in virulence, though it is obviously the most important in laboratory animals. When we turn to typhoid prophylaxis in man, however, we are plunged into almost total ignorance.

In spite of the fact that typhoid vaccination in man was first tried about 1897 it was not until the publication by the Yugoslav Typhoid Commission (1957, 1962) of the results of the field trials run in Yugoslavia by the Commission and the World Health Organization that we knew unequivocally that at least some typhoid vaccines protected man against natural infection. Unfortunately these trials and the extensive laboratory testing that went hand in hand with them showed that we have no satisfactory laboratory method for the assay of potency of typhoid vaccines to be used in man (Ikic, 1956; Edsall, Carlson, Formal & Benenson, 1959; Standfast, 1960a). Later field trials in

British Guiana and in Poland, Yugoslavia and Russia have confirmed the first results, and we now have samples of several vaccines which have been proved to be effective in protecting man against natural infections of typhoid in areas where typhoid is endemic.

It is not difficult to immunize mice with a variety of vaccines against various challenges with *Salmonella typhi*. The difficulty is to find a method of assay which will arrange vaccines in the same order of potency as is found in field trials in man. This difficulty suggests that perhaps infection, and certainly the mechanism of immunity, are different in mice and men. An illustration of the difficulties in the laboratory is shown by some recent tests on two vaccines coded 'K' and 'L'. Vaccine 'K' is known to be significantly better than 'L' in field trials in man. In the laboratory 'K' was better than 'L' in a test in which the vaccines were given intraperitoneally to mice and the challenge given seven days later by the same route. When the vaccines were given subcutaneously to the mice which were then challenged 7 days later with the same intraperitoneal challenge dose, vaccine 'L' was significantly better than 'K'. If this was a single observation one would be inclined to wonder what had happened, but this test has been repeated in the U.S.A., England, Denmark, and the U.S.S.R., so one is therefore forced to the conclusion that it may be true. But until some idea of the mechanisms of immunity involved are obtained one cannot be certain that the first method is a satisfactory method of assay.

Cholera

The fourth example is cholera, where our knowledge is even more patchy. Anticholera vaccination was first attempted in Spain in 1885; and now, 80 years later, it is hoped that it will be possible in 1964–65, with the collaboration of the World Health Organization and the Indian and Pakistan Governments, to carry out a proper field trial which will answer the question—is cholera vaccine any use in man?

If a vaccine is found which protects man against a natural infection with *Vibrio cholerae* we shall know that this is possible and that the vaccine does contain a 'protective' antigen. The next step will be to develop a laboratory assay to detect and assay the good vaccines, for the one thing we know from some of the field trials of the past is that it is possible to have a bad vaccine. In the laboratory, cholera vaccine poses a problem different from those with other vaccines. It is not difficult to kill mice with *V. cholerae* suspensions, although the symptoms in mice do not resemble those of cholera in man; nor is it difficult to protect mice against such a challenge. This can be done with almost any cholera vaccine; in fact it is fair to say that it is more difficult to find a cholera vaccine that will not protect mice than one that will. Indeed if men were as easily protected as mice there would be no problem in cholera prophylaxis. This confusion suggests that we know very little about the behaviour in laboratory animals of the antigen (if there is one) which protects men and that we, also, know very little about the true virulence factor in laboratory animals or about the symptoms in any laboratory animal which correspond to cholera in man, though the work of De and his colleagues may change this (De, 1961).

Conclusions

This brief account of the phenomena *in vivo* and *in vitro* in four human infections is enough to show that our knowledge about 'protective' antigens is limited. We are nowhere near obtaining purified antigens nor proper tests for them *in vitro*; these may follow each other closely, for it is much easier to evolve a test *in vitro* from a single pure antigen than from the complicated mixtures of antigens of which most organisms are composed.

SUMMARY AND CONCLUSIONS

From this brief survey of a few examples of host–parasite relations we must come to the conclusion that very few generalizations can be made. Organism by organism, host by host, and even route by route, each case must be treated on its own. Two questions were asked on p. 70. These were: has the analysis *in vitro* satisfactorily explained the phenomena *in vivo*, or has the approach *in vitro* led us astray, either by oversimplification or by stressing the wrong correlations? I am doubtful whether we should, in the present state of our knowledge, attempt to answer them. Nevertheless, there is enough evidence, here and there, with pneumococcal polysaccharides and the Vi-antigen in mice for example, to suggest that explanations from experiments *in vitro* will be found for phenomena occurring *in vivo*.

We must distinguish between the natural host and the artificial laboratory host, and it seems to me from our experiences in the past that we would do far better to consider that experiments carried out in laboratory animals are '*in vitro*' (i.e. artificial) rather than '*in vivo*'. *In vitro* in the sense that laboratory animals are convenient self-incubating self-adjusting test tubes. Whichever way we regard laboratory animals they must not be considered as synonymous with natural hosts until careful experiments show that it is safe to consider them to be so. Our knowledge of the prophylaxis of typhoid must have been put back for

30 years by the belief that one could extrapolate from mice to men. When the reaction in the laboratory animal is different from the reaction in the true host it is preferable to have a clear understanding about the mechanisms of the former, if not of both.

An analysis of the causes and effects of immunity in man must go from man to laboratory animal and from laboratory animal to test tube. The analysis of virulence is more difficult. Direct observation in man is difficult if not impossible and experiments in animals are often difficult to design and carry out. I think the real danger here is that the investigation gets lost in the intriguing by-ways of experimental pathology. Does the experiment in which 2 ml. of sterile broth are injected intraperitoneally into a mouse, followed the next day by ten thousand million living organisms, so that samples of peritoneal exudate can be examined hourly, bear any relation to the events which follow a natural infection? Can such an experiment contribute to the analysis of the host–parasite relationship? Nevertheless, as we have seen, sometimes we can find a reason for virulence, we can demonstrate and even measure this by tests *in vitro* and this gives some hope for the future.

REFERENCES

AMIES, C. R. (1954). The pathogenesis of diphtheria. *J. Path. Bact.* **67**, 25.

BACON, G. A., BURROWS, T. W. & YATES, M. (1951). The effects of biochemical mutation on the virulence of *Bacterium typhosum*: the loss of virulence of certain mutants. *Brit. J. exp. Path.* **32**, 85.

BEUMER, J. (1938). L'action du serum de cheval normal sur la staphylotoxine. *Ann. Inst. Pasteur*, **61**, 54.

BURNET, F. M. & TIMMINS, C. (1937). Experimental infection with *Haemophilus pertussis* in the mouse by intranasal inoculation. *Brit. J. exp. Path.* **18**, 83.

COLE, R. (1917). The neutralization of anti pneumococcus immune bodies by infected exudates and sera. *J. exp. Med.* **26**, 453.

DE, S. N. (1961). *Cholera, its Pathology and Pathogenesis*. Edinburgh: Oliver and Boyd, Ltd.

DE, S. N. & CHATTERJE, D. N. (1953). An experimental study of the mechanism of action of *Vibrio cholerae* on the intestinal mucous membrane. *J. Path. Bact.* **66**, 559.

DOLBY, J. M. & STANDFAST, A. F. B. (1958). A comparison of passive protection tests against intranasal and intracerebral challenges with *Bordetella pertussis*. *Immunology*, **1**, 144.

DOWNIE, A. W. (1937). The specific capsular polysaccharide of pneumococcus type 1 and the immunity which it induces in mice. *J. Path. Bact.* **45**, 131.

EDSALL, G., CARLSON, M. C., FORMAL, S. B. & BENENSON, A. S. (1959). Laboratory tests of typhoid vaccines used in a controlled field study. *Bull. Wld Hlth Org.* **20**, 1017.

ENDERS, J. F., SHAFFER, M. F. & WU, C. J. (1936). Studies on natural immunity to pneumococcus type III. 3. Correlation of the behaviour *in vivo* of pneumococci type III varying in their virulence for rabbits with certain differences observed *in vitro*. *J. exp. Med.* **64**, 307.

EVANS, D. G. (1940). The production of Pertussis antitoxin in rabbits and the neutralisation of Pertussis, Para-pertussis and Bronchiseptica toxins. *J. Path. Bact.* **51**, 49.

EVANS, D. G. (1947). The failure of whooping-cough and adult sera to neutralise Pertussis toxin. *J. Path. Bact.* **59**, 341.

EVANS, D. G. & MAITLAND, H. B. (1937). The preparation of the toxin of *H. pertussis*: its properties and relation to immunity. *J. Path. Bact.* **45**, 715.

FELIX, A. & PITT, R. M. (1934). A new antigen of *B. typhosa*. Its relation to virulence and to active and passive immunisation. *Lancet*, ii, 186.

FINDLAY, H. T. (1951). Mouse virulence of strains of *Salmonella typhi* from a mild and a severe outbreak of typhoid fever. *J. Hyg., Camb.* **49**, 111.

FRISCH, A. W., TUPP, J. T., BARRETT, C. D. & PIDGEON, B. E. (1942). The specific polysaccharide content of pneumonic lungs. *J. exp. Med.* **76**, 505.

GAINES, S., TULLY, J. G. & TIGGERT, W. D. (1961). Enhancement of mouse virulence of non Vi variant of *Salmonella typhosa* by Vi-antigen. *J. Immunol.* **86**, 543.

GERWE, E. G. (1961). Tests for non-specific toxicity of pertussis vaccine. *Proc. VIIth Int. Congr. Microbiol. Stand.* p. 229.

HOYLE, L. (1942). The lipid antigens of *C. diphtheriae* and *C. hofmannii*. *J. Hyg., Camb.* **42**, 416.

IKIC, D. (1956). Problems of standardization and control of typhoid fever vaccines. *Atti del secondo congresso internazionale de standardizzazione immunomicrobiologica*, p. 311.

KAUFFMANN, F. (1951). *Enterobacteriaceae*. Copenhagen: E. Munksgaard.

KENDRICK, P. L., ELDENING, G., DIXON, M. K. & MISNER, J. (1947). Mouse protection tests in the study of pertussis vaccines. *Amer. J. publ. Hlth*, **37**, 803.

LANDY, M., GAINES, S. & SPRINZ, H. (1957). Studies on intra-cerebral typhoid infection in mice. 1. Characteristics of the infection. *Brit. J. exp. Path.* **38**, 15.

LANDY, M. & WEBSTER, M. E. (1952). Studies on Vi-antigen. 3. Immunological properties of purified Vi-antigen derived from *Escherichia coli* 5396/38. *J. Immunol.* **69**, 143.

MCLEOD, J. W. (1943). The types *mitis, intermedius* and *gravis* of *Corynebacterium diphtheriae*. A review of observations during the past ten years. *Bact. Rev.* **7**, 1.

MCLEOD, C. M. & KRAUSS, M. R. (1950). Relation of virulence of pneumococcal strains for mice to the quantity of capsular polysaccharide formed *in vitro*. *J. exp. Med.* **92**, 1.

MEDICAL RESEARCH COUNCIL (1951). The prevention of whooping cough by vaccination. *Brit. med. J.* i, 1464.

MEDICAL RESEARCH COUNCIL (1956). Vaccination against whooping cough. Relation between protection in children and results of laboratory tests. *Brit. med. J.* ii, 454.

MEDICAL RESEARCH COUNCIL (1959). Vaccination against whooping cough. The final report. *Brit. med. J.* i, 994.

MILES, A. A. (1951). The mouse pathogenicity and toxicity of *Proteus vulgaris*. *J. gen. Microbiol.* **5**, 307.

MILES, A. A. (1955). The meaning of pathogenicity. In *Mechanisms of Microbial Pathogenicity, Symp. Soc. gen. Microbiol.* **5**, 1.

MUELLER, J. H. (1941). Toxin production as related to the clinical severity of diphtheria. *J. Immunol.* **42**, 353.

NIKONOV, A. G., YEVSEYEVA, V. I., BIBIKOVA, P. D. & BICHUL, K. G. (1958). Cultivation of *Vibrio comma* in the small intestine of guinea pigs. *J. Microbiol., Moscow*, **51–53**, 1925.

NORTH, E. A., ANDERSON, G. & GRAYDON, J. J. (1941). Active immunization in experimental pertussis. *Med. J. Aust.* **2**, 589.

NORTON, J. F. & DINGLE, J. H. (1935). Virulence tests for typhoid bacilli and antibody relationships in antityphoid sera. *Amer. J. publ. Hlth*, **37**, 803.

NYE, R. N. & HARRIS, A. H. (1937). Viable pneumococci and pneumococci specific soluble substance in the lungs from cases of lobar pneumonia. *Amer. J. Path.* **13**, 749.

PIERSMA, H. D., BINGHAM, F. W., BOLYN, A. E., BRUECKNER, A. H., DETTWILER, H. A., FRANKHOUSER, W. W., MARSHALL, D. M. & NEWMAN, C. (1962). Recent laboratory experience in the U.S.A. with the pertussis toxicity test. *Round Table Conference on Pertussis Immunization, Prague*, **1**, 111.

PILLEMER, L., BLUM, L. & LEPOW, I. H. (1954). Protective antigen of *Haemophilus pertussis. Lancet*, i, 1257.

ROWLEY, D. (1954). The virulence of strains of *Bacterium coli. Brit. J. exp. Path.* **35**, 528.

ROWLEY, D. & JENKINS, C. R. (1962). Antigenic cross-reaction between host and parasite as a probable cause of pathogenicity. *Nature, Lond.* **193**, 151.

SCHIEBEL, I. F. (1943). Hereditary differences in the capacity of guinea-pigs for the production of diphtheria antitoxin. *Acta path. microbiol. scand.* **20**, 464.

SMITH, H. & KEPPIE, J. (1955). Studies on the chemical basis of the pathogenicity of *Bacillus anthracis* using organisms grown *in vivo*. In *Mechanisms of Microbiol. Pathogenicity, Symp. Soc. gen. Microbiol.* **5**, 126.

SMITH, H., WILLIAMS, A. E., PEARCE, J. H., KEPPIE, J., HARRIS-SMITH, P. W., FITZ-GEORGE, R. B. & WITT, K. (1962). Foetal erythritol: a cause of the localization of *Brucella abortus* in bovine contagious abortion. *Nature, Lond.* **193**, 47.

STANDFAST, A. F. B. (1951). The virulence of *Haemophilus pertussis* for mice by the intranasal route. *J. gen. Microbiol.* **5**, 250.

STANDFAST, A. F. B. (1958). The comparison between field trials and mouse protection tests against intranasal and intra-cerebral challenges with *Bordetella pertussis. Immunology*, **1**, 135.

STANDFAST, A. F. B. (1960a). A report on the laboratory assays carried out at the Lister Institute of Preventive Medicine on the typhoid vaccines used in the field study in Yugoslavia. *Bull. Wld Hlth Org.* **23**, 37.

STANDFAST, A. F. B. (1960b). Experiments with Vi-negative strains of *Salmonella typhi. Bull. Wld Hlth Org.* **23**, 47.

STOCKER, B. A. D. (1959). Bacterial genetics and infectious disease. *J. med. Educ.* **34**, 354.

TAYLOR, J., MALTBY, M. P. & PAYNE, J. M. (1958). Factors influencing the response of ligated rabbit-gut segments to injected *Escherichia coli. J. path. Bact.* **76**, 491.

VAN HEYNINGEN, W. E. (1950). *Bacterial Toxins*. Oxford: Blackwell Scientific Publications.

VAN HEYNINGEN, W. E. (1955). The role of toxins in pathology. In *Mechanisms of Microbial Pathogenicity, Symp. Soc. gen. Microbiol.* **5**, 17.

WILLIAMS, A. E., KEPPIE, J. & SMITH, H. (1963). Erythritol, its effect on the growth of various strains of *Brucella abortus* and its possible significance in infections with *Brucella melitensis* and *Brucella suis. J. gen. Microbiol.* **31**, xxiii.

WINTER, J. L. (1953). Development of antibodies in children convalescent from whooping cough. *Proc. Soc. exp. Biol., N.Y.* **83**, 866.

YUGOSLAV TYPHOID COMMISSION (1957). Field and laboratory studies with typhoid vaccines. A preliminary report. *Bull. Wld Hlth Org.* **16**, 897.

YUGOSLAV TYPHOID COMMISSION (1962). A controlled field trial of the effectiveness of phenol and alcohol typhoid vaccines. *Bull. Wld Hlth Org.* **26**, 357.

SAPROPHYTIC AND PARASITIC MORPHO-
LOGY OF PATHOGENIC FUNGI

F. MARIAT

Institut Pasteur, Paris

INTRODUCTION

The study of microfungi pathogenic for man and animals is based largely on their morphology. Historically, it was problems of form which first occupied research workers, and the main reason for this was the relatively large size of pathogenic fungi. At present, the physiology of these micro-organisms is more particularly studied, but the question of their morphology still gives rise to numerous interesting investigations. Moreover the morphology of a fungus is closely bound up with its physiological properties, and is conditioned by biochemical reactions the significance of which is only surmise (Nickerson, 1958; Ward, 1958).

One of the interesting aspects of medical mycology, but one which is often disconcerting, is the great morphological variability among pathogenic fungi. This variability depends not only on genetic properties but also on the environment. The conditions encountered by a pathogenic fungus in the soil (or, by analogy, on a culture medium) where it develops in the saprophytic state, are different from those in the parasitized host. This often results in a morphology which is different in the two situations. Frequently, the morphology of pathogenic fungi in their natural habitats is not well known, and we possess a better knowledge of their 'saprophytic' morphology observed *in vitro* on laboratory culture media. However, it is possible to compare the parasitic and saprophytic aspects of some pathogenic fungi by studying pathological materials such as pus and infected tissue sampled *in vivo*, and specimens sampled after growth *in vitro* on culture media.

The parasitic morphology of pathogenic fungi is generally simpler than their saprophytic morphology. This simplification is explained by the absence of sporulation *in vivo* except in some rare cases, for example *Aspergillus fumigatus* in aspergillar bronchitis. The simplification can also be explained by a reduction from a multicellular mycelial state to a unicellular state as in fungi causing deep mycoses. Sometimes fungal forms *in vivo* are more complex than *in vitro*. This is the case in lesions where the parasite aggregates in specific parasitic granules when, however, it is more a question of colony structure than of actual cellular form.

A knowledge of saprophytic and parasitic morphology is important, not only from an academic point of view, but also from a practical one. It is essential in medical mycology that diagnosis is established by the examination of pathological specimens (direct examination or histological examination) or by the study of cultures obtained. For epidemiological research the saprophytic morphology, particularly in the soil, is of primary importance. Finally, a knowledge of the physiological factors governing the morphology of fungi is interesting on a theoretical level; such factors can also be used to obtain *in vitro* a parasitic form for the production of more potent antigens or vaccines.

Examples of morphological duality *in vivo* and *in vitro* are many and varied and it is not possible within the limits of this review to examine them all. We will not linger, therefore, over those cases in which the fungal form *in vivo* is a barely modified reflexion of the form *in vitro* (e.g. in aspergillosis, mucormycosis and cryptococcosis); neither will we linger over the interesting cases of the dermatophytes which *in vitro* present a morphology of varying richness according to the species, and *in vivo* a reduction to fragmented filaments.

We will limit this study to some examples taken from: (*a*) pathogenic fungi or actinomycetes forming granules and clubs in the parasitic state; and (*b*) dimorphic fungi taking on a spherule form or a unicellular form within the host.

FUNGI-FORMING GRANULES

The word granules is generally used to mean an agglomeration of varying density of fungal or actinomycotic filaments within parasitized tissue. In a broad sense this can be applied to, for example, the fungal masses of bronchial aspergillomas. However, it is often given a more restricted meaning to define the small agglomerations encountered in mycetomas. *In vivo* it is not the true form of a single fungal or actinomycotic cell that one sees, but that of a group of cells. It is therefore more correct to compare the morphology of granules to those of colonies obtained *in vitro*. The granules of mycetomas are not simple, disorganized masses of filaments; they have constant specific form, composition and colour. The mere examination of granules in a histological section in most cases permits diagnosis of the species of the parasite in question (Mackinnon, 1951; Abbott, 1954; Mackinnon & Artagaveytia-Allende, 1956; Segretain & Mariat, 1958; Rey, 1961). This is true for actinomycotic mycetomas caused by *Streptomyces madurae*, *S. pelletieri* and *S. soma-*

liensis, and in the case of fungal mycetomas caused by *Madurella mycetomi* (Brumpt, 1906; Camain, Segretain & Nazimoff, 1957), *Leptosphaeria senegalensis* (Baylet, Camain & Segretain, 1959), *Pyrenochaeta romeroi* (Borelli, 1959), *Neotestudina rosatii* (Rosati *et al.* 1961), *M. grisea* (Mackinnon, Ferrada-Urzúa & Montemayor, 1949*a*). On the other hand, it is sometimes difficult to distinguish vesicular granules of *M. mycetomi* from those of *L. senegalensis*; it is hardly possible to differentiate in section a mycetoma caused by *Monosporium apiospermum* (*Allescheria boydii*) from one caused by a *Cephalosporium*; finally, it is not

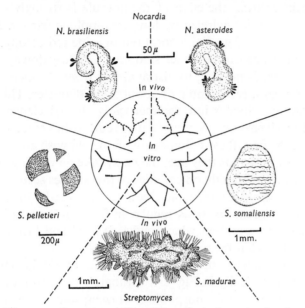

Fig. 1. Diagrammatic representation of the principal agents of actinomycotic mycetomas *in vitro* and *in vivo*.

possible to distinguish the parasitic granules of *Nocardia brasiliensis* from those of *N. asteroides* (Destombes, Mariat, Nazimoff & Satre, 1961). In the latter cases, diagnosis necessarily requires the examination of the saprophytic form *in vitro*.

The following examples of fungal behaviour *in vivo* and *in vitro* illustrate the contrast of properties in the two environments.

A diagrammatic representation of the main agents causing actinomycotic mycetomas is shown in Fig. 1. The microscopic study of actinomycetes from culture *in vitro* provides little information; *Streptomyces* can be distinguished from *Nocardia*, but with difficulty (González-Ochoa & Sandoval, 1955; Mackinnon & Artagaveytia-Allende, 1956; Mariat, 1962, 1963). A study of their macroscopic and morphological characters and their physiological properties is essential for diagnosis of the species (Mackinnon & Artagaveytia-Allende, 1956; Mariat, 1958). *S. madurae* appears in culture as a white to pinkish colony. The surface of the colony is glossy, smooth or slightly downy and furrowed with folds of varying depth. Microscopic examination shows only branched, non-fragmented filaments, 1μ diam. and sometimes (González-Ochoa & Sandoval, 1955) forming chains of spores characteristic of the species. *In vivo*, *S. madurae* forms characteristic, cartographic granules measuring from 0·5 to 5 mm. In sections stained with haemalum (haematoxylin–eosin–saffron) the edge of the granule is strongly stained with haematoxylin and is encircled by long eosinophilic fringes. *S. pelletieri* gives, *in vitro* on acid media, a coral red or wine red colony. The finely cerebriform surface can be either smooth or slightly downy. Under the microscope there is nothing to distinguish it from *S. madurae*. *In vivo*, *S. pelletieri* gives a red granule of about 500μ diameter. This originally spherical granule divides into geometrically shaped elements; haematoxylin colours it strongly. *S. somaliensis* in culture has colonies of varying appearance. At first, these colonies are often yellowish and covered with a white or grey down; as they age, they may take on varying shades from white to black. Frequently on colonies of this species sectoral variations are seen. There is no microscopic characteristic worthy of note. *In vivo* the granule of *S. somaliensis* is regular in form, 1–2 mm. diameter and impregnated with a cement which makes it hard and brittle. It can be stained only by eosin from haemalum.

The principal agents of fungal mycetomas or maduromycoses are shown diagrammatically in Fig. 2. Fortunately these fungi often have a characteristic microscopic morphology *in vitro*; some of them even have a sexual type of reproduction (e.g. *Leptosphaeria senegalensis, Allescheria boydii* and *Neotestudina rosatii*). Hence, to the specific characteristics shown *in vivo* and observed in histological section, the characteristics shown *in vitro* can be added for diagnosis (Destombes & Segretain, 1963). *In vivo*, some of the granules may be impregnated with cement which may be coloured by a pigment synthesized by the fungus (*Madurella mycetomi, L. senegalensis* and *Pyrenochaeta romeroi*); others are soft and not coloured (*A. boydii, N. rosatii* and *Cephalosporium* sp.; Mackinnon, 1951). *In vitro* each of these fungi is distinguished by characteristic conidial elements. *M. mycetomi*, which is the most widespread of the mycetoma fungi, shows sclerotia of 1 mm. diameter and on certain media, conidia isolated in short chains or grouped in small masses at the end of a phialide (Abbott, 1954; Borelli, 1957; Segretain, 1958). *M. grisea* is less widespread (Mackinnon, Ferrada-

Urzúa & Montemayor, 1949 b) and does not sporulate *in vitro*. *L. senegalensis*, found in West Africa as an agent of black grain mycetoma (Segretain, Baylet, Darasse & Camain, 1959), produces no asexual spores but has a sexual sporulation with perithecia, ascus and ascospores. *P. romeroi*, studied by Borelli (1959), causes black grain mycetomas and produces, *in vitro*, filaments and pycnidia containing a multitude of small asexual spores. *Monosporium apiospermum*, the

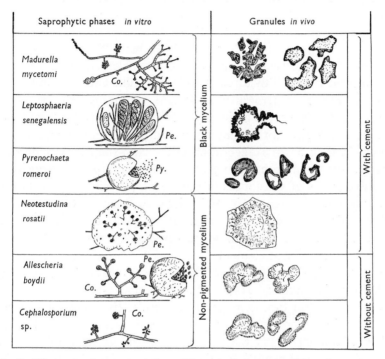

Fig. 2. Diagrammatic representation of the principal agents of fungal mycetomas *in vitro* and *in vivo*. (*Co.*, conidium; *Pe.*, perithecium; *Py.*, pycnidium).

cause of a white grain mycetoma (Shear, 1922), was studied by Emmons (1944). This fungus produces conidia generally at the extremity or laterally on a conidiophore. It has a sexual state (*A. boydii*) characterized by small black perithecia containing very ephemeral asci enclosing ascospores. Some white grain mycetomas are caused by species of *Cephalosporium*, filamentous fungi bearing phialides producing conidia which remain grouped together in accumulations by means of mucus.

The ability of these micro-organisms to form granules *in vivo* is curious. It is probably related to development within reacting tissues, both the host and parasite participating in their elaboration. The study of the morphological evolution of mycetoma granules would be simpler if it were easier to obtain them experimentally in animals. Recently some workers have obtained experimental granules with *Allescheria boydii* (Vanbreuseghem & Benaerts, 1955), *Madurella mycetomi* (Borelli, 1957; Murray, Spooner & Walker, 1960) and with *Nocardia brasiliensis* and *N. asteroides* (Destombes *et al.* 1961; Macotela-Ruiz & Mariat, 1963). One can hope for favourable developments from this work.

FUNGI FORMING CLUBS

In actinomycosis caused by anaerobic actinomycetes the granules formed *in vivo* are surrounded by a rosette of acidophilic clubs. Similar formations are found *in vivo* around fungal, actinomycotic or bacterial structures. *In vitro* some actinomycetes show enlarged extremities and these morphological forms may be related to the club *in vivo*. Thus, the club would then be an exaggeration of the club-shaped extremity of the filament, the membrane being thickened *in vivo* by apposition of successive layers of substances of a mineral, protein, polysaccharide or other nature. The recent investigations of Pine & Overman (1963) confirm this hypothesis. This explanation could be equally valid for *Nocardia* mycetomas but it is less acceptable for clubs of bacterial or fungal origin. Clubbed elements may be formed from tubercle or para-tubercle bacilli (Meyer & Mayer, 1927) or from true bacteria, for example staphylococci (Magrou, 1914, 1919). Fungi may give rise to clubs which develop *in vivo* on the periphery of spherules (in coccidioidomycosis), filaments (in aspergillosis) or yeast-like elements (*Candida*, *Sporotrichum*). Similar clubs may be formed round deposits of foreign inorganic bodies (Levaditi & Dimancesco-Nicolau, 1926) when introduced into an animal. Whatever their origin these clubs always have a similar appearance and arrangement. True clubs are only very rarely found *in vitro*.

As the thickening of the membrane of a filament cannot always be involved, various other hypotheses for the formation of clubs have been put forward (Moore, 1946). Some of these hypotheses are unlikely, others admissible. Magrou (1914, 1919) thought that the formation of clubs could take place only *in vivo* at the centre of a chronic lesion, and that the clubs resulted from a sort of balance between the host cells and the parasite (Magrou & Mariat, 1954). Magrou's experiments on staphylococcal granules, and the observations we have carried out on the asteroid bodies of *Sporotrichum schenckii* (Mariat & Drouhet, 1954), show that the clubs have on their edge eosinophilic substance which, *in vivo*, surrounds the staphylococcal granules or the cells of *Sporotrichum*. This coating of living micro-organisms in a semi-fluid matrix also occurs when killed tubercle bacilli are used; granules with well differentiated clubs are obtained and, besides these, bacilli enclosed in a non-differentiated eosinophilic jelly (Meyer & Mayer, 1927). A similar phenonomenon is seen round deposits of intramuscular tellurium (Levaditi & Dimancesco-Nicolau, 1926); this would also seem to be the situation in many cases observed in routine examinations of histological sections of mycoses. The formation of clubs could take place on the periphery of the semi-fluid droplet of colloidal nature which envelops the various parasites or foreign 'actinomycogenous' bodies owing to a physical phenomenon. In this connexion, an experiment of Leduc's (1906) is worth mentioning; he observed beautiful clubbed microscopic forms on the edge of a drop of indian ink in physiological saline solution. It is probable that not all clubs or asteroid formations observed *in vivo* depend on a single phenomenon. This interesting problem deserves closer study.

DIMORPHIC FUNGI

Dimorphic pathogenic fungi have *in vitro* a mycelial saprophytic phase, generally sporulating, and *in vivo* a distinct parasitic phase: either budding yeast, sclerotic cell or spherule. These dimorphic fungi are the cause of serious visceral mycoses. They can be arranged in four categories: (1) *Candida albicans* and other pathogenic *Candida* species forming *in vitro* just as *in vivo*, filaments and yeasts at one and the same time; (2) fungi forming *in vivo* spherules or similar structures (*Coccidioides immitis*, *Emmonsia crescens*); (3) fungi producing *in vivo* sclerotic cells (agents of chromomycosis); and (4) fungi with a purely yeast-like parasitic phase (*Histoplasma capsulatum*, *Blastomyces dermatitidis*, *B. brasiliensis* and *Sporotrichum schenckii*). These fungi are of special interest to medical mycologists. Now more knowledge is available on the morphology of the two forms due to progress in microscopy, investigations are being directed into the conversion *in vitro* of the saprophytic to the parasitic form. The factors causing the conversion of saprophytic phase to parasitic phase, which are optimum in the parasitized host, are now being studied, and explanations of their action at a cellular level are being sought. Recent reviews on dimorphic fungi are those of Nickerson (1951, 1954), Scherr & Weaver (1953), Pine (1962), Howard (1962).

Candida albicans

This fungus can cause infections of the mucous membranes on which it is normally present. *In vivo* it appears both as mycelial and yeast-like forms and thereby constitutes a peculiar case of dimorphism. The yeast cells are round, budding, with thin walls, and $2 \cdot 5$–4μ diameter. The filaments are elongated cells joined end to end and have a constriction at the point of the articulations (pseudomycelium). True cylindrical filaments are also present. Some authors believe the yeast-like form is the invasive form of *Candida* (Mackinnon, 1940; Scherr & Weaver, 1953); others, on the contrary, think the infective element *in vivo* is the filament (Young, 1958; Whittle & Gresham, 1960; Gresham & Whittle, 1961).

In vitro likewise, *Candida* has both budding cells and pseudomycelium. The latter bears numerous blastospores arising by budding at the points of junction of the elongated elements of the pseudomycelium. *Candida* is the only pathogenic yeast exhibiting filamentation *in vitro*. *C. albicans* forms, in addition, chlamydospores which are large, rounded, refractive, thick-walled cells. For mycological diagnosis it is important to obtain *in vitro* the formation of filaments and blastospores characteristic of the genus *Candida* and the chlamydospores specific for *C. albicans*.

Some study of the factors conditioning the morphology of *Candida albicans* was recorded in the work of Linossier & Roux (1890), Talice (1930) and Langeron & Guerra (1939–1940). Nickerson and his pupils have made a careful study of the dimorphism of this fungus (Ward, 1958). In a medium with a potato base and favourable for the culture of the mycelial phase of *C. albicans*, cysteine (10^{-2} M) inhibited the filamentous form but allowed a good growth of *C. albicans* and *C. tropicalis* in the yeast form (Nickerson & Van Rij, 1949). A certain specificity of active sulphydryl compounds for producing yeast forms was noted, for example, cysteine > glutathione > sodium thioglycollate. Penicillin and cobalt acetate (10^{-3} M) favoured the conversion of yeast form to mycelium, a conversion inhibited by cysteine. Nickerson (1951) and Nickerson & Mankowsky (1953) obtained an excellent mycelial growth of *C. albicans* with chlamydospores on a synthetic medium containing purified starch. If glucose or cysteine was added to the starch medium then the yeast phase was obtained. Glucose, an easily metabolized sugar, would allow the maintenance of a high content of –SH intracellular groups which, by favouring cellular division, forced *C. albicans* to develop in the yeast form. Various investigations by Nickerson and his school, for example, the isolation of complex compounds involved in the phenomena of cellular division (Nickerson & Falcone, 1956), confirm their initial hypothesis of the importance of –SH groups in the formation of the yeast form.

Other workers have been interested in the filamentation of *Candida* either from a theoretical (Scherr & Weaver, 1953) or from a practical viewpoint. In this connexion the morphology of *C. albicans* and the formation of chlamydospores has been the subject of numerous communications (Drouhet, 1956; Pavlatou & Marcelou, 1956; Gordon & Little, 1963).

Fungi causing chromomycosis

Chromomycosis is a verrucose dermatitis caused by various fungi with a complex saprophytic morphology (Carrion & Silva, 1947). Among the principal agents, three are of particular interest to us: *Phialophora pedrosoi*, *P. verrucosa* and *Cladosporium carrionii*. These three fungi which are dematiaceous show different types of sporulation *in vitro* (see below) but *in vivo* they are alike morphologically (Silva, 1958, 1960) and appear as single, brown, rounded, generally thick-walled cells, $6–12\mu$ or more in diameter (so-called sclerotic cells or *cellules fumagoïdes*) (see Fig. 3). These elements do not bud, but multiply by producing septa which may be at right angles. Often, but not invariably, the sclerotic cells meet inside giant cells *in vivo*. In skin scrapings, and more rarely in the tissue itself, filamentation of the fungus can be seen.

In vitro these sclerotic cells can give rise to various fungi of which three are now described. *Phialophora verrucosa* is characterized by brown, septate filaments $2–6\mu$ diameter, bearing terminal or lateral conidiophores, $5–12\mu$ long. The conidiophore, in the form of a swollen phialide, ends in a small cup from which spring, by budding, endogenous spores which remain agglutinated by mucus. In one strain of this fungus Ajello & Runyon (1953) have observed aborted perithecia. The most common cause of chromomycosis is *P. pedrosoi*. On brown black filaments, it bears three types of sporulation: (1) a rare phialide type, similar to that of *P. verrucosa*; (2) an *Acrotheca* type in which there are filaments of varying length whose surfaces bristle with spicules bearing spores, usually isolated; and (3) a short *Hormodendrum* type, characterized by a filament bearing on its tip two or three cells from which emerge conidia isolated, or in chains and sometimes branched. *Cladosporium carrionii* (Trejos, 1954) has as its only form of sporulation, a long, branched chain of spores; the oldest spores are nearest the short, swollen filament which bears them; they are also larger than the more distal. This fungus is difficult to distinguish from *C. trichoides*, the agent of a cerebral mycosis, which in the brain occurs as filaments and vesicles (Bindford, Thompson, Gorham & Emmons, 1952; Segretain, Mariat & Drouhet, 1955; Trejos, 1954).

The parasitic phase of several agents of chromomycosis has been obtained *in vitro* (Silva, 1957) on different media; the most satisfactory

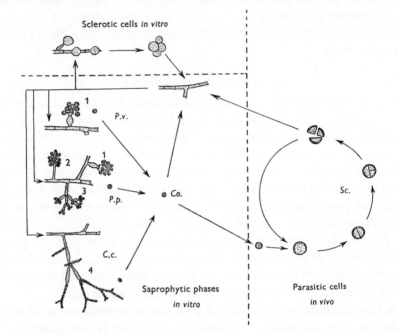

Fig. 3. Diagrammatic representation of the cycles of three agents of chromomycosis *in vitro* (saprophytic and sclerotic phases) and *in vivo*. (*P.v.*, *Phialophora verrucosa*; *P.p.*, *Phialophora pedrosoi*; *C.c.*, *Cladosporium carrionii*. 1, Phialide form; 2, *Acrotheca* form; 3, short *Hormodendrum* form; 4, long *Hormodendrum* form; *Co.*, conidium; *Sc.*, sclerotic cells.)

results were obtained on Francis's glucose cystine blood agar at 37°. Under these conditions the sclerotic cells develop like intercalary chlamydospores. They can also rise from a conidia. These sclerotic cells may develop septa in the manner of cells *in vivo*.

Fungi forming spherules

Coccidioides immitis

This fungus (Fig. 4) is the agent of an important mycosis limited to certain geographical areas (Fiese, 1958). *In vitro Coccidioides immitis* has branched, septate filaments, and at intervals, elements similar to arthrospores. These are rectangular or oval spores ($2\cdot5$–$3\mu \times 3$–4μ) which have thick walls and constitute the method of dissemination of the disease.

In vivo the fungus has a round structure 10–80μ diameter, sometimes more, and is bounded by a thick-walled membrane. This spherule (or sporangium) contains round endospores 2–8μ diameter. When mature, they are released by the rupture of the membrane. On rare occasions

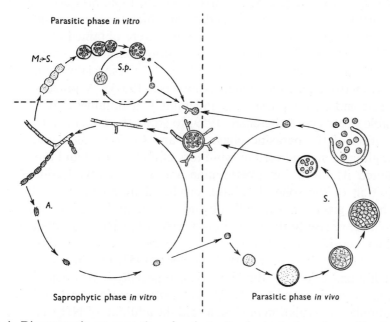

Fig. 4. Diagrammatic representation of cycle *in vitro* (saprophytic and parasitic phases) and *in vivo* of *Coccidioides immitis*. (*A*, arthrospores; *S*, spherule; *S.p.*, spherule phase; *M.→S.*, mycelium-spherule conversion.)

filaments are observed *in vivo* (Fiese, 1958). The endospore released into the tissues, or the infecting exogenous arthrospore, grows progressively larger and nuclear divisions take place (Emmons, 1942; Baker, Mrak & Smith, 1943; O'Hern & Henry, 1956). The cytoplasm of this spherule then breaks up and the endospores form round nuclei. When arthrospore-bearing filaments are inoculated into mice (Tarbet, Wright & Newcomer, 1952), some of the arthospores enlarge rapidly, become round, and their membrane thickens. They then take on the appearance of spherules which may be attached to one another. Endospores form, ripen and are released into the tissues.

When endospores or immature spherules are deposited on a culture medium *in vitro*, they normally develop into an arthrospore-bearing mycelium. However the spherule-endospore parasitic phase can also be obtained *in vitro*. As early as 1914, MacNeal & Taylor observed the development and temporary multiplication of parasitic forms by spreading a complex medium with pus containing spherules. These results were confirmed by some but not by others. Later a more or less pure spherule phase was obtained but always on complex media (Lack, 1938; Baker & Mrak, 1941; Burke, 1951; Conant & Vogel, 1954; Lubarsky & Plunkett, 1955). Converse (1955, 1956) produced spherules on a chemically defined liquid medium in which the glucose and mineral salt concentration was reduced; spherules were obtained after shaking for 72 hr. at 34°. No spherules were produced at less than 26°, and the optimum number was produced at 33–34° especially if an anionic surface active agent was added (Converse, 1957). A practically pure spherule-endospore phase was indefinitely maintained. It was then possible (Converse & Besemer, 1959) to study the factors favouring the formation of the parasitic phase (Na and Ca ions, oleic and linoleic acid, glutathione and certain inhibitors of metabolism) and those which inhibited its formation (methionine and sodium thioglycollate). An inhibition of the spherule phase, produced by compounds acting on sulphydryl groups was annulled by adding glutathione. The addition of cysteine gave rise to the formation of large abortive spherules and with sodium fluoride the spherules had dimensions similar to those of the sporangia seen *in vivo*. Henry & O'Hern (1957) obtained an excellent parasitic phase in a synthetic medium especially if this contained 4 % glucose. Cytosine, and to a lesser extent other purines and pyrimidines, encouraged spherulation *in vitro*, as did the addition of NH_4Cl (0.1–1 %) and an atmosphere containing up to 10 % CO_2 (Henry & O'Hern, 1957; Lones & Peacock, 1960 *a*, *b*). The favourable action of CO_2 was not noted by Converse & Besemer (1959).

Emmonsia crescens

Emmonsia in the lungs of the rodents appears as a large spore or multinuclear spherule up to 450μ or more in diameter. This adiaspore (Emmons & Jellison, 1960) is surrounded by a very thick wall. The species *E. parva* produces uninucleate adiaspores, smaller than those of *E. crescens*. In contrast to the spherule of *Coccidioides immitis*, the adiaspore of *E. crescens* cannot multiply *in vivo*.

In vitro the adiaspore germinates at 20–30° producing septate filaments 0.5–2μ diameter bearing short conidiophores. These conidiophores may be branched and bear conidia (2–$4\mu \times 2.5$–4.5μ), either isolated or on rare occasions grouped in chains of 2 or 3 elements. The parasitic adiaspore form of *Emmonsia* can easily be produced *in vitro* (Dowding, 1947; Carmichael, 1951; Emmons & Jellison, 1960) at 37° on any rich medium (e.g. glucose blood agar).

Rhinosporidium seeberi

This fungus, which cannot be grown artificially, causes an infection in some mucous membranes. *In vivo*, the sporangia appear as huge spherules (350 μ) containing an enormous number of endospores 6–7 μ diameter. The endospores are released into the tissue by the rupture of the sporangium membrane at a pore. Ashworth (1923) observed *in vivo* the evolution of spores into sporangia, which involved simultaneous nuclear divisions and the thickening of the wall during the development of the spherule. It is interesting to note the analogy between this fungus and *Coccidioides immitis*, and also *Ichthyophonus hoferi*, a phycomycete parasite of fish (Dorier & Degrange, 1960–61).

Yeast-like dimorphic fungi

Blastomyces

A diagrammatic representation of the cycles of these fungi is shown in Fig. 5. *Blastomyces dermatitidis*, the agent of North American blasto-mycosis (Gilchrist's disease), and *B. brasiliensis*, the agent of South American blastomycosis (the disease of Lutz & Splendore, 1908), have a similar mycelial morphology but slightly different parasitic morphology. *In vitro* at room temperature, they exhibit a regularly septate mycelium and round, oval or pear-shaped conidia, about 4 μ diameter. The spores are sessile or borne on a short lateral branch. In old cultures numerous intercalary chlamydospores can be seen. *In vivo*, *B. dermatitidis* takes on a spherical yeast form, 8–15 μ diameter bounded by a thick-walled membrane. When this cell buds, it is reunited to the daughter cell by a broad isthmus and takes on the appearance of a figure 8. *B. brasiliensis*, *in vivo* is a spherical yeast with thick walls, generally larger than *B. dermatitidis* (10–60 μ diameter). These yeast-like elements produce either one or several daughter cells of varying dimensions (2–3 μ × 8–10 μ). The multi-budded cells are characteristic of the species; they sometimes present a peculiar 'rueda de timon' appearance (Mackinnon, 1950).

Conant & Howell (1942) and Howard & Herndon (1960) have studied the morphology of *Blastomyces* during the conversion from the sapro-phytic to the parasitic phase *in vivo*. The mycelium is modified by the condensing of the cytoplasm into a few cells which become round and then multiply by budding. Howard & Herndon (1960) have observed this conversion from single filamentous elements and Carmichael (1951) from conidia.

The culture of the parasitic phase *in vitro* is relatively easy and was observed early (Ricketts, 1901). Hamburger (1907) demonstrated the importance of temperature in the conversion of *Blastomyces dermatitidis* from mycelium to yeast. De Monbreun (1935), Kelley (1939) and Conant (1939) obtained the yeast phase after incubation on blood medium at 37°. Conant & Howell (1942) cultivated the parasitic form of

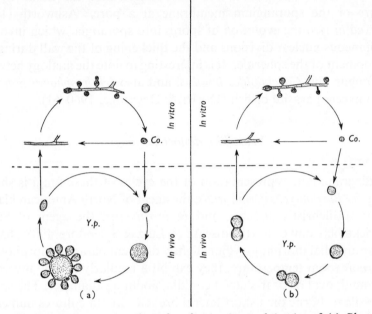

Fig. 5. Diagrammatic representation of cycles *in vitro* and *in vivo* of (*a*) *Blastomyces brasiliensis* and (*b*) *B. dermatitidis*. (*Co*, Conidium; *Y.p.*, yeast phase.)

both species on Sabouraud's medium at 37°. It was soon shown (Levine & Ordal, 1946; Salvin, 1949 *b*) that, whatever the medium, temperature is the factor determining conversion. This is the 'thermal dimorphism' of Nickerson & Edwards (1949), a phenomenon which is reversible. The conversion *in vitro* of *B. brasiliensis* from mycelium to yeast takes place under the same conditions as that of *B. dermatitidis* with, however, a tendency towards an aborted mycelial phase. Some authors note that different compounds are favourable to the respiration or growth of one or other of the dimorphic phases (Bernheim, 1942; Drouhet, 1954; Holliday & McCoy, 1955). Nickerson & Edwards (1949) showed that for both species the oxygen consumption (endogenous respiration) of the yeast phase was 5 to 6 times greater than that of the mycelial phase.

Histoplasma capsulatum *and* H. duboisii

A diagrammatic representation of the cycles of these fungi is shown in Fig. 6. *Histoplasma capsulatum* is present in soil (Emmons, 1949) and causes histoplasmosis or Darling's disease (Sweany, 1960), a systemic mycosis characterized by the presence of small yeasts in the cells of the reticulo-endothelial system. An African form of this disease has been observed recently (Dubois, Janssens, Brutsaert & Vanbreuseghem, 1952;

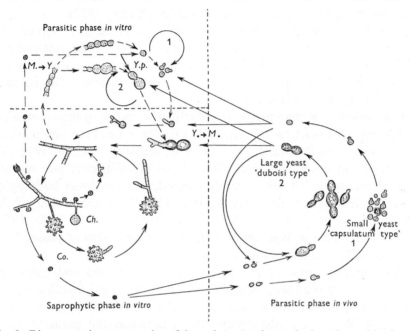

Fig. 6. Diagrammatic representation of the cycle *in vitro* (saprophytic and parasitic phases) and *in vivo* of *Histoplasma capsulatum* (1) and *H. duboisii* (2). (*Y.*, yeast; *Y.p.*, yeast phase; *M.→Y.*, mycelial-yeast conversion; *Y.→M.*, yeast-mycelial conversion; *Co.*, conidium; *Ch.*, chlamydospores.)

Drouhet, 1962); the fungus responsible, *H. duboisii* (or *H. capsulatum* var. *duboisii*), has a larger parasitic yeast form.

In vitro, at room temperature on the usual media, these *Histoplasma* are similar in appearance (Pine, 1960). A septate mycelium ($1-4\mu$ diameter), produces two types of conidia: (*a*) spherical or pyriform macroconidia ($2-5\mu$), either sessile or pedicellate; and (*b*) large ($10-25\mu$) macroconidia, round or pyriform with smooth or tuberculate walls; some tubercles can reach a length of 5μ. The tuberculate macroconidia (or chlamydospores) are characteristic of both *H. capsulatum* and *H. duboisii*. The growth and metabolism of the mycelial saprophytic phase of *H. capsulatum* has been studied, together with the effect of various factors such as amino acids, various sources of carbon and vitamins (Salvin, 1949a; Scheff, 1950; Rowley & Pine, 1955).

In vivo, Histoplasma capsulatum forms small (1–4μ) spherical, intracellular yeast cells which cause tissue reaction of the histiocytic type (Levaditi, Drouhet, Segretain & Mariat, 1959). The yeast forms may bud and are surrounded by a halo which is an artifact. Larger fungal cells are also encountered (Schwarz & Drouhet, 1957) and filamentous forms have sometimes been observed in vivo, particularly in certain areas of necrosis in experimentally inoculated mice (Haley, 1952). In lesions, H. duboisii is found in giant cells and appears as large lemon-like yeasts (6–15μ) surrounded by a thickened double-walled membrane. Lipid droplets can be seen inside. The yeasts bud at their extremities and remain joined in small chains of two or three elements. The isthmus produced in budding is narrow.

The evolution of the yeast in vivo during the course of the conversion of the mycelial (M.) to the parasitic (Y.) form has been summarized in a recent review (Howard, 1962). Pine & Webster (1962) followed this evolution and observed the swelling of the mycelial cells into moniliform elements, which subsequently budded, and the direct budding of mycelial elements to give blastospores. The direct conversion of the microconidia into yeasts has also been observed (Dowding, 1948, 1950; Milne, 1957; Pine & Webster, 1962), but the formation of yeasts from macroconidia is less well understood.

The parasitic phase of H. capsulatum was obtained and cultivated in vitro. This allowed the study of factors permitting the mycelium-yeast (M.→Y.) conversion, and those affecting the growth and metabolism of the yeast phase. De Monbreun (1934) obtained the parasitic yeast phase of H. capsulatum in vitro at 37° in a serum or a rabbit blood medium in sealed tubes. Campbell (1947) obtained M.→Y. conversion at 37° on blood agar containing glucose and cysteine. The sulphydril compounds were essential for the conversion M.→Y. in vitro (Salvin, 1949a) and favoured the maintenance of the yeast form at 25° (Scherr, 1957). Carbon dioxide (15–30 %) stimulated the conversion (Bullen, 1949), as did also chelating agents by exhausting the Ca^{2+} ions (Pine & Peacock, 1958). A temperature of 37° is generally required for the M.→Y. conversion, but with media containing certain compounds lower temperatures are equally effective (Pine, 1957). Chemically defined media on which the M.→Y. conversion is optimum have been formulated (Pine & Peacock, 1958; Pine & Drouhet, 1963). Certain compounds are favourable to the growth of the yeast phase, for example Zn^{2+} and Mg^{2+} ions, glucose (Pine, 1954); citric and α-keto glutaric acids (Pine &

Peacock, 1958); cystine, cysteine and methionine (Salvin, 1949 a; Pine & Peacock, 1955); and a mixture of glutamic acid, aspartic acid and cysteine (Pine, 1954). Some of these materials may not be the same as those permitting $M. \rightarrow Y.$ conversion.

Sporotrichum schenckii

This dimorphic fungus (see Fig. 7) causes sporotrichosis, a subacute or chronic subcutaneous infection, which rarely affects viscera. The mycelial form is seen in nature and *Sporotrichum schenckii* has been isolated from plants (de Beurmann & Gougerot, 1911; Brown, Wein-

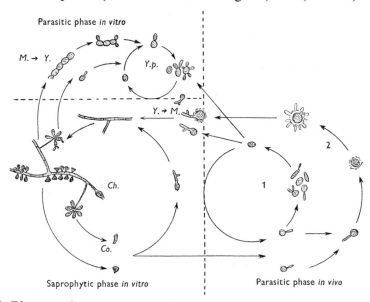

Fig. 7. Diagrammatic representation of the cycle *in vitro* (mycelial and yeast phases) and *in vivo* of *Sporotrichum schenckii* (*Y.p.*, yeast phase; *M.→Y.*, mycelial-yeast conversion; *Y.→M.* yeast-mycelial conversion; *Co.*, conidium; *Ch.*, chlamydospore; 1, 'cigar bodies'; 2, asteroid bodies).

troub & Simpson, 1947). *In vitro* at normal temperatures, *S. schenckii* has septate hyphae bearing pedicellate round or oval conidia ($1 \cdot 5 - 3 \mu \times 2 \cdot 5 - 5 \mu$), usually with thin walls. The conidia are united in a cluster at the tip of an upright conidiophore or in a sheath round the filaments. Cone-shaped conidia are also encountered (Mariat, Lavalle & Destombes, 1962).

Although in spontaneous human lesions the fungal parasite is barely visible, in experimental animals it is frequently an elongated element ($1 - 3 \mu \times 2 - 5 \mu$) capable of budding. These 'conidies-levures' or 'cigar bodies' are surrounded by halo-like artifacts giving the impression of a capsule. In areas of necrosis, asteroid bodies (Mariat & Drouhet,

Table 1. *Morphology in vivo and in vitro of the principal fungi pathogenic for man and animals*

Pathogenic fungi	Mycosis	Morphology in vivo						Morphology in vitro					
		Sclerotic cells	Granules	Spores	Spherules	Yeasts	Filaments	Yeasts	Sexual sporulation	Internal spores	Conidia	Arthrospores	Filaments
*Nocardia asteroides**	Nocardiosis	—	—	—	—	—	+	—	—	—	—	+	+
*N. asteroides**	Actinomycotic mycetomas	—	+	—	—	—	—	—	—	—	—	+	+
*N. brasiliensis**	Actinomycotic mycetomas	—	+	—	—	—	—	—	—	—	—	+	+
Pathogenic streptomyces*	Actinomycotic mycetomas	—	+	—	—	—	—	—	—	—	+	—	+
Madurella mycetomi	Maduromycosis	—	+	—	—	—	—	—	—	—	+	—	+
Leptosphaeria senegalensis	Maduromycosis	—	+	—	—	—	—	—	+	—	+	—	+
Allescheria boydii	Maduromycosis	—	+	—	—	—	—	—	+	—	+	—	+
Various dermatophytosis	Ring Worm (dermatophytosis)	—	—	+	—	—	+	—	+†	—	+	—	+
Candida	Candidiasis	—	—	—	—	+	+	+	—	—	—	—	+
Cryptococcus neoformans	Cryptococcosis	—	—	—	—	+	—	+	—	—	—	—	—
Pathogenic aspergilli	Aspergillosis	—	—	—	—	—	+	—	+†	—	+	—	+
Pathogenic aspergilli	Aspergilloma	—	+	—	—	—	—	—	+†	—	+	—	+
Pathogenic aspergilli	Aspergillar bronchitis	—	—	—	—	—	+	—	+†	—	+	—	+
Pathogenic Mucoraceae	Mucormycosis	—	—	+	—	—	+	—	—	+	—	—	+
Basidiobolus ranarum	Phycomycosis	—	—	—	—	—	+	—	+	+	—	—	+
Histoplasma capsulatum	Histoplasmosis	—	—	—	—	+	—	—	—	—	+	—	+
Blastomyces dermatitidis	Blastomycosis	—	—	—	—	+	—	—	—	—	+	—	+
B. brasiliensis	Blastomycosis	—	—	—	—	+	—	—	—	—	+	—	+
Sporotrichum schenckii	Sporotrichosis	—	—	—	—	+	—	—	—	—	+	—	+
Coccidioides immitis	Coccidioidomycosis	—	—	—	+	—	—	—	—	—	—	+	+
Emmonsia crescens	Adiaspiromycosis	—	—	—	+	—	—	—	—	—	+	—	+
Phialophora verrucosa	Chromomycosis	+	—	—	—	—	—	—	—	—	+	—	+
P. pedrosoi	Chromomycosis	+	—	—	—	—	—	—	—	—	+	—	+
Cladosporium carrionii	Chromomycosis	+	—	—	—	—	—	—	—	—	+	—	+
C. trichoides	Cladosporiosis	—	—	—	—	—	+	—	—	—	+	—	+

* Actinomycete and not fungus.

† Possible in some cases.

Table 2. *Factors required by dimorphic fungi for their conversion in vitro from saprophytic to parasitic morphological forms*

Fungi	Factors producing conversion when used with a medium adequate for growth of the saprophytic form			
	Temperature 35–37°	CO_2	–SH or –S–S compounds	Others, including more complex media
Candida albicans	–	–	×	
Fungi of chromomycosis	×	–	×	Francis's, cystine, blood agar medium
Coccidioides immitis	×	×	–	Decreased nutrient concentration, sodium aryl sulphonic acid
Emmonsia crescens	×	–	–	Glucose, blood agar medium
Blastomyces dermatitidis	×	–	–	
B. brasiliensis	×	–	–	
Histoplasma capsulatum	×	×	×	Chelating agents, thiamin
H. farciminosum	×	×	–	Complex media
Sporotrichum schenckii	×	×	–	Biotin
Penicillium marneffei	×	–	–	Amino acids, maltose

1954) may be encountered. These have a circular form with double walls surrounded by a corona of acidophilic substance, and at the centre of this substance varying degrees of club formation occur. Some asteroid bodies of *Sporotrichum schenckii* can reach a diameter of $20–30\mu$. In areas of necrosis, filaments occasionally occur (Okudaira, Ono, Araki & Fukushiro, 1959; Okudaira, Tsubura & Schwarz, 1961; Mariat *et al.* 1962).

The parasitic yeast form (*Y.*) was obtained *in vitro* by Lutz & Splendore (1908). The mycelium-yeast (*M.→Y.*) conversion has been observed at 37° on agar media when covered with pus (De Beurmann & Gougerot, 1912) or paraffin oil (Davis, 1913) and on other complex media (Campbell, 1945; Salvin, 1947). Drouhet & Mariat (1951) obtained the yeast phase on a chemically defined medium containing casein hydrolysate or amino acids, together with glucose, thiamine and biotin. Mechanical agitation was essential and a temperature of 37° was optimum for *M.→Y.* conversion. Below 28°, only the mycelial phase developed. Some amino acids, arginine and glycine, were as good as casein hydrolysate in this medium. Various sulphydril compounds had no effect on conversion (Drouhet & Mariat, 1952) but biotin was favourable (Mariat & Drouhet, 1953). A mixture of air and CO_2 facilitated conversion at *M.→Y.* at 37° in a complex medium (Bullen, 1949) and in either solid or liquid chemically defined medium containing an inorganic source of nitrogen (Mariat & Drouhet, 1952; Drouhet & Mariat, 1952). The yeast form obtained *in vitro* differs slightly from that observed *in vivo*. It is larger and more rounded, sometimes with multiple budding. In some cases, a hardly modified conidial form is found.

The yeast form can be produced from a hypha, the cells of which swell, become oval, and develop into a budding moniliform element. Conidia can also directly produce budding yeast forms. In tissue cultures of mouse peritoneal exudate cells, Howard (1961) observed two kinds of *M.→Y.* evolution: the formation of clubbed, budding element sat the tip of the hyphae or on the lateral branches; and the formation of chains of oidia (*sic*) which were isolated in budding unicellular forms.

Physiological studies on yeast and mycelial phases *in vitro* have shown that the nucleic acid composition of the two phases differs considerably (Mariat, 1959, 1960*a*). The RNA/DNA ratios of the mycelial phase vary from 17·3 to 6·01; those of the yeast phase from 1·40 to 0·71. The respiratory metabolism of the two phases is also different; the oxygen consumption of the yeast phase is about half that of the mycelial phase. Although CO_2 is a growth factor for each culture phase, it is required in greater quantity by the yeast phase (Mariat, 1960*b*).

Penicillium marneffei

This curious *Penicillium* (Segretain, 1959) causes mycosis in rodents in Viet-Nam. Cultivated on normal media at 25–30°, it takes on the characteristic form of *Penicillium*. *In vivo* it is found in the cells of the reticulo-endothelial system as is *Histoplasma capsulatum*. The invasive form of *H. capsulatum* is a yeast whereas the invasive form *in vivo* of *P. marneffei* is an arthrospore which multiplies by schizogenesis. *In vitro* the parasitic arthrospore phase is easily obtained in a liquid medium containing hydrolysate and maltose, agitated at 37°, or even on the normal agar media at the same temperature.

CONCLUSION

This study is intentionally limited to certain examples selected to demonstrate the dual aspect of the morphology of fungi in their saprophytic and parasitic states; these examples are summarized in Table 1. When possible the conditions, which *in vitro* permit the passage from the saprophytic to the parasitic form, have been indicated. This conversion is relatively well known for the dimorphic fungi (see Table 2). However, some parasitic states are still unexplained. Nothing is known of the factors conditioning the specific morphology of the granules in fungal or actinomycotic mycetomas. No explanation is available for the dual parasitic morphology, granules or isolated filaments as the case may be, which a single strain of *Nocardia asteroides* can show. It is not known why some agents of chromomycosis exhibit sclerotic cells or filaments. There is also the interesting question of the dermatophytes which, in passing from the saprophytic to the parasitic state, lose their usual mycological characteristics. They attack the hair according to a constant and special pattern. If the various forms of attack on hair by dermatophytes is well known, conditions causing the peculiar arrangement of the arthrospores in the parasitized hair are completely unknown. Numerous problems remain to be studied.

REFERENCES

ABBOTT, P. H. (1954). Mycetoma, a clinical and epidemiological study. Thesis for M.D. degree of the University of Cambridge.

AJELLO, L. & RUNYON, L. (1953). Abortive 'perithecial' production by *Phialophora verrucosa*. *Mycologia*, **45**, 947.

ASHWORTH, J. H. (1923). On *Rhinosporidium seeberi* with special reference to its sporulation and affinities. *Trans. roy. Soc. Edinb.* **53**, 301.

BAKER, E. E. & MRAK, E. M. (1941). Spherule formation in culture by *Coccidioides immitis*. *Amer. J. trop. Med.* **21**, 589.

BAKER, E. E., MRAK, E. M. & SMITH, C. E. (1943). The morphology, taxonomy and distribution of *Coccidioides immitis*. *Farlowia*, 1, 199.

BAYLET, J., CAMAIN, R. & SEGRETAIN, G. (1959). Identification des agents des maduromycoses du Sénégal et de la Mauritanie. Description d'une espèce nouvelle. *Bull. Soc. Pat. exot.* 52, 448.

BERNHEIM, F. (1942). The effect of various substances on the oxygen uptake of *Blastomyces dermatitidis*. *J. Bact.* 44, 533.

DE BEURMANN, L. & GOUGEROT, H. (1911). Les Sporotrichum pathogènes. Classification botanique. *Arch. Parasit.*, Paris, 15, 5.

DE BEURMANN, L. & GOUGEROT, H. (1912). *Les Sporotrichoses*. Paris: Félix Alcan.

BINDFORD, C. H., THOMPSON, R. K., GORHAM, M. E. & EMMONS, C. W. (1952). Mycotic brain abscess due to *Cladosporium trichoides*, a new species. *Amer. J. clin. Path.* 22, 535.

BORELLI, D. (1957). *Madurella mycetomi*: Fialides, fialospores, inoculación al ratón. *Bol. venez. Lab. clin.* 2, 1.

BORELLI, D. (1959). *Pyrenochaeta romeroi* n.sp. *Rev. derm. venezol.* 1, no. 4.

BROWN, R., WEINTROUB, D. & SIMPSON, M. W. (1947). Timber as a source of sporotrichosis infection. In *Sporotrichosis infections in mines of the Witwatersrand*, p. 5. Johannesburg.

BRUMPT, E. (1906). Les mycétomes. *Arch. Parasit.*, Paris, 10, 489.

BULLEN, J. J. (1949). The yeast like form of *Cryptococcus farciminosus* (*Histoplasma farciminosum*). *J. Path. Bact.* 61, 117.

BURKE, R. C. (1951). *In vitro* cultivation of the parasitic phase of *Coccidioides immitis* (18481). *Proc. Soc. exp. Biol., N.Y.* 76, 332.

CAMAIN, R., SEGRETAIN, G. & NAZIMOFF, O. (1957). Etude histopathologique des mycétomes du Sénégal. *Sem. Hôp. (P & B), Paris,* 33, 923.

CAMPBELL, C. C. (1945). Use of Francis' glucose cystine blood agar in the isolation and cultivation of *Sporotrichum schenckii*. *J. Bact.* 50, 233.

CAMPBELL, C. C. (1947). Reverting *Histoplasma capsulatum* to the yeast phase. *J. Bact.* 54, 263.

CARMICHAEL, J. W. (1951). The pulmonary fungus *Haplosporangium parvum*. II. Strain and generic relationships. *Mycologia,* 43, 605.

CARRION, A. L. & SILVA, M. (1947). Chromoblastomycosis and its etiologic fungi. In *Biology of Pathogenic Fungi*, p. 20. Waltham, Mass.: Chronica Botanica.

CONANT, N. F. (1939). Laboratory study of *Blastomyces dermatitidis*. *South Pacif. Congr.* 5, 853.

CONANT, N. F. & HOWELL, A., Jr. (1942). The similarity of the fungi causing South American blastomycosis (Paracoccidioidal granuloma) and North American blastomycosis (Gilchrist's disease). *J. invest. Derm.* 5, 353.

CONANT, N. F. & VOGEL, R. A. (1954). The parasitic growth phase of *Coccidioides immitis* in culture. *Mycologia,* 46, 157.

CONVERSE, J. L. (1955). Growth of spherules of *Coccidioides immitis* in a chemically defined liquid medium. *Proc. Soc. exp. Biol., N.Y.* 90, 709.

CONVERSE, J. L. (1956). The effect of physico-chemical environment on spherulation of *Coccidioides immitis* in a chemically defined medium. *J. Bact.* 72, 784.

CONVERSE, J. L. (1957). Effect of surface active agents on endosporulation of *Coccidioides immitis* in a chemically defined medium. *J. Bact.* 74, 106.

CONVERSE, J. L. & BESEMER, A. R. (1959). Nutrition of the parasitic phase of *Coccidioides immitis* in a chemically defined liquid medium. *J. Bact.* 78, 231.

DAVIS, D. J. (1913). The morphology of *Sporotrichum schenckii* in tissues and in artificial media. *J. infect. Dis.* 12, 453.

DESTOMBES, P., MARIAT, F., NAZIMOFF, O. & SATRE, J. (1961). A propos des mycétomes à *Nocardia*. *Sabouraudia,* 1, 161.

DESTOMBES, P. & SEGRETAIN, G. (1963). Les mycétomes fongiques. Caractères histologiques et culturaux. *Arch. Inst. Pasteur Tunis*, **39**, 273.

DORIER, A. & DEGRANGE, C. (1960–61). L'évolution de l'*Ichthyosporidium* (*Ichthyophonus*) *hoferi* chez les salmonides d'élevage. *Trav. Lab. Hydrobiol. Grenoble*, **52–53**, 7.

DOWDING, E. S. (1947). The pulmonary fungus *Haplosporangium parvum* and its relationship with some human pathogens. *Canad. J. Res.* E, **25**, 195.

DOWDING, E. S. (1948). The spores of *Histoplasma*. *Canad. J. Res.* E, **26**, 265.

DOWDING, E. S. (1950). *Histoplasma* and Brazilian *Blastomyces*. *Mycologia*, **42**, 668.

DROUHET, E. (1954). Action de certaines vitamines sur les champignons dimorphiques. *C.R. VIIIe Congr. Int. Bot.*, sect. 19, 62.

DROUHET, E. (1956). Contribution à l'étude des *Candida*. Thesis for M.D. degree of University of Paris.

DROUHET, E. (1962). Sur les histoplasmoses à petites et à grandes formes. *Arch. Inst. Pasteur Tunis*, **39**, 291.

DROUHET, E. & MARIAT, F. (1951). Sur le déterminisme du développement *in vitro* de *S. schenckii* sous la forme levure. *C.R. Acad. Sci., Paris*, **233**, 433.

DROUHET, E. & MARIAT, F. (1952). Etude des facteurs déterminant le développement de la phase levure de *Sporotrichum schenckii*. *Ann. Inst. Pasteur*, **83**, 506.

DUBOIS, A., JANSSENS, P. G., BRUTSAERT, P. & VANBREUSEGHEM, R. (1952). Un cas d'histoplasmose africaine. *Ann. Soc. belge Méd. trop.* **32**, 559.

EMMONS, C. W. (1942). Coccidioidomycosis. *Mycologia*, **34**, 452.

EMMONS, C. W. (1944). *Allescheria boydii* and *Monosporium apiospermum*. *Mycologia*, **36**, 188.

EMMONS, C. W. (1949). Isolation of *Histoplasma capsulatum* from soil. *Publ. Hlth Rep., Wash.* **64**, 892.

EMMONS, C. W. & JELLISON, W. L. (1960). *Emmonsia crescens* sp.n. and Adiaspiromycosis (Haplomycosis) in mammals. *Ann. N.Y. Acad. Sci.* **89**, 91.

FIESE, M. J. (1958). *Coccidioidomycosis*. Springfield: C. C. Thomas.

GONZÁLEZ-OCHOA, A. & SANDOVAL, M. L. A. (1955). Características de los Actinomicetes patógenos más comunes. *Rev. Inst. Salubr. Enferm. trop., Méx.* **15**, 149.

GORDON, M. A. & LITTLE, G. N. (1963). Effective dehydrated media with surfactants for identification of *Candida albicans*. *Sabouraudia*, **2**, 171.

GRESHAM, G. H. & WHITTLE, C. H. (1961). Studies of the invasive, mycelial form of *Candida albicans*. *Sabouraudia*, **1**, 30.

HALEY, L. D. (1952). 'Saprophytic form' of *Histoplasma capsulatum in vivo*. *Yale J. Biol. Med.* **24**, 381.

HAMBURGER, W. W. (1907). A comparative study of four strains of organisms isolated from four cases of generalized blastomycosis. *J. infect. Dis.* **4**, 201.

HENRY, B. S. & O'HERN, E. M. (1957). The production of spherules in a synthetic medium by *Coccidioides immitis*. *U.S. Publ. Hlth Rep.* **575**, 183.

HOLLIDAY, W. J. & McCOY, E. (1955). Biotin as a growth requirement for *Blastomyces dermatitidis*. *J. Bact.* **70**, 464.

HOWARD, D. H. (1961). Dimorphism of *Sporotrichum schenckii*. *J. Bact.* **81**, 464.

HOWARD, D. H. (1962). The morphogenesis of the parasitic form of dimorphic fungi. *Mycopathol. Mycol. appl.* **18**, 127.

HOWARD, D. H. & HERNDON, R. L. (1960). Tissue cultures of mouse peritoneal exudates inoculated with *Blastomyces dermatitidis*. *J. Bact.* **80**, 522.

KELLEY, W. H. (1939). A study of the cell and colony variations of *Blastomyces dermatitidis*. *J. infect. Dis.* **64**, 293.

LACK, A. R. (1938). Spherule formation and endosporulation of the fungus *Coccidioides in vitro*. *Proc. Soc. exp. Biol., N.Y.* **38**, 907.

LANGERON, M. & GUERRA, P. (1939–40). Valeur et nature des variations et dissociations de colonies chez les champignons levuriformes. *Ann. Parasit. hum. comp.* **17**, 447.

LEDUC, S. (1906). *Les bases physiques de la vie.* Paris: Masson.

LEVADITI, C. & DIMANCESCO-NICOLAU, O. (1926). Formations astéroïdes autour des dépôts telluriques. *C.R. Soc. Biol., Paris,* **95**, 531.

LEVADITI, J., DROUHET, E., SEGRETAIN, G. & MARIAT, F. (1959). Sur le caractère histiocytaire de l'histoplasmose à petites formes et le caractère giganto-cellulaire de l'histoplasmose à grandes formes. *Ann. Inst. Pasteur,* **96**, 659.

LEVINE, S. & ORDAL, Z. J. (1946). Factors influencing the morphology of *Blastomyces dermatitidis. J. Bact.* **52**, 687.

LINOSSIER, G. & ROUX, G. (1890). Recherches biologiques sur le champignon du muguet. *Arch. Méd. exp.* **2**, 222.

LONES, G. W. & PEACOCK, C. L. (1960a). Role of carbon dioxide in the dimorphism of *Coccidioides immitis. J. Bact.* **79**, 308.

LONES, G. W. & PEACOCK, C. L. (1960b). Studies of the growth and metabolism of *Coccidioides immitis. Ann. N.Y. Acad. Sci.* **89**, 102.

LUBARSKY, R. & PLUNKETT, O. A. (1955). *In vitro* production of the spherule phase of *Coccidioides immitis. J. Bact.* **70**, 182.

LUTZ, A. & SPLENDORE, A. (1908). Ueber eine bei Menschen und Ratten beobachtete mykose. Ein Beitrag zur Kenntnis der sogenanten Sporotrichosen. *Zbl. Bakt.* **45**, 631.

MACKINNON, J. E. (1940). Dissociation in *Candida albicans. J. infect. Dis.* **66**, 59.

MACKINNON, J. E. (1950). Naturaleza o significado de la forma en 'rueda de timon' de *Paracoccidioides braziliensis. An. Fac. Med. Montevideo,* **35**, 653.

MACKINNON, J. E. (1951). Los agentes de maduromicosis de los generos *Monosporium, Allescheria, Cephalosporium* y otros de dudosa identitad. *An. Fac. Med. Montevideo,* **36**, 153.

MACKINNON, J. E. & ARTAGAVEYTIA-ALLENDE, R. C. (1956). The main species of pathogenic actinomycetes causing mycetomas. *Trans. R. Soc. trop. Med. Hyg.* **50**, 31.

MACKINNON, J. E., FERRADA-URZÚA, L. V. & MONTEMAYOR, L. (1949a). *Madurella grisea* n.sp. *Mycopathol. Mycol. appl.* **4**, 384.

MACKINNON, J. E., FERRADA-URZÚA, L. V. & MONTEMAYOR, L. (1949b). Investigaciones sobre las maduromicosis y sus agentes. *An. Fac. Med. Montevideo,* **34**, 231.

MACNEAL, W. J. & TAYLOR, R. M. (1914). *Coccidioides immitis* and coccidioidal granuloma. *J. med. Res.* **30**, 261.

MACOTELA-RUIZ, E. & MARIAT, F. (1963). Sur la production de mycétomes expérimentaux par *Nocardia brasiliensis* et *Nocardia asteroides. Bull. Soc. Pat. exot.* **56**, 46.

MAGROU, J. (1914). *Les grains botryomycotiques. Leur signification en pathologie et en biologie générales.* Paris: Institut Pasteur.

MAGROU, J. (1919). Les formes actinomycosiques du staphylocoque. *Ann. Inst. Pasteur,* **33**, 344.

MAGROU, J. & MARIAT, F. (1954). Sur la genèse des massues dans les grains radiés d'origine fongique. *C.R. VIIIe Congr. Int. Bot.* Sect. 19, 69.

MARIAT, F. (1958). Physiologie des Actinomycètes aérobies pathogènes. *Mycopathol. Mycol. appl.* **9**, 111.

MARIAT, F. (1959). Sur la teneur en acides nucléiques des formes levure et mycélienne de *Sporotrichum schenckii. C.R. Acad. Sci., Paris,* **248**, 3468.

MARIAT, F. (1960a). Sur la composition en acides nucléiques de la phase mycélienne de *Sporotrichum schenckii. C.R. Acad. Sci., Paris,* **250**, 3368.

MARIAT, F. (1960b). Action de l'anhydride carbonique sur la croissance de *Sporotrichum schenckii*. *C.R. Acad. Sci.*, *Paris*, **250**, 3503.

MARIAT, F. (1962). Critères de détermination des principales espèces d'Actinomycètes aérobies pathogènes. *Ann. Soc. belge Méd. trop.* **4**, 651.

MARIAT, F. (1963). Des Nocardioses et des mycétomes à *Nocardia asteroides* et *Nocardia brasiliensis*. *Arch. Inst. Pasteur Tunis*, **39**, 309.

MARIAT, F. & DROUHET, E. (1952). Rôle de l'anhydride carbonique dans le développement de la phase levure de *Sporotrichum schenckii*. *C.R. Acad. Sci.*, *Paris*, **234**, 2554.

MARIAT, F. & DROUHET, E. (1953). Action de la biotine sur la croissance de *Sporotrichum schenckii*. *Ann. Inst. Pasteur*, **84**, 659.

MARIAT, F. & DROUHET, E. (1954). Sporotrichose expérimentale du hamster—Observation de formes astéroïdes de *Sporotrichum*. *Ann. Inst. Pasteur*, **86**, 485.

MARIAT, F., LAVALLE, P. & DESTOMBES, P. (1962). Recherches sur la sporotrichose. Etude mycologique et pouvoir pathogène de souches mexicaines. *Sabouraudia*, **2**, 60.

MEYER, K. & MAYER, K. (1927). Kolbenkranzbildung um tote tuberkelbacillen als reaktion des Wirtsorganismus. Ein Beitrag zur genese der Actinomyces formen. *Z. Hyg. InfektKr.* **108**, 38.

MILNE, H. A. (1957). The morphology and cytochemistry of *Histoplasma capsulatum*. *J. med. Lab. Tech.* **14**, 142.

DE MONBREUN, W. A. (1934). The cultivation and cultural characteristics of Darling's *Histoplasma capsulatum*. *Amer. J. trop. Med.* **14**, 93.

DE MONBREUN, W. A. (1935). Experimental chronic cutaneous blastomycosis in monkeys. A study of the etiological agents. *Arch. Derm. Syph.*, *N.Y.* **31**, 831.

MOORE, M. (1946). Radiate formation on pathogenic fungi in human tissue. *Arch. Path.* (*Lab. Med.*), **42**, 113.

MURRAY, I. G., SPOONER, E. T. C. & WALKER, J. (1960). Experimental infection of mice with *Madurella mycetomi*. *Trans. R. Soc. trop. Med. Hyg.* **54**, 335.

NICKERSON, W. J. (1951). Physiological bases of morphogenesis in animal disease fungi. *Trans. N.Y. Acad. Sci.* **13**, 140.

NICKERSON, W. J. (1954). Experimental control of morphogenesis in microorganisms. *Ann. N.Y. Acad. Sci.* **60**, 50.

NICKERSON, W. J. (1958). Biochemistry of morphogenesis. A report on symposium. VI. *Trans. IVth Int. Congr. Biochem.* **14**, 191.

NICKERSON, W. J. & EDWARDS, G. A. (1949). Studies on the physiological bases of morphogenesis in fungi. I. The respiratory metabolism of dimorphic pathogenic fungi. *J. gen. Physiol.* **33**, 41.

NICKERSON, W. J. & FALCONE, G. (1956). Identification of protein disulphide reductase as a cellular division enzyme in yeasts. *Science*, **124**, 722.

NICKERSON, W. J. & MANKOWSKY, Z. (1953). A polysaccharide medium of known composition favoring chlamydospore formation in *Candida albicans*. *J. infect. Dis.* **92**, 20.

NICKERSON, W. J. & VAN RIJ, N. J. W. (1949). The effect of sulfhydril compounds, penicillin and cobalt on the cell division mechanism of yeasts. *Biochim. biophys. Acta*, **3**, 461.

O'HERN, E. M. & HENRY, B. S. (1956). The cytological study of *Coccidioides immitis* by electron microscopy. *J. Bact.* **72**, 632.

OKUDAIRA, M., ONO, T., ARAKI, T. & FUKUSHIRO, R. (1959). Sporotrichosis with hyphal element in tissue; report of a biopsy case. *Trans. Jap. path. Soc.* **48**, 254.

OKUDAIRA, M., TUSBURA, E. & SCHWARZ, J. (1961). A histopathological study of experimental murine sporotrichosis. *Mycopathol. Mycol. appl.* **14**, 284.

Pavlatou, M. & Marcelou, U. (1956). Milieu favorisant la formation des chlamydospores de *Candida albicans*. *Ann. Inst. Pasteur*, **91**, 410.

Pine, L. (1954). Studies on the growth of *Histoplasma capsulatum*. I. Growth of the yeast phase in liquid medium. *J. Bact.* **68**, 671.

Pine, L. (1957). Studies on the growth of *Histoplasma capsulatum*. III. Effect of thiamin and other vitamins on the growth of the yeast and mycelial phases of *Histoplasma capsulatum*. *J. Bact.* **74**, 239.

Pine, L. (1960). Morphological and physiological characteristics of *Histoplasma capsulatum*. In *Histoplasmosis*, p. 40. Springfield: C. C. Thomas.

Pine, L. (1962). Nutritional determinants of fungous morphology. In *Fungi and Fungous Diseases*, p. 84. Springfield: C. C. Thomas.

Pine, L. & Drouhet, E. (1963). Sur l'obtention et la conservation de la phase levure d'*Histoplasma capsulatum* et *H. duboisii* en milieu chimiquement défini. *Ann. Inst. Pasteur*, **105**, 798.

Pine, L. & Overman, J. R. (1963). Determination of the structures and composition of the 'sulfur granules' of *Actinomyces bovis*. *J. gen. Microbiol.* **31**, 888.

Pine, L. & Peacock, C. L. (1955). Reaction of fumaric acid with cysteine. *J. Amer. chem. Soc.* **77**, 3153.

Pine, L. & Peacock, C. L. (1958). Studies on the growth of *Histoplasma capsulatum*. IV. Factors influencing conversion of the mycelial phase to the yeast phase. *J. Bact.* **75**, 167.

Pine, L. & Webster, R. E. (1962). Conversion in strains of *Histoplasma capsulatum*. *J. Bact.* **83**, 149.

Rey, M. (1961). Les mycétomes dans l'ouest Africain. Thesis for M.D. degree of the University of Paris. Paris: Foulon.

Ricketts, H. T. (1901). Oidiomycosis of skin. *J. med. Res.* **6**, 373.

Rosati, L., Destombes, P., Segretain, G., Nazimoff, O. & Arcouteil, A. (1961). Sur un nouvel agent de mycétome isolé en Somalia. *Bull. Soc. Pat. exot.* **54**, 1265.

Rowley, D. A. & Pine, L. (1955). Some nutritional factors influencing growth of yeast cells of *Histoplasma capsulatum* to mycelial colonies. *J. Bact.* **69**, 695.

Salvin, S. B. (1947). Cultural studies on the yeastlike phase of *Histoplasma capsulatum* Darling. *J. Bact.* **54**, 655.

Salvin, S. B. (1949a). Cysteine and related compounds in the growth of the yeast-like phase of *Histoplasma capsulatum*. *J. infect. Dis.* **84**, 275.

Salvin, S. B. (1949b). Phase determining factors in *Blastomyces dermatitidis*. *Mycologia*, **41**, 311.

Scheff, G. S. (1950). Biochemical and immunological properties of *Histoplasma capsulatum* (no. 650). *Yale J. Biol. Med.* **18**, 41.

Scherr, G. H. (1957). Studies on the dimorphism of *Histoplasma capsulatum*. I. The role of –SH groups and incubation temperature. *Exp. Cell Res.* **12**, 92.

Scherr, G. H. & Weaver, R. H. (1953). The dimorphism phenomenon in yeasts. *Bact. Rev.* **17**, 51.

Schwarz, J. & Drouhet, E. (1957). Morphologic features of an African strain of *Histoplasma* in hamsters and mice. *Arch. Path.* (*Lab. Med.*), **64**, 409.

Segretain, G. (1958). Sur les phialides et phialospores produites par *Madurella mycetomi*. *C.R. Acad. Sci., Paris*, **247**, 130.

Segretain, G. (1959). *Penicillium marneffei* n.sp. Agent d'une mycose du système réticulo-endothélial. *Mycopathol. Mycol. appl.* **11**, 327.

Segretain, G., Baylet, J., Darasse, H. & Camain, R. (1959). *Leptosphaeria senegalensis* n.sp. Agent de mycétomes à grains noirs. *C.R. Acad. Sci., Paris*, **248**, 3730.

Segretain, G. & Mariat, F. (1958). Contribution à l'étude de la mycologie et de la bactériologie des mycétomes du Tchad et de la Côte des Somalis. *Bull. Soc. Pat. exot.* **51**, 833.

Segretain, G., Mariat, F. & Drouhet, E. (1955). Sur *Cladosporium trichoides* isolé d'une mycose cérébrale. *Ann. Inst. Pasteur*, **89**, 465.

Shear, C. L. (1922). Life history of an undescribed ascomycete from a granular mycetoma of man. *Mycologia*, **14**, 239.

Silva, M. (1957). The parasitic phase of the fungi of chromoblastomycosis. Development of sclerotic cells *in vitro* and *in vivo*. *Mycologia*, **49**, 318.

Silva, M. (1958). The saprophytic phase of the fungi of chromoblastomycosis. Effect of nutrients and temperature upon growth and morphology. *Trans. N.Y. Acad. Sci.* **21**, 46.

Silva, M. (1960). Growth characteristics of the fungi of chromoblastomycosis. *Ann. N.Y. Acad. Sci.* **89**, 17.

Sweany, H. C. (1960). *Histoplasmosis*. Springfield: C. C. Thomas.

Talice, R. V. (1930). Sur la filamentisation des *Monilia*. *Ann. Parasit. hum. comp.* **8**, 394.

Tarbet, J. E., Wright, E. T. & Newcomer, V. D. (1952). Experimental coccidioidal granuloma; developmental stages of sporangia in mice. *Amer. J. Path.* **28**, 901.

Trejos, A. (1954). *Cladosporium carrionii* n.sp. and the problem of cladosporia isolated from chromoblastomycosis. *Rev. Biol. trop. Univ. Costa Rica*, **2**, 75.

Vanbreuseghem, R. & Bernaerts, J. P. (1955). Production expérimentale de grains maduromycosiques par *Monosporium apiospermum* et *Allescheria boydii*. *Ann. Soc. belge Méd. trop.* **35**, 451.

Ward, J. M. (1958). Biochemical system involved in differentiation of fungi. *Trans. IVth Int. Congr. Biochem.* **14**, 33.

Whittle, C. H. & Gresham, G. A. (1960). *Candida in vitro* and *in vivo*. *Mycopathol. Mycol. appl.* **12**, 207.

Young, G. (1958). The process of invasion and persistence of *Candida albicans* infected intraperitoneally into mice. *J. infect. Dis.* **102**, 114.

THE REACTION OF TRYPANOSOMES TO THEIR ENVIRONMENT

B. WEITZ

The Lister Institute of Preventive Medicine, Elstree, Hertfordshire

INTRODUCTION

Unlike many other kinds of protozoa the haemoflagellates are strictly parasitic, having no free living stage outside the body during their complicated life history. Thus the African trypanosomes are continually under the influence of living processes in the mammalian hosts or the insect vectors. These parasites therefore can be characterized in nature only by the observable effects they have on the living host, but as such effects are not very informative and cannot always be specifically attributed to the trypanosomes the latter must usually be removed from their normal habitat for an adequate study of their characters. Such manipulations normally entail techniques *in vitro*; and the study of the behaviour of the parasites in laboratory animals must be regarded as an experiment *in vivo* with connotations *in vitro*. It is impracticable to distinguish clearly between procedures in the laboratory *in vivo* and those *in vitro*; even simple operations like syringe inoculations between animal passages have the flavour of an experiment *in vitro*. It is more useful to distinguish the two types of study by relating them to their environment; that is, to distinguish 'natural environmental conditions' and 'artificial environmental conditions' under which the parasites may be characterized.

NATURAL ENVIRONMENTAL CONDITIONS

The natural state of trypanosomes is, by definition, one in which the intentional and direct interference by man is absent, and in which techniques *in vitro* are only used for the necessary examination of the state of the parasites in the hosts at any given time.

It is thus limited to naturally infected vertebrate hosts and to the vectors capable of transmitting the organisms to and from these hosts. Both these biological environments are themselves subjected to various external influences which may induce alterations of the environment of the parasite.

Table 1. *The environments of trypanosomes*

Natural environment *in vivo*	Artificial environment *in vivo*
A. Hosts	
1. Natural hosts: reservoir hosts	1. Artificial hosts: laboratory animals
2. Adventitious natural hosts: domestic animals or man	2. Natural hosts: drug-treated animals
B. Vector	
1. Cyclical transmission: *Glossina* spp.	
2. Mechanical transmission: tabanids, haematopota etc.	
	In vitro Cultivated forms Laboratory procedures

Natural hosts of trypanosomes

A large number of species of African mammals are known to be infected with trypanosomes and act as the main reservoir of the parasites. Little is known about the course of parasitization of natural reservoir hosts and about the behaviour of the parasites in them, but it may be assumed that the hosts rarely suffer from disease as a result. It is difficult to isolate the organisms from wild hosts, and because of the low-grade parasitaemia in these animals microscopic examination of the blood is an unreliable means of diagnosis. Moreover, if the behaviour of the parasite in reservoir hosts is similar to that in domestic animals, parasitaemia in the natural disease is probably limited to certain periods during the infection. It follows that the data for infection rates of different wild mammals indicates only minimum rates. The course of infection with African trypanosomes in domestic animals and man has been more precisely studied, and these hosts, which may be called 'adventitious natural hosts', are presumably only temporary reservoirs of infection, although as such they may contribute to a local intensification of the natural infection under certain ecological conditions.

Natural vectors of parasites

The parasites come into intimate relation with the tsetse fly, the main vector, and the insect influences the course of the cyclical development of the parasite. There is a less intimate relationship between the vectors and parasites in blood-sucking flies of various species, for example tabanids and haematopota, which mechanically transmit infection without cyclical development. The incidence of mechanical transmission

8

of certain species of trypanosomes in natural hosts is not known but undoubtedly plays some part in the maintenance of infection, particularly in gregarious species of host.

The development of trypanosomes in the invertebrate hosts also depends on a number of ecological factors. There are about 22 species of *Glossina*, all of which have their own peculiar ecology and habits. The chances of infection of the fly at the time of feeding may depend initially on a number of factors concerning the behaviour of the parasites in the host, or possibly on the species of host on which the fly has fed. It is known that each species of tsetse fly has its own preferences for different species of hosts, irrespective of the availability of different animals in the natural state (Weitz, 1963c). Thus *G. longipennis* restricts its feeding mainly to rhinoceros, elephant and buffalo, and *G. pallidipes* chiefly to bushbuck or bushpig. When, however, these two may exist in the same area, transmission may occur between the distinct groups of hosts they feed on by a third species of tsetse fly like *G. morsitans* which feeds on some animals in both groups. In this way a trypanosome infection maintained in bushbuck by *G. pallidipes* could become established in the rhinoceros through an intermediate infection of *G. morsitans*. This example is over-simplified, taking no account of factors like the movement of game into different areas, which facilitate the spread of trypanosome infections into different ecological niches. These problems are particularly evident in the study of the epidemiology of *Trypanosoma rhodesiense*, which is maintained in game animals and occasionally in man as the result of an endemic prevalence of human infection. Unfortunately, little is known about the infection rate of the natural host reservoirs and the epidemiology of the disease is to that extent obscure. Some species of wild host are known to resist infection, and others to maintain it without harm to themselves (Ashcroft, Burtt & Fairbairn, 1959). A notable exception is the transmission of *T. gambiense* in man by *G. palpalis*. No other host of *T. gambiense* is known and man, although severely ill from such chronic infection, is an effective reservoir of the parasite. Wild animal hosts might likewise be affected clinically by parasitic infection without necessarily becoming extinct or severely reduced in numbers, and still act as an effective reservoir.

ARTIFICIAL ENVIRONMENTAL CONDITIONS

These conditions obtain when the environment of the parasites is subject to artificial interference either in the field or in the laboratory, and include both procedures *in vitro* and *in vivo*. Processes *in vivo* include interference with the normal behaviour of the natural or adventitious hosts as, for example, the treatment of domestic hosts with drugs, and the study of the parasite in the laboratory animals which are 'unnatural' hosts; and methods *in vitro* include the growth of trypanosomes in culture or the maintenance of the parasites outside the body for laboratory purposes.

Trypanosome antigens in laboratory animals

Nature of trypanosome antigens

A number of characters of trypanosomes vary under natural or artificial environmental conditions. Variations of the antigens of the parasites readily occur, and have an obvious bearing on the immunology of trypanosome infection. For the understanding of such variations, an accurate picture of the antigenic constitution of trypanosomes is required. In view of the low parasitaemia occurring in natural hosts, which precludes harvesting sufficient numbers of trypanosomes for study, the antigenic analysis is necessarily based on laboratory experiments, both *in vitro* and *in vivo* with trypanosomes from an artificial environment. The relevance of such findings to the natural environment is determined at a later stage under field conditions. Most antigenic studies have started from trypanosomes *in vivo*, and the species or strains of trypanosomes which can be studied immunologically in the laboratory are limited to those which cause a sufficiently intense parasitaemia in the laboratory animals to provide enough antigenic material.

The antigens of *Trypanosoma rhodesiense* obtained from disintegrated whole trypanosomes from laboratory animals were shown by Brown & Williamson (1962) to be protein in nature, and not associated with nucleotides, lipoproteins or polysaccharides. Antigenic variants of a strain, or different strains of the same species, were however chemically similar, so that no chemical distinction of variants was possible. It is nevertheless possible that chemical analysis of cultivated trypanosomes (see later), which are antigenically, morphologically and biologically different from the blood forms, would reveal differences due to the growth of the organisms in a different metabolic environment.

Immunological characters of trypanosomes

The antigens of trypanosomes are identifiable by agglutination tests (Soltys, 1957; Cunningham & Vickerman, 1962; Gray, 1962; Weitz, 1962), the precipitin reaction (Gray, 1960, 1961; Weitz, 1960) and fluorescent antibody techniques (Weitz, 1963b). The differences in sensitivity and specificity of these tests and in the nature of the antigen-antibody reaction involved may be reflected in the contradictory results obtained by the different techniques. Thus, the agglutination reaction with living whole trypanosomes is a complex antigen-antibody system and, although highly sensitive, is not suitable for the determination of antigenic components of trypanosomal extracts except in conjunction with absorption experiments. Extracts may be characterized by precipitin tests. The protective action of antibodies to different antigenic constituents may be tested in actively immunized animals, or in animals passively immunized by antisera to the extracts.

The properties of soluble antigens, called exoantigens (Weitz, 1960) occurring in the serum of animals suffering from a heavy parasitaemia with *Trypanosoma brucei* and *T. vivax* were investigated by this means, and differed immunologically from the antigens extracted from whole trypanosomes. The exoantigens were species specific. Rats or mice immunized with these soluble antigens were protected against infection with the homologous strains only, and antibodies from rats immunized with the exoantigens contained in the serum of infected rats, precipitated only the exoantigens of the homologous species of trypanosome (Weitz, 1963*a*), and agglutinated only the homologous living trypanosomes. Similarly, fluorescent antibodies from conjugated antisera against the exoantigens of *T. brucei* and *T. vivax* stained each species specifically (Weitz, 1963*b*).

The 'bound antigens' released after the disintegration of whole cells were not species specific, but common to both species of trypanosome. This difference was confirmed by Dodin & Fromentin (1962), and Seed & Weinman (1963) distinguished a precipitin and a protective antibody to the exoantigen of *Trypanosoma rhodesiense*. The distinction between the exoantigen and bound antigens is perhaps arbitrary but it is useful in considering antigenic variations of trypanosomes, since the antigens correspond roughly to species, or strain, and generic antigens respectively.

The influence of the host on the antigens of trypanosomes

The characterization of the antigens of different strains of a single species of trypanosome is confused by the prolific variation of antigens which occurs in the chronically infected host.

The change of the antigens of the infecting trypanosomes during the course of chronic infection of rabbits has been known for many years (Ritz, 1916; Russell, 1936). Using a precise agglutination test, Gray (1962) showed that each successive antigen induced antibodies specific for itself so that the serum of an infected rabbit came to contain agglutinins for each variant that had arisen during the course of infection, and for those variants only. When the variant prevailing at a given period during the infection was passed to another animal, initially it induced antibodies specific for itself; later antibodies to subsequent variants developed in the recipient rabbit. Each new antigenic variant was presumably produced as the result of suppression by antibody of the preceding antigenic variants. Gray (1962) demonstrated this antigenic selection by passive immunization. When rabbits passively immunized with antibodies to a number of variants were infected with the original strain, none of the variants that developed corresponded to the antibodies passively given. This 'antigenic defence' of the organism against environmental antibody seems to be mainly associated with changes in the exoantigen, since the agglutinins found in the serum of chronically infected rabbits can be specifically absorbed by the corresponding homologous exoantigens (Miller, private communication).

Another manifestation of exoantigen activity and its variation was described by Dodin, Fromentin & Gleye (1962) who immunized mice with the exoantigen in cell-free plasma of mice infected with strains of *Trypanosoma gambiense*. The peritoneal macrophages obtained from these immunized mice adhered *in vitro* to trypanosomes of the homologous strain of *T. gambiense* only, whereas variants of this strain, or other strains of the same species, failed to adhere to the macrophages. These macrophages also passively protected mice against the homologous strains. The extent to which the appearance of agglutinating and other antibodies correspond to the early sensitization of the macrophages to trypanosomal antigen needs to be established; by analogy with the response to many other microbial antigens it is probable that cellular sensitization would precede antibody formation.

All these investigations stress the ability of trypansomal antigens to change *in vivo* as a result of variations in the immune status of the host.

The influence of the vector on trypanosomal antigens

The vector of the trypanosomes also influences the antigenic constitution of the parasites. Although the identity of a strain alters continually in the vertebrate host, it is likely that some stabilization of the strain occurs in the course of the life history of the organism, and Broom & Brown (1940) suggested that variant strains reverted to their original antigenic state after passage through the tsetse fly. Gray (1963) infected tsetse flies with different variants arising in the host. The first variant to develop in animals newly infected by these tsetse flies had the antigen of the original, parent strains. It is not clear at what stage during the development of metacyclic forms the reversion to the parent type occurs, nor is the nature of the antigens known at the time of infection of the host from the tsetse fly, but Gray showed that the antigens did not vary in the infected fly. Animals which were bitten by the same infective tsetse flies at intervals up to 18 days became infected with serologically identical strains. During a similar period in the untreated host, five or more different serological variants would have been produced.

Artificial factors influencing the parasites in the vector

The physiology of the tsetse fly maintained under artificial conditions differs considerably from that in the natural state. For example, tsetse flies maintained in a cage produce less than the normal amount of fat (Jackson, 1937). Again, *Glossina* species in captivity take about 4 days to digest the proteins of blood meals, but when freshly fed flies are released into the bush, the blood meal disappears from the gut in about 16–20 hr. (Weitz & Buxton, 1953). Even artificial stimulation of the pupal stage may have considerable effect on the parasitic infection of the adult fly. Burtt (1946) showed that cyclical infection is almost three times as common in flies from pupae incubated at 30° as in normally bred flies. It will be evident that the results of experiments in artificially kept tsetse flies must be applied with caution to field conditions.

Studies of the parasites in the vector have not been very fruitful, the technical difficulties being too great. However, with techniques now available such as the fluorescent antibody technique, it should be possible to characterize the antigens in the tsetse fly.

The antigens of trypanosomes cultivated in vitro

The cultivation of trypanosomes in artificial media may provide a source of antigenic material for the study of the antigens of trypanosomes in the vector. Trypanosomes cultivated *in vitro* resemble the gut forms of trypanosomes in tsetse flies and are not usually infective to the mammalian host. Weinman (1957) cultivated infective trypanosomes by adding trehalose to the medium and suggested (Weinman, 1960) that the trehalose present in the salivary glands of *Glossina* species may have a similar function in inducing virulence for the mammalian host. Tissue cultures of tsetse fly salivary glands in the presence of normal sheep serum supported the growth of *Trypanosoma vivax*; the culture was infective for sheep (Trager, 1959). Gordon & Miller (1961) artificially infected tsetse flies with cultivated forms of *T. rhodesiense* and the flies developed metacyclic forms capable of infecting susceptible hosts.

The relationship between the blood forms, artificially cultivated forms and the tsetse fly forms of trypanosomes is instructive. Cultures from the infected gut of the tsetse fly can be established easily, but not those from the blood of vertebrate hosts. Furthermore, cultivated trypanosomes will not infect the mammalian host unless first passed through the tsetse fly (Fig. 1). This suggests that the forms in the gut of the tsetse fly and the cultivated forms are similar, but the vector is necessary for the development of trypanosomes to the infective stage. The cultivated forms are antigenically distinct from the blood-infecting forms. In other methods of transmission by tsetse flies of species such as *Trypano-*

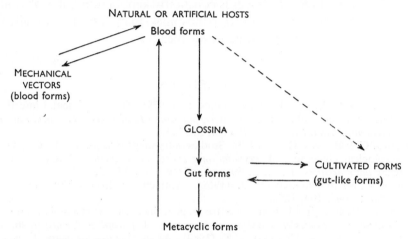

Fig. 1. The transmission of trypanosomes *in vivo* and *in vitro*.

soma grayi, *T. congolense* and *T. vivax*, where the infective forms of the parasite are not transmitted by the salivary glands, other mechanism may operate. The transmission of the trypanosomes by mechanical means, as for example the transmission of *T. evansi* in camels, must be considered on yet another basis. Here the blood form appears merely to be transfused into the new host, and whether in such conditions any antigenic variation occurs is not known.

CONCLUSIONS AND DISCUSSION

In the development of trypanosomes, as in most biological problems, the relationship of procedures *in vitro* and *in vivo* is not a simple one. Basically, all investigations outside the natural environment must be regarded with some caution; however, little would be known about the natural behaviour of trypanosomes without the aid of artificial experiments. At all stages in the development of trypanosomes the examination of the organism is a procedure *in vitro*, even when the material is obtained directly from its natural state. Similarly, the artificial behaviour of trypanosomes *in vivo* or *in vitro* has clarified the complex mechanisms of the development and transmission of trypanosomes in the field. However, the laboratory investigator is constantly in danger of evolving principles of behaviour of the organisms which are unique to artificial environment and it is, therefore, most important to relate any findings in the laboratory to field conditions at every stage.

Thus, problems of the field are influenced by a large number of factors which cannot always be imitated in the laboratory and unless the natural history of trypanosomes is constantly kept in mind, it is obvious that the laboratory study of trypanosomes may lead to unrealistic conclusions, however interesting academically.

REFERENCES

Ashcroft, M. T., Burtt, E. & Fairbairn, H. (1959). The experimental infection of some African wild animals with *Trypanosoma rhodesiense, T. brucei* and *T. congolense. Ann. trop. Med. Parasit.* **53**, 147.

Broom, J. C. & Brown, H. C. (1940). Studies in trypanosomiasis. IV: Notes on the serological characters of *Trypanosoma brucei* after cyclical development in *Glossina morsitans. Trans. R. Soc. trop. Med. Hyg.* **34**, 53.

Brown, K. N. & Williamson, J. (1962). Antigens of Brucei Trypanosomes. *Nature, Lond.* **194**, 1253.

Burtt, E. (1946). Incubation of tsetse pupae: increased transmission rate of *Trypanosoma rhodesiense* in *Glossina morsitans. Ann. trop. Med. Parasit.* **40**, 18.

Cunningham, M. P. & Vickerman, K. (1962). Antigenic analysis in the *Trypanosoma brucei* group, using the agglutination reaction. *Trans. R. Soc. trop. Med. Hyg.* **56**, 48.

Dodin, A. & Fromentin, H. (1962). Mise en évidence d'un antigène vaccinant dans le plasma de souris expérimentalement infectées par *Trypanosoma gambiense* et par *Trypanosoma congolense. Bull. Soc. Pat. exot.* **55**, 128.

Dodin, A., Fromentin, H. & Gleye, M. (1962). Mise en évidence d'un antigène dans le plasma de souris infectées par diverses espèces de trypanosomes. *Bull. Soc. Pat. exot.* **55**, 291.

Gordon, R. M. & Miller, J. K. (1961). Cyclical infection of *Glossina morsitans* with culture forms of *Trypanosoma rhodesiense. Nature, Lond.* **191**, 1317.

Gray, A. R. (1960). Precipitating antibody in trypansomiasis of cattle and other animals. *Nature, Lond.* **186**, 1058.

Gray, A. R. (1961). Soluble antigens of *Trypanosoma vivax* and of other trypanosomes. *Immunology*, **4**, 253.

Gray, A. R. (1962). The influence of antibody on serological variation in *Trypanosoma brucei. Ann. trop. Med. Parasit.* **56**, 4.

Gray, A. R. (1963). Antigenic variation in a fly-transmitted strain of *Trypanosoma brucei*. In *Report of 9th Meeting of the Commission for Technical Cooperation in Africa South of the Sahara, International Scientific Committee for Trypanosomiasis Research* (in the Press).

Jackson, C. H. N. (1937). Some new methods in the study of *Glossina morsitans. Proc. zool. Soc. Lond.* 1936, p. 811.

Ritz, H. (1916). Über Rezidive bei experimenteller Trypanosomiasis, II Mitteilung. *Arch. Schiffs- u. Trophenhyg.* **20**, 397.

RUSSELL, H. (1936). Observations on immunity in relapsing fever and trypanoso-miasis. *Trans. R. soc. Med. Hyg.* **30**, 179.

SEED, R. J. & WEINMAN, D. (1963). Characterization of antigens from *Trypanosoma rhodesiense*. *Nature, Lond.* **198**, 197.

SOLTYS, M. A. (1957). Immunity in trypanosomiasis. II. Agglutination reaction with African trypanosomes. *Parasitology*, **47**, 390.

TRAGER, W. (1959). Tsetse fly tissue culture and the development of trypanosomes to the infective stage. *Ann. trop. Med. Parasit.* **53**, 473.

WEINMAN, D. (1957). Cultivation of trypanosomes. Correspondence. *Trans. R. Soc. trop. Med. Hyg.* **51**, 560.

WEINMAN, D. (1960). Trehalose metabolism of trypanosomes. *Nature, Lond.* **186**, 166.

WEITZ, B. (1960). The properties of some antigens of *Trypanosoma brucei*. *J. gen. Microbiol.* **23**, 589.

WEITZ, B. (1962). Immunity in Trypanosomiasis. In *Drug, Parasites and Hosts. Biological Council Symposium*, p. 180. Ed. L. G. Goodwin & R. H. Nimmo-Smith. London: Churchill.

WEITZ, B. (1963 a). Immunological relationships between African trypanosomes and their hosts. *Ann. N.Y. Acad. Sci.* (in the Press).

WEITZ, B. (1963 b). The specificity of trypanosomal antigens by immuno-fluorescence. *J. gen. Microbiol.* **32**, 145.

WEITZ, B. (1963 c). The feeding habits of Glossina. *Bull. World Hlth Org.* **28**, 711.

WEITZ, B. & BUXTON, P. A. (1953). The rate of digestion of blood meals of various haematophagous arthropods as determined by the precipitin test. *Bull. ent. Res.* **44**, 445.

SOME FACTORS INFLUENCING THE RESPONSE OF YOUNG DOMESTICATED ANIMALS TO *ESCHERICHIA COLI*

P. L. INGRAM

Department of Pathology, Royal Veterinary College, London

Shortly after the isolation of the colon bacillus, *Escherichia coli*, by Escherich in 1885, Jensen (1893) attributed a septicaemic disease in young calves (Kälberruhr or Calf Scours) to infection by this organism. His observations were confirmed by other workers (e.g. Joest, 1903) but later, between 1920 and 1930, Theobald Smith and his colleagues made their classical studies on calves showing that non-septicaemic (infection localized to the intestine) and septicaemic forms of the disease were recognizable; the type of the disease depended on whether colostrum, the first milk from the dam containing maternal antibodies, was fed to the calf or not. *E. coli* are part of the normal intestinal flora of all animals and many antigenic types are known. The first indication that special strains were responsible for the disease in calves was obtained when Lovell (1937) found, by means of a precipitin reaction, that the majority of strains isolated from such cases could be assigned to one or other of eight serological types. Subsequent investigations have confirmed that certain serological types are more frequently associated with disease than others, and in an epidemic the same serotype of *E. coli* can usually be isolated from consecutive deaths.

It was not until 1945 that an association between special strains of *Escherichia coli* and gastro-enteritis in babies was established (Bray, 1945), although earlier Adam (1927) had claimed that certain fermentative types of the organism were frequently found in the disease. Since then it has been confirmed that particular serotypes of *E. coli* are frequently isolated from the intestinal contents of such cases and they predominate over other strains in the gut (Giles & Sangster, 1948; Taylor & Charter, 1952; Thomson, 1955a, b).

Recently a similar intestinal disorder has become common amongst young piglets; also cases of coli-septicaemia have been seen frequently in chickens. In other species of domesticated animals the disease is not unknown but serious outbreaks are infrequently encountered.

In all these diseases, which mainly take the form of a diarrhoea, profuse growth of one or sometimes two predominant serotypes of

Escherichia coli can be cultured from the blood or intestinal contents. *E. coli* are found universally in the alimentary tracts of man and animals in normal circumstances; the large numbers of *E. coli* which are present in the faeces during the first few weeks of life gradually diminish to lower levels with advancing age (Smith, 1961). Whereas some years ago *E. coli* was considered to be only a harmless inhabitant of the alimentary tract, and probably beneficial to the host, the association of particular strains with disease is now firmly established.

The reproduction of colibacillosis* in normal animals, however, has been less certain. Some workers (Jensen, 1893; Joest, 1903; Dunne, Glanz, Hokanson & Bortree, 1956; Schoenaers & Kaeckenbeeck, 1958, 1960*a*; Fey & Margadant, 1962; Fey, Lanz, Margadant & Nicolet, 1962) reproduced the condition in newborn calves by feeding cultures of *Escherichia coli* isolated from diseased animals, but others (Smith & Little, 1927; McEwen, 1950; Moll & Ingalsbe, 1955) have had less success. The disease has been reproduced successfully, however, without the deliberate administrations of *E. coli*, by withholding colostrum and maintaining the calves under special conditions (Aschaffenburg *et al.* 1949*a*). The reproduction of gastro-enteritis in naturally reared piglets (Saunders, Stevens, Spence & Sojka, 1963) and in pathogen-free piglets (Saunders, Stevens, Spence & Betts, 1963) by oral dosing with cultures of *E. coli* has been reported recently; in these experiments the piglets were either deprived of colostrum or dosed before their first feed of colostrum. Dosing piglets with a supposedly non-pathogenic strain either failed to induce the disease or produced mild illness only. Sojka & Carnaghan (1961) have reproduced typical coli-septicaemia by the intravenous injection of young chickens with *E. coli* strains isolated from diseased birds, but were unable to do so with either a strain isolated from a normal fowl or a strain isolated from disease in pigs.

For the establishment of *Escherichia coli* infections in animals the environment and resistance of the host and the qualities and potentialities of the bacteria are important factors. Some of these factors, which have been investigated by experiments *in vivo* and *in vitro*, will be discussed to illustrate the theme of the Symposium. Most observations described under 'Host factors' were made *in vivo*. These are epidemiological observations and their support by experiments in the field; and experiments made in animals which clearly demonstrate the importance of antibody-containing colostrum for controlling these infections, and the influence on the latter of post-colostrum nutrition. Only a few

---

* In this paper the term 'colibacillosis' is restricted to septicaemic infection and enteric disease of young animals associated with *E. coli*.

attempts have been made *in vitro* to demonstrate the mode of action of the colostrum and these are described. Under 'Bacterial factors', the demonstration that certain serological types of *E. coli* are associated with infection is followed by a description of experiments made *in vitro* and *in vivo* to recognize properties of these organisms which might be responsible for, or associated with, their pathogenicity. The interaction of the host and bacterial factors is discussed under 'General Considerations', especially in relation to the conditions influencing the growth of *E. coli* in the mixed culture of the gut contents, and this involves a description of experiments on mixed infections *in vivo*.

HOST FACTORS

Age incidence

Generalized infections with *Escherichia coli* organisms are largely confined to young animals; in older animals localized infections are not uncommon but they rarely have a fatal outcome. Calves, pigs, lambs and foals are usually affected in the first few days or weeks of life, and the age of chickens most frequently affected is 6–10 weeks. In addition to the gastro-enteritis of young piglets, another condition also attributed to *E. coli*, known as oedema disease or bowel oedema, is seen in pigs of 8–12 weeks, i.e. in the early post-weaning period. This disease is characterized by the formation of marked oedema in the stomach wall and in the mesentery of the intestine, and heavy growths of certain *E. coli* serotypes can be cultured from the alimentary tract.

The characteristic age incidence of these conditions can be only partly explained in the light of our present knowledge. The passive immunity transferred from the dam to young via the colostrum plays an indisputable role, particularly in preventing the septicaemic condition. To what extent immaturity of the natural defence mechanisms of the host contribute—detoxication processes, responsiveness of the reticulo-endothelial system and its capacity to produce antibodies—is largely speculative. It is interesting to note, however, that the human infant—which receives immunity from the mother by another route, and is more physically immature than the newborn domesticated animals—is susceptible to intestinal infection with *Escherichia coli* up to about 12 months.

Breed incidence

Little is known of the incidence of *Escherichia coli* infections in different breeds of animal species, except in calves. Animals of the beef breeds are less frequently affected than those of dairy breeds, and of the latter some types appear more susceptible under farm conditions than others (Withers, 1952); some of these differences, however, have not been confirmed under experimental conditions (Aschaffenburg *et al.* 1952). Whether the different incidence of calf scours amongst various breeds of calves is a true breed characteristic is debatable. Dairy calves are usually reared under more artificial conditions than are beef calves and a difference of environment may be the true explanation of the apparent breed incidence.

Environmental factors

Surveys of the mortality rates due to calf scours indicate that climatic conditions have an influence on the occurrence of the disease, though methods of management may also vary in different climates. Comparison of the mortality rates in England and Wales (Lovell & Hill, 1940; Withers, 1952) with those in Scotland (Jordan, 1933; Smith, 1934; Lovell & Hill, 1940; Withers, 1952) suggest that the incidence increases farther north. Withers (1952) concluded that subjection of animals to wide fluctuations of temperature was a more important predisposing factor for calf scours than a constant suboptimal temperature. The greatest mortality occurs in late winter and spring, which are not only the seasons when the environmental temperature fluctuates widely but also when there is a greater number of young calves on a farm. Experiments made with calves at the National Institute for Research in Dairying have confirmed that epidemics of *Escherichia coli* infection can be established by bringing young susceptible animals together in one building. These epidemics were entirely spontaneous, and the calves were in no way 'infected' artificially. Newborn calves, before they had sucked their dams, were introduced into a calfhouse containing individual pens and in which constant environmental conditions were maintained. In the first day of life some calves were fed a standard amount of bulked deep-frozen colostrum whilst others were deprived of colostrum, and the calves were then reared for 3 weeks either on whole milk or a variety of semi-synthetic diets. When a calf died or reached 3 weeks of age it was removed, the pen was disinfected, and another newborn calf was introduced. After a period when the calfhouse had not been in use, the first calves brought in grew well and many, although deprived of colostrum, survived for 3 weeks. Later, however, as more calves passed through the pens, those deprived of colostrum died with a septicaemic *E. coli* infection, and those that had received colostrum at their first feed scoured. Still later, at the height of the epidemic, many of the colostrum-fed calves succumbed after an illness characterized by severe diarrhoea which was typical of non-septicaemic calf scours; these deaths occurred only after the incidence of diarrhoea had risen to a high level.

The outbreaks of spontaneous disease in calves observed under these conditions bear a close resemblance to the epidemics established in mice by Topley with *Salmonella typhimurium* infections; it is probable that the same factors operate under natural farm conditions, except that the numbers of susceptible animals in the community are not so great; they may in part account for the higher incidence of calf scours towards the end of the normal calving season. The recent increased frequency of colibacillosis in poultry has coincided with the development of the broiler chicken industry where large numbers of young birds are herded together in broiler houses. Likewise, studies on *Escherichia coli* scours in lambs under natural conditions have indicated that the lambing of flocks of ewes in sheds, together with severe weather conditions, were important predisposing causes for epidemics of the disease (Marsh & Tunnicliff, 1938).

Nutritional factors

Colostrum feeding

Unlike the human infant which acquires maternal antibodies mainly *in utero*, the newborn domesticated farm animals rely entirely upon the first milk or colostrum from their dams for passive protection against micro-organisms in their environment (McGirr, 1947).

The importance of colostrum to the newborn calf was recognized by Jensen (1905) who found that calves fed boiled milk during the first 24 hr. of life usually died of an enteritis, but it was not until the classical studies of Howe, Theobald Smith, and their colleagues, that the reason for this became clear. They showed that the blood of a newborn calf contained no euglobulin, pseudoglobulin I or antibodies until after the ingestion of colostrum on the first day of life (Howe, 1921, 1922; Little & Orcutt, 1922). Nelson (1924) demonstrated a similar absorption of agglutinins against *Escherichia coli* from colostrum by the newborn calf. In the light of these results, Smith & Little (1922) examined the effects of depriving calves of colostrum; calves that received colostrum usually survived the first few weeks of life, whilst most of those that received only milk died of an *E. coli* septicaemia. They concluded that a function of colostrum was protective against miscellaneous bacteria which were harmless to older animals.

Subsequently, a divergence of opinion arose as to the relative importance of the antibody and vitamin content of colostrum for the survival and normal growth of the young calf. In experiments on calves made at the National Institute for Research in Dairying, it was shown that the protective value of colostrum resided in its lactoglobulin content, that as little as 80 ml. of the non-fatty fraction of colostrum was sufficient to protect calves from invasion by *Escherichia coli* and that supplements of vitamins were no replacement for colostrum (Aschaffenburg *et al.* 1949*a*, *b*, 1951*a*). Similarly, the performance of calves that received mammary secretion from parturient cows that had been pre-partum milked was no better than that of calves receiving milk only, and no *E. coli* antibodies could be demonstrated in this secretion (Aschaffenburg *et al.* 1951*b*).

Investigation of the strains of *Escherichia coli* isolated from these calves at death, and samples of colostrum that had been fed to some of the calves, revealed that the presence of agglutinins against the 'K' antigens in the colostral whey protected mice against infection with the corresponding strain of *E. coli* (Briggs, 1951). It was also shown that colostrum given to calves which subsequently died with septicaemia did not contain agglutinins against the 'K' antigens of the strain of *E. coli* isolated from the dead calves; and agglutinins against strains that killed contemporary calves in the same environment were usually present in the colostrum given to calves that survived (Briggs *et al.* 1951; Ingram *et al.* 1953; Ingram *et al.* 1956). A relationship between the feeding of colostrum containing specific antibody and the type of *E. coli* infection at death was clearly demonstrated (Wood, 1955; Ingram, 1962); septicaemia usually followed the withholding of colostrum and in animals that died even after receiving colostrum the infection was usually non-septicaemic (Table 1).

Table 1. *Relationship between the feeding of colostrum and the type of* Escherichia coli *infection*

Diet	Number of calves	Septicaemic infection			Non-septicaemic infection (chronic)
		Acute	Subacute	Total	
Deprived of colostrum	131	74	49	123	8
Colostrum-fed	153	15	7	22	131

(From Ingram & Lovell, 1960)

Thus, by virtue of its specific antibody content, colostrum protects calves against invasion by pathogenic *Escherichia coli* in their environment. The importance of allowing a cow to calve in the same environment in which she has been for some weeks prior to parturition, and the dangers of transporting young calves from one farm to another, can be readily appreciated.

The precise mechanisms by which calves are protected against septicaemic *Escherichia coli* infection after they have received colostrum containing the homologous 'K' antibodies have not been fully elucidated. Some information has been obtained from studies *in vitro* on the bactericidal activity of sera and colostral whey for *E. coli* organisms. Glanz, Dunne, Heist & Hokanson (1959) showed that the bactericidal activity of sera from colostrum-deprived calves was limited to two of seven strains tested, whereas sera from calves receiving colostrum showed activity for six of the seven strains as did all sera from adult cows. Samples of colostral whey had no bactericidal activity. A heat-labile substance in serum was a necessary participant in this bactericidal activity; the substance was not complement and Glanz *et al.* (1959) suggested a similarity between it and properdin. These results have been confirmed and extended by Smith (1962) who found that two substances were involved in the bactericidal action of serum against *E. coli*—a heat-labile non-specific factor (possibly properdin) and a heat-stable factor (antibody). All strains of *E. coli* cultured from the blood of diseased colostrum-deprived calves grew well in precolostral calf serum, whilst few of the strains isolated from the intestines of diseased and healthy colostrum-fed calves grew in the post-colostral serum of the calves from which they were isolated.

Post-colostrum nutrition

Following the period of colostrum feeding many calves, especially those reared on an intensive system and also those in experiments, receive a semi-synthetic diet. Comparisons of animals having whole milk, pasteurized milk and a semi-synthetic diet have shown that their growth rate was less and the incidence of scouring and mortality due to *Escherichia coli* infection was significantly greater with the artificial diet than with raw or pasteurized milk (Shillam, Dawson & Roy, 1960; Shillam, Roy & Ingram, 1962*a*). Investigation of the components of the semi-synthetic diet demonstrated that the dried skim-milk was the detrimental factor when the method of preparing the milk powder involved high temperatures, or moderate temperatures for long periods. The feeding value of the dried skim-milk fell as the degree of denaturation of the milk proteins increased (Shillam *et al.* 1962*a*; Shillam, Roy & Ingram, 1962*b*; Shillam, Roy & Ingram, 1962*c*).

Apart from the actual constituents of the diet, sudden changes of diet and environment may influence the bacterial flora of the intestine.

This has been shown recently in pigs. The strains of *Escherichia coli* that are commonly associated with diarrhoea and oedema disease are frequently haemolytic, and a small proportion of the *E. coli* organisms in the faeces of normal animals are often haemolytic. Buxton & Thomlinson (1961) determined the proportions of haemolytic and non-haemolytic *E. coli* in rectal swabs from a litter of 8-week-old pigs, which were sampled at intervals for 28 days. During this period the pigs had two changes of husbandry; first they were moved to another farm, mixed with other pigs, and their diet was changed from dry meal to a wet mash, then secondly, the diluted whey used to prepare the wet mash was replaced by undiluted whey. No clinical illness was observed with

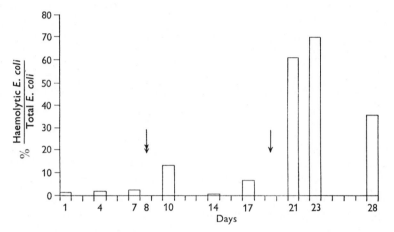

Fig. 1. The relationship of environment to sudden changes in the proportion of haemolytic to total *Escherichia coli* in rectal swabs from pigs (redrawn from Buxton & Thomlinson, 1961). $\twoheadrightarrow$ indicates a change in environment, $\rightarrow$ indicates a change in diet; for details of these changes see text.

these changes of husbandry, but 48 hr. after each change there was a marked increase in the average percentage of haemolytic *E. coli* (O 138) in the rectal swabs (Fig. 1).

The adverse influence of the artificial feeding of calves as opposed to natural feeding was stressed many years ago (Lovell & Hill, 1940), and over-distension of the digestive tract with too much milk at infrequent intervals may be one of the factors concerned (Sheehy, 1934; Hewison, 1939). Under both farm and experimental conditions attention to details of management, feeding routine and composition of the diet has reduced the incidence of calf scours. Withers (1952) concluded that a personal factor in the husbandry of calves, e.g. the greater care of the calves which is often provided by female rather than male attendants, also had an important bearing on the well-being of calves.

It is evident from the studies that have been made under natural and experimental conditions that a variety of factors predispose the host to infections with *Escherichia coli*, and that at least one of these factors—the antibody content of colostrum—influences the type of infection seen at death.

BACTERIAL FACTORS

The concept that strains of *Escherichia coli* differ in their pathogenicity to calves first put forward by Jensen (1893) was supported by Smith & Orcutt (1925) and Lovell (1937). The results of subsequent experimental reproduction of *E. coli* infections in piglets, fowls and calves, together with the greater frequency with which certain serological types of *E. coli* are isolated from diseased animals have supported the view that differences exist between strains of the organism in their ability to produce disease.

Strains of Escherichia coli *associated with disease in young animals*

Strains of *Escherichia coli* of many of the recognized O groups have been associated with disease in young animals on the basis of their isolation from the blood in septicaemic infections, or their predominance in the intestinal contents of animals suffering from enteric disorders. Organisms of certain O groups are isolated more frequently than others, and those most frequently involved in these diseases in different species of animals are given in Table 2; the number of groups concerned

Table 2. *O groups of* Escherichia coli *most frequently associated with disease in various animals*

O group	Calves	Lambs	Babies	Pigs	Fowls
1	+	.	.	.	+ +
2	+	.	.	+	+ +
8	+	+	.	+ +	+
15	+ +	.	.	.	+
26	+ +	.	+ +	+	.
55	+	.	+ +	.	+
78	+ +	+ +	.	+	+ +
111	+	.	+ +	.	+
115	+ +	.	.	+	.
117	+ +	.	.	+	.
119	+	.	+ +	+	.
127	.	.	+ +	.	.
128	+	.	+ +	.	.
138	.	.	.	+ +	.
139	.	.	.	+ +	+
141	+	.	.	+ +	+

+ + indicates frequently isolated; + isolations have been made but not with great frequency; no symbol (.) infrequently isolated or isolation not recorded.

is relatively limited. Apart from groups 26 and 78, those isolated repeatedly from disease in one species of animal are apparently less frequently associated with disease in other species.

The restricted number of O groups associated with disease in lambs may be due to the few cases in which the strains of *Escherichia coli* concerned have been examined serologically. Most of the O groups have been found in outbreaks of scours in calves, whereas the serotypes associated with scouring in pigs are more restricted; Sojka, Lloyd & Sweeney (1960) were able to classify 58·5 % of 549 strains with three sera only. Unlike the strains isolated from other species of animals, most, 74·0 %, of those isolated from gastro-enteritis in piglets were haemolytic. The strains of *E. coli* isolated from the intestinal contents and

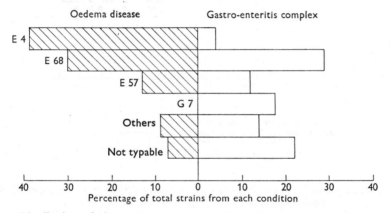

Fig. 2. Distribution of the common serotypes of *Escherichia coli* in oedema disease and gastro-enteritis of pigs (redrawn from Sojka *et al.* 1960). E4 = O139:K82 (B). E68 = O141:K85 (B). E57 = O138:K81 (B). G7 = O8:K87 (B), K88 (L).

mesenteric lymph nodes of pigs with oedema disease were also restricted; Sojka *et al.* (1960) were able to type 82 % of 411 strains using three sera only. Ninety-two % were haemolytic, but the distribution of serotypes in the two conditions was different (Fig. 2). Similarly, in fowls with coli-septicaemia, the *E. coli* strains that were most frequently encountered belonged to a limited number of O groups. Sojka & Carnaghan (1961) were able to type 61·3 % of 212 strains with only three O sera, and over 40 % belonged to O2.

Although *Escherichia coli* of a particular O group are encountered in acute outbreaks of disease in any one herd or flock, in prolonged epidemics, such as those in experimental calves, different serotypes come into prominence at different stages of the epidemic (Fig. 3).

Many investigations have been made in attempts to find some

property of the organisms or a method by which potentially pathogenic-strains can be distinguished in the laboratory from the non-pathogenic strains of the normal gut flora.

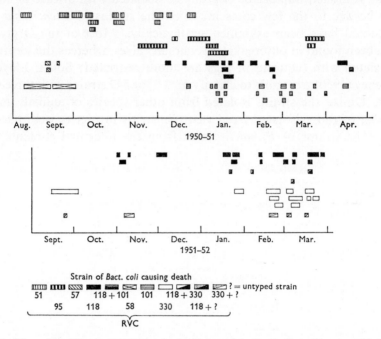

Fig. 3. The epidemiology of white scours (calf scours) (redrawn from Ingram *et al.* 1953). Each rectangular symbol represents the life of one calf.

Methods in vitro

Colony morphology and biochemical reactions of strains have provided no features differentiating between pathogenic and non-pathogenic *Escherichia coli*. Haemolysin production is not helpful, although most *E. coli* isolated from piglet scours and oedema disease of pigs are haemolytic. Ross & Dawes (1954) compared strains isolated from healthy children and those from cases of infantile diarrhoea and found no differences between the strains with regard to their ability to utilize certain amino acids as the sole source of nitrogen, pH range favourable to growth, pH limits of growth, and ability to grow in extracts of faeces from normal breast-fed and artificially fed infants, and infants with gastro-enteritis. Likewise, *E. coli* having the K1 antigen contain sialic acid, but their content of this acid bears no relation to virulence as measured by the intraperitoneal injection of mice (Forbes & Kuck, 1961).

No universal property responsible for pathogenicity has yet been determined, but some correlation between pathogenicity and serine-inhibition and mucinase production has been indicated. Rowley (1953) compared strains of *Escherichia coli* isolated from babies suffering from gastro-enteritis with coliform strains picked at random, for inhibition of growth by certain amino acids when these were added to cultures on a simple salt medium containing glucose and ammonia as the sole source of nitrogen; 65 % of the pathogenic strains and only 6 % of the random coliforms were inhibited by serine. Ross (1959) showed that a significantly larger proportion of *E. coli* strains associated with infantile gastro-enteritis produced mucinase than strains of *E. coli* isolated from healthy children. Further investigations on these lines may show some difference of metabolism or enzyme production between pathogenic and non-pathogenic strains.

The chief toxic component of smooth strains of *Escherichia coli* is an endotoxin considered to be identical with the O antigen, and many investigations have demonstrated the profound physiological disturbances which follow injection of these endotoxins into experimental animals. Broth cultures of strains of *E. coli* cultured from the intestines of scouring calves (Smith & Little, 1927) or antigen prepared by Boivin's method from similar strains (Lovell, unpublished) were highly toxic for calves when given intravenously. They produced dyspnoea, pyrexia and diarrhoea. Chemical analyses and toxicity tests made on lipopolysaccharide extracts from pathogenic and non-pathogenic strains of *E. coli* from calves have failed to show any consistent differences, and the pathogenicity of *E. coli* strains did not appear to be associated with a qualitative or quantitative variation of the endotoxin extracted from them (Harvey & Carne, 1960). Other reports also indicate that the toxicity of heat-killed organisms for mice is approximately the same regardless of the antigenic structure of the strains, and whether they were associated with colibacillosis or isolated from normal animals (Rowley, 1954; Forbes & Kuck, 1961).

Methods in vivo

Mouse virulence tests

Although variations exist between strains of *Escherichia coli* with regard to their virulence for mice (Rowley, 1954; Forbes & Kuck, 1961) no correlation has been found between virulence for mice and association with disease conditions. Taylor, Powell & Wright (1949) examined two strains isolated from calf scours, one from infantile gastro-enteritis, one from a subcutaneous abscess, and one from a healthy child; no marked difference in virulence could be detected. Similarly, Rowley (1954) estimated the mouse LD 50 of 13 strains of *E. coli* from cases of infantile gastro-enteritis and of 6 normal faecal coliform organisms; there was no relationship between the LD 50 and the origin of the strain.

Oral dosing of weaned and suckling mice with living cultures of *Escherichia coli* has been equally unsatisfactory for differentiating pathogenic from non-pathogenic strains. Moll & Ingalsbe (1955) found neither death nor illness in weaned mice dosed with broth cultures of 9 strains isolated from normal calves, and from cases of calf scours and infantile enteritis. Two of the 4 strains associated with scouring calves caused death of all suckling mice up to the age of 8 days, but the suckling mice became completely resistant to the same two strains by the 10th or 12th day.

Shwartzman reaction

For many years it has been known that *Escherichia coli* organisms produce potent tissue-sensitizing substances that can stimulate a phenomenon of tissue reactivity, the Shwartzman reaction. In some cases of infantile gastro-enteritis the appearances post mortem resemble those seen in rabbits in which this reaction has been elicited. Lindberg & Young (1956) compared some enteropathogenic strains of *E. coli* with non-pathogenic strains as regards their ability to sensitize for and evoke the cutaneous Shwartzman reaction. A significant difference was demonstrated between the two groups, and the preparations from enteropathogenic strains were even more potent in this respect when the cultures were maintained under conditions which favoured the production of 'K' antigen. Recently specific substances, resembling endotoxins, and resulting from the proliferation of *E. coli* in the intestine, have been detected in the blood of some babies suffering from gastro-enteritis, by inhibition of haemagglutination reactions (Young, Sochard, Gillem & Ross, 1960) and sensitization of the red cells *in vivo* (Young, Gillem & Akeroyd, 1962).

Although these findings do not necessarily implicate the Shwartzman reaction in the pathogenesis of diarrhoea due to *Escherichia coli*, the difference between enteropathogenic and non-pathogenic strains in eliciting the reaction may be of significance, and warrants further study and consideration.

Rabbit gut technique

A technique originally described by Violle & Crendiropoulo (1915) and rediscovered by De & Chatterje (1953) to study the pathogenicity of *Vibrio cholerae* organisms has been used to investigate *Escherichia coli* strains. The method consists of injecting cultures of organisms into segments of rabbit intestine isolated by ligatures, and killing the rabbit 24 hr. later to examine the injected loops of bowel. De, Bhattacharya & Sarkar (1956) found that all three strains of E. coli from cases of infantile gastro-enteritis, and many of those from cases of acute or chronic enteritis in adults gave positive reactions, which were seen as severe congestion of the injected intestine and distension of the loop with greyish or haemorrhagic fluid; the majority of strains of E. coli from healthy people produced neither congestion nor fluid accumulations in the injected loops. These findings were confirmed by McNaught & Roberts (1958) and by Taylor, Maltby & Payne (1958); the latter found that reproducible results could be obtained with living cultures of E. coli, provided standard rabbits fed on their customary diet were used; furthermore heat-killed cultures, filtrates of cultures or extracts of organisms failed to induce positive reactions.

Taylor, Wilkins & Payne (1961) demonstrated a close correlation between the ability of a strain of Escherichia coli to cause a positive rabbit gut reaction and the association of the strain with diarrhoea in babies. This correlation was not so close with strains of E. coli isolated from calves, but many of these strains had been kept in the laboratory for some years before being tested and it was shown that the ability of strains to produce a positive reaction in rabbit's intestines may be reduced and finally lost under these conditions. E. coli strains isolated from gastro-enteritis and oedema disease in pigs were also tested by the rabbit gut technique by Taylor et al. (1961); all failed to produce dilatation of the injected loop of bowel. This is in contrast with the results obtained by Namioka & Murata (1962), who report positive results in the rabbit intestine with each of seven strains from cases of oedema disease and gastro-enteritis in pigs; seven strains isolated from normal pig faeces gave negative results.

The absence of any reaction in the rabbit gut injected with killed organisms and extracts of organisms suggests that the pathogenicity of Escherichia coli strains is a property intimately connected with the living cell in vivo and not solely with the toxic products it can produce in vitro. To produce a positive rabbit gut reaction the broth cultures must be moderately alkaline (pH 8·4), and no reactions were produced when the pH of the broth culture was 7·4 (De et al. 1956). Milk media were found by Taylor et al. (1961) to enhance the pathogenic effects in the intestines of rabbits, whilst hog gastric mucin reduced the effect. Thus various subsidiary factors were concerned in the elicitation of positive reactions by pathogenic strains in this technique; it would be of some interest to know the relationship between strong mucinase activity of strains and the rabbit gut reaction produced by them, also the results of using the gut of pigs and other animals in the technique.

The results of experiments by Taylor *et al.* (1961) led them to believe that positive reactions in the rabbit gut indicated enteropathogenicity and not pathogenicity in general, since strains of *Escherichia coli* isolated from urinary infections failed to produce reactions in rabbits. Although it has been established that certain serological types of *E. coli* are commonly associated with infantile gastro-enteritis and calf scours, the relationship between antigenic structure and enteropathogenicity as indicated by the isolated loop of rabbit intestine is not absolute; the agent causing dilatation of the rabbit gut may be present or absent in strains of the same serotype (Taylor *et al.* 1961).

Chick embryos

Recently, attempts have been made to distinguish between strains of *Escherichia coli* which predominate in calves affected with colibacillosis and those isolated from healthy calves, by their virulence in 13-day-old chick embryos when infected by way of the chorio-allantoic membrane (Hughes & Lovell, 1962; Miss B. A. Hughes, unpublished work). By estimating the LD 50 for a number of strains a gradation of virulence for chick embryos was demonstrated with the strains tested. Those that were highly virulent for chick embryos were nearly all isolated from diseased calves and were serological types recognized to be frequently associated with colibacillosis. Strains of intermediate virulence included many isolated from calves with colibacillosis and some from healthy cattle; whilst the relatively avirulent, non-capsulated strains were largely derived from healthy cattle. An exception to this relationship between virulence of *E. coli* strains and association with disease was the group of strains which proved the least virulent for the chick embryo; practically all in this group, irrespective of their source, were capsulated or mucoid strains. Strains of the same serological type had the same order of virulence for chick embryos whether they had been isolated from scouring or from healthy animals, and dissimilar but related antigenic types had different degrees of virulence.

Both virulent and avirulent strains which were non-capsulated multiplied rapidly on the chorio-allantoic membrane and produced a bacteraemia within 12 hr. of inoculation. The virulent strains then invaded the tissues of the embryo where they continued to multiply and produced death. The avirulent strains, on the other hand, were usually removed from the blood and no tissue invasion occurred at this stage;

waves of tissue invasion and clearance subsequently occurred but there were few deaths. It was considered that the continued viability of most embryos infected with avirulent organisms may be influenced by increasing age of the embryo. Smear and histological examinations of the exudates and tissues of the embryos suggested that virulence of the strains was associated with their reduced susceptibility to phagocytosis.

Capsulated strains of *Escherichia coli* initially invaded the blood and tissues of the embryo, but soon became localized in the extraembryonic tissue; when re-invasion of the embryonic tissues occurred later the embryos, being older, were possibly more resistant to endotoxins and thus usually survived the infection. At first the capsulated strains were not particularly sensitive to phagocytosis, but susceptibility increased, resulting in removal of the bacteria from the blood-stream. The anomalous behaviour of capsulated strains, which failed to conform with established concepts for capsulated bacteria, indicated that the effects of different strains in chick embryos must be interpreted with caution; conditions within the embryo and the behaviour of strains in this medium may be quite different from those existing in the young animal.

GENERAL CONSIDERATIONS

Escherichia coli form part of the normal intestinal flora of animals and, in healthy individuals, host and bacteria exist together harmoniously. Some serological types of *E. coli* are associated with enteric disease in young animals and certain strains have greater potentialities for producing disease or tissue reactions than others. Different species of host have various susceptibilities to infection by different strains of one bacterial genus, for some serotypes are more frequently isolated from disease in one species of animal than others. The same serological types of *E. coli* which are grown in profusion from cases of calf scours can be isolated in small numbers from rectal swabs of some normal healthy calves (Wramby, 1948; Wood, 1955; Schoenaers & Kaeckenbeeck, 1960b), and this has also been found in other animal species. It implies that healthy animals carry potentially pathogenic strains of *E. coli* in their intestines, and thus provide a source of infection for clinical disease when the appropriate predisposing factors are present. With the development of clinical disease, the excretion of large numbers of these strains in the faeces would supply more opportunity for transmission of greater numbers of pathogenic organisms to in-contact animals. Whether strains can become enhanced in virulence by passage is still uncertain; the experiments of Miss B. A. Hughes (unpublished) suggest that non-capsulated strains that are relatively avirulent in chick embryos cannot be made more virulent by passage through mice, but chick embryo virulent strains may be reduced in virulence over a period of time by serial transfer in laboratory media, and then increased in virulence by mouse passage.

Predisposing factors relating to the resistance of the host are of vital importance with regard to the establishment and character of *Escherichia coli* infections of young animals. If calves deprived of colostrum die they usually show a septicaemic *E. coli* infection, whereas in colostrum-fed calves the tissues are not usually invaded and *E. coli* organisms are confined to the lumen of the alimentary tract; the presence of specific antibodies in the colostrum protects the calves' tissues against bacterial invasion. The precise way in which adverse environmental conditions, and poor methods of management and feeding, including quality of the food, predispose the host to *E. coli* infections is still unknown. The oral administration of substances such as strong intestinal antiseptics and endotoxins, which may damage or interfere with the normal functioning of the alimentary mucosa, can produce colisepticaemias experimentally in colostrum-deprived calves (Jensen, 1893; Schoenaers & Kaeckenbeeck, 1958).

Shortly after a calf is born the lumen of the alimentary tract becomes populated with bacteria which include *Escherichia coli*. In clinically normal calves the number of viable *E. coli* organisms increases rapidly during the first 2 days of life and large numbers can be cultured from the duodenum as well as the lower parts of the small intestine and colon. Thereafter the numbers decrease and by 2 or 3 weeks of age large numbers are found only in the large intestine (Ingram, 1962). The factors responsible for these adjustments in the intestinal flora during the early days of life and those that subsequently maintain an equilibrium between the various species of bacteria in the flora are largely unknown.

The recent work of Dubos and his colleagues (Schaedler & Dubos, 1962; Dubos & Schaedler, 1962; Dubos & Kessler, 1963) is pertinent to these considerations. A colony of mice was established which was free of many of the common mouse pathogens. Maintained under proper conditions the faecal flora of these mice contained very large numbers of lactobacilli but no *Escherichia coli*, *Proteus* or *Pseudomonas*. Small alterations to the diet fed to these mice, the administration of antibiotics by mouth, or injections of endotoxin produced profound changes in the type and numbers of bacteria in the stools. The administration of penicillin in the drinking water for one week, or the single injection of endotoxin, for example, caused a disappearance of all viable lacto-bacilli in the faecal flora and an enormous increase of enterococci and coliform bacilli, including *E. coli*; the changes were more marked when the mice were fed on a synthetic diet in place of their normal pellet food. Weather conditions and crowding in the mouse colony were some of the other factors mentioned which disturbed the balance of the faecal flora. The results of these experiments in laboratory animals provide some clearer understanding of the way in which dietary and environmental factors may exert an influence on the well-being of the host, as the increase of enterococci and Gram-negative bacilli in the faecal flora was accompanied by a loss of body weight of the mice. If potentially pathogenic strains of *E. coli* are present amongst the alimentary flora, with their ability to proliferate more readily in the intestinal lumen and to multiply in living tissues, the disturbance of the equilibrium of the gut flora by predisposing factors may be the initiating factor for the precipitation of clinical disease.

Colibacillosis of young animals is one of many diseases where pre-disposing factors together with the presence of the exciting agent are necessary for the initiation of active infection. In some diseases the predisposing causes are more fully understood and the exciting agents have been more amenable to investigation. For example, in black disease of sheep the spores of *Clostridium oedematiens* remain latent in the liver until stimulated by migrating liver flukes. Also in transit fever of cattle, *Pasteurella septica* and possibly also a virus reside harmlessly in the respiratory tract, and only under conditions of stress when the animals are transported by rail or road do they proliferate and invade the host's tissues, resulting in pneumonia. The predisposing factors concerned in *Escherichia coli* infections appear to be more diverse and complex, and different strains of the bacterium vary in their potentialities for producing disease. Most of the known serotypes appear to be able to produce disease when given the right conditions. There are clearly many other strains of *E. coli* not yet characterized serologically which have the same capacity, but whether all strains are capable or have the necessary qualities to produce disease is purely speculative. In no other disease is the host–parasite relationship of greater fundamental importance.

CONCLUSIONS

Certain serological types of *Escherichia coli* are associated with an enteric disease which occurs mainly in young animals, especially when they are crowded together or reared in an inclement environment. Antibodies in the colostrum play a major role in protection against the septicaemic forms of these infections which are often fatal in the absence of colostral feeding. Even when colostrum is fed enteric disease may occur, which is usually non-septicaemic and may be fatal. Post-colostrum nutrition also influences the prevalence of these infections.

Investigations of the basis for the enteropathogenicity of these types of *Escherichia coli* have been made *in vitro* and *in vivo*. No test *in vitro* or property of the organisms has been found to distinguish potentially pathogenic and supposedly non-pathogenic types, although some features such as increased mucinase production and inhibition by serine are more frequent attributes of strains isolated from cases of colibacillosis. As yet there is no evidence of any toxin production by pathogenic strains other than the lipopolysaccharide complex which is common to both types; the studies which have been reported, however, have been confined to strains grown *in vitro*, and the possibility of additional toxin formation which occurs only *in vivo* deserves investigation. Of the techniques *in vivo* virulence tests in mice and guinea-pigs provide no means of differentiating strains with respect to their enteropathogenicity. In this method and also in the chick embryo virulence tests, the bacteria are exposed to the natural defence mechanisms of the host—phagocytosis and serum factors—which is unlikely to be the case in natural non-septicaemic infections in which the *E. coli* are localized to the lumen of the intestine. However, the results of chick embryo virulence tests and the isolated rabbit gut technique show promise as means of distinguishing enteropathogenicity of strains and as useful tools for further studies.

I wish to thank Miss B. A. Hughes for permission to quote from her unpublished work and allowing me access to her collected data; also Professor R. Lovell for his helpful criticisms during the preparation of this paper, and the editors of Research in Veterinary Science and the Veterinary Record, and the Nationaal Comité van den Internationalen Zuivelbond for permission to reproduce figures and tables already published.

REFERENCES

ADAM, A. (1927). Dyspepsiekoli. Zur Frage der bacteriellen Ätiologie der sogennanten alimentären Intoxikation. *Jb. Kinderheilk.* **116**, 8.

ASCHAFFENBURG, R., BARTLETT, S., KON, S. K., ROY, J. H. B., BRIGGS, C. & LOVELL, R. (1951 *a*). The nutritive value of colostrum for the calf. 4. The effect of small quantities of colostral whey, dialysed whey and 'immune lactoglobulins'. *Brit. J. Nutr.* **5**, 171.

ASCHAFFENBURG, R., BARTLETT, S., KON, S. K., ROY, J. H. B., SEARS, H. J., INGRAM, P. L., LOVELL, R. & WOOD, P. C. (1952). The nutritive value of colostrum for the calf. VIII. The performance of Friesian and Shorthorn calves deprived of colostrum. *J. comp. Path.* **62**, 80.

ASCHAFFENBURG, R., BARTLETT, S., KON, S. K., ROY, J. H. B., WALKER, D. M., BRIGGS, C. & LOVELL, R. (1951*b*). The nutritive value of colostrum for the calf. 5. The effect of prepartum milking. *Brit. J. Nutr.* **5**, 343.

ASCHAFFENBURG, R., BARTLETT, S., KON, S. K., TERRY, P., THOMPSON, S. Y., WALKER, D. M., BRIGGS, C., COTCHIN, E. & LOVELL, R. (1949*a*). The nutritive value of colostrum for the calf. 1. The effect of different fractions of colostrum. *Brit. J. Nutr.* **3**, 187.

ASCHAFFENBURG, R., BARTLETT, S., KON, S. K., WALKER, D. M., BRIGGS, C., COTCHIN, E. & LOVELL, R. (1949*b*). The nutritive value of colostrum for the calf. 2. The effect of small quantities of the non-fatty fraction. *Brit. J. Nutr.* **3**, 196.

BRAY, J. (1945). Isolation of antigenically homogeneous strains of *Bact. coli neopolitanum* from summer diarrhoea of infants. *J. Path. Bact.* **57**, 239.

BRIGGS, C. (1951). The nutritive value of colostrum for the calf. 6. The 'K' antigens of *Bacterium coli*. *Brit. J. Nutr.* **5**, 349.

BRIGGS, C., LOVELL, R., ASCHAFFENBURG, R., BARTLETT, S., KON, S. K., ROY, J. H. B., THOMPSON, S. Y. & WALKER, D. M. (1951). The nutritive value of colostrum for the calf. 7. Observations on the nature of the protective properties of colostrum. *Brit. J. Nutr.* **5**, 356.

BUXTON, A. & THOMLINSON, J. R. (1961). The detection of tissue-sensitizing antibodies to *Escherichia coli* in oedema disease, haemorrhagic gastro-enteritis and in normal pigs. *Res. vet. Sci.* **2**, 73.

DE, S. N., BHATTACHARYA, K. & SARKAR, J. K. (1956). A study of strains of *Bacterium coli* from acute and chronic enteritis. *J. Path. Bact.* **71**, 201.

DE, S. N. & CHATTERJE, D. N. (1953). An experimental study of the mechanism of action of *Vibrio cholerae* on the intestinal mucous membrane. *J. Path. Bact.* **66**, 559.

DUBOS, R. J. & KESSLER, A. (1963). Integrative and disintegrative factors in symbiotic associations. In *Symbiotic Associations, Symp. Soc. gen. Microbiol.* **13**, 1.

DUBOS, R. J. & SCHAEDLER, R. W. (1962). The effect of diet on the fecal bacterial flora of mice and on their resistance to infection. *J. exp. Med.* **115**, 1161.

DUNNE, H. W., GLANZ, P. J., HOKANSON, J. F. & BORTREE, A. L. (1956). *Escherichia coli* as a cause of diarrhoea in calves. *Ann. N.Y. Acad. Sci.* **66**, 129.

FEY, H., LANZ, E., MARGADANT, A. & NICOLET, J. (1962). Zur Pathogenese der Kälbercolisepsis. VI. Experimentelle Infektion zum Beweis der parenteralen Genese. *Dtsch. tierärztl. Wschr.* **69**, 581.

FEY, H. & MARGADANT, A. (1962). Zur Pathogenese der Kälbercolisepsis. V. Versuche zur künstlichen Infektion neugeborener Kälber mit dem Colityp 78:80B. *Zbl. Vet.-Med.* **9**, 767.

FORBES, M. & KUCK, N. A. (1961). Lack of relationship between sialic acid content, toxicity, and lethality of *Escherichia coli*. *Proc. Soc. exp. Biol., N.Y.* **108**, 34.

GILES, C. & SANGSTER, G. (1948). An outbreak of infantile gastro-enteritis in Aberdeen. The association of a special type of *Bact. coli* with the infection. *J. Hyg., Camb.* **46**, 1.

GLANZ, P. J., DUNNE, H. W., HEIST, C. E. & HOKANSON, J. F. (1959). Bacteriological and serological studies of *Escherichia coli* serotypes associated with calf scours. *Bull. Pa agric. Exp. Sta.* no. 645, 1.

HARVEY, D. G. & CARNE, P. (1960). Studies on some chemical aspects of the pathological activities of strains of *Escherichia coli* of bovine origin. *J. comp. Path.* **70**, 84.

HEWISON, N. V. (1939). Calf rearing. *J. Minist. Agric.* **46**, 18.

Howe, P. E. (1921). An effect of the ingestion of colostrum upon the composition of the blood of new-born calves. *J. biol. Chem.* **49**, 115.

Howe, P. E. (1922). The relation between age and the concentration of protein fractions in the blood of the calf and cow. *J. biol. Chem.* **53**, 479.

Hughes, B. A. & Lovell, R. (1962). Virulence of *Escherichia coli* for chick embryos. *Abstr. VIIIth Int. Congr. Microbiol., Montreal*, no. 114.

Ingram, P. L. (1962). Observations on the pathology and pathogenesis of experimental colibacillosis in calves. Thesis for Ph.D. degree of the University of London.

Ingram, P. L. & Lovell, R. (1960). Infection by *Escherichia coli* and salmonella. *Vet. Rec.* **72**, 1183.

Ingram, P. L., Lovell, R., Wood, P. C., Aschaffenburg, R., Bartlett, S., Kon, S. K., Palmer, J., Roy, J. H. B. & Shillam, K. W. G. (1956). *Bacterium coli* antibodies in colostrum and their relation to calf survival. *J. Path. Bact.* **72**, 561.

Ingram, P. L., Lovell, R., Wood, P. C., Aschaffenburg, R., Bartlett, S., Kon, S. K., Roy, J. H. B. & Sears, H. J. (1953). Further observations of the significance of colostrum for the calf. I. White scours of calves in an enclosed community. *Proc. XIIIth Int. Dairy Congr. The Hague*, **3**, 1365.

Jensen, C. O. (1893). Über die Kälberruhr und deren Aetiologie. *Mh. prakt. Tierheilk.* **4**, 97.

Jensen, C. O. (1905). Über Kälberruhr und deren Verhütung durch Serum-injectionen. *Z. Tiermed.* **9**, 321.

Joest, E. (1903). Untersuchungen über Kälberruhr. *Z. Tiermed.* **7**, 377.

Jordan, L. (1933). Diseases of young calves. *Vet. J.* **89**, 202.

Lindberg, R. B. & Young, V. M. (1956). Observations on enteropathogenic *Escherichia coli. Ann. N.Y. Acad. Sci.* **66**, 100.

Little, R. B. & Orcutt, M. L. (1922). The transmission of agglutinins of *Bacillus abortus* from cow to calf in the colostrum. *J. exp. Med.* **35**, 161.

Lovell, R. (1937). Classification of *Bacterium coli* from diseased calves. *J. Path. Bact.* **44**, 125.

Lovell, R. & Hill, A. B. (1940). A study of the mortality rates of calves in 335 herds in England and Wales (together with some limited observations for Scotland). *J. Dairy Res.* **11**, 225.

Marsh, H. & Tunnicliff, E. A. (1938). Dysentery of new-born lambs. *Bull. Mont. agric. Exp. Sta.* no. 361, 1.

McEwen, A. D. (1950). The resistance of the young calf to disease. *Vet. Rec.* **62**, 83.

McGirr, J. L. (1947). Colostral transmission of antibody substances from mother to offspring. A review. *Vet. J.* **103**, 345.

McNaught, W. & Roberts, G. B. S. (1958). Enteropathogenic effects of strains of *Bacterium coli* isolated from cases of gastro-enteritis. *J. Path. Bact.* **76**, 115.

Moll, T. & Ingalsbe, C. K. (1955). The pathogenicity of certain strains of *Escherichia coli* for young mice and calves. *Amer. J. vet. Res.* **16**, 337.

Namioka, S. & Murata, M. (1962). Studies on the pathogenicity of *Escherichia coli*. II. The effect of a substance obtained from pseudoeosinophils of rabbits on the organism. *Cornell Vet.* **52**, 289.

Nelson, J. B. (1924). Normal immunity reactions of the cow and the calf with reference to antibody transmission in the colostrum. *Res. Bull. Mo. agric. Exp. Sta.* no. 68, 1.

Ross, C. A. C. (1959). Mucinase activity of intestinal organisms. *J. Path. Bact.* **77**, 642.

Ross, C. A. C. & Dawes, E. A. (1954). Resistance of the breast-fed infant to gastro-enteritis. *Lancet*, i, 994.

ROWLEY, D. (1953). Interrelationships between amino-acids in the growth of coliform organisms. *J. gen. Microbiol.* **9**, 37.

ROWLEY, D. (1954). The virulence of strains of *Bacterium coli* for mice. *Brit. J. exp. Path.* **35**, 528.

SAUNDERS, C. N., STEVENS, A. J., SPENCE, J. B. & BETTS, A. O. (1963). *Escherichia coli* infection: reproduction of the disease in 'pathogen-free' piglets. *Res. vet. Sci.* **4**, 347.

SAUNDERS, C. N., STEVENS, A. J., SPENCE, J. B. & SOJKA, W. (1963). *Escherichia coli* infection: reproduction of the disease in naturally-reared piglets. *Res. vet. Sci.* **4**, 333.

SCHAEDLER, R. W. & DUBOS, R. J. (1962). The fecal flora of various strains of mice. Its bearing on their susceptibilities to endotoxin. *J. exp. Med.* **115**, 1149.

SCHOENAERS, F. & KAECKENBEECK, A. (1958). Études sur la colibacillose du veau. I. Réalisation expérimentale de la maladie. *Ann. Méd. vét.* **102**, 211.

SCHOENAERS, F. & KAECKENBEECK, A. (1960a). Études sur la colibacillose du veau. III. Essai d'immunoprophylaxie. *Ann. Méd. vét.* **104**, 117.

SCHOENAERS, F. & KAECKENBEECK, A. (1960b). Les colibacilles entéropathogènes de la flore intestinale du veau normal. *Ann. Méd. vét.* **104**, 240.

SHEEHY, E. J. (1934). Derangement of the digestive processes in the milk-fed calf due to abnormal curd formation in the fourth stomach. *Sci. Proc. R. Dublin Soc.* **21**, 73.

SHILLAM, K. W. G., DAWSON, D. A. & ROY, J. H. B. (1960). The effect of heat treatment on the nutritive value of milk for the young calf. The effect of ultra-high-temperature treatment and of pasteurization. *Brit. J. Nutr.* **14**, 403.

SHILLAM, K. W. G., ROY, J. H. B. & INGRAM, P. L. (1962a). The effect of heat treatment on the nutritive value of milk for the young calf. 2. The factor in a milk substitute associated with a high incidence of scouring and mortality. *Brit. J. Nutr.* **16**, 267.

SHILLAM, K. W. G., ROY, J. H. B. & INGRAM, P. L. (1962b). The effect of heat treatment on the nutritive value of milk for the young calf. 3. The effect of the preheating treatment of spray-dried skim milk and a study of the effect of ultra-high-temperature treatment of separated milk. *Brit. J. Nutr.* **16**, 585.

SHILLAM, K. W. G., ROY, J. H. B. & INGRAM, P. L. (1962c). The effect of heat treatment on the nutritive value of milk for the young calf. 4. Further studies on the effects of the preheating treatment of spray-dried skim milk and of ultra-high-temperature treatment. *Brit. J. Nutr.* **16**, 593.

SMITH, H. W. (1961). The development of the bacterial flora of the faeces of animals and man: the changes that occur during ageing. *J. appl. Bact.* **24**, 235.

SMITH, H. W. (1962). Observations on the aetiology of neonatal diarrhoea (scours) in calves. *J. Path. Bact.* **84**, 147.

SMITH, R. McD. (1934). White scour and allied diseases in calves. *Vet. Rec.* **14**, 1004.

SMITH, T. & LITTLE, R. B. (1922). The significance of colostrum to the new-born calf. *J. exp. Med.* **36**, 181.

SMITH, T. & LITTLE, R. B. (1927). Studies on pathogenic *B. coli* from bovine sources. I. The pathogenic action of culture filtrates. *J. exp. Med.* **46**, 123.

SMITH, T. & ORCUTT, M. L. (1925). The bacteriology of the intestinal tract of young calves with special reference to early diarrhoea (scours). *J. exp. Med.* **41**, 89.

SOJKA, W. J. & CARNAGHAN, R. B. A. (1961). *Escherichia coli* infection in poultry. *Res. vet. Sci.* **2**, 340.

SOJKA, W. J., LLOYD, M. K. & SWEENEY, E. J. (1960). *Escherichia coli* serotypes associated with certain pig diseases. *Res. vet. Sci.* **1**, 17.

TAYLOR, J. & CHARTER, R. E. (1952). The isolation of serological types of *Bact. coli* in two residential nurseries and their relation to infantile gastro-enteritis. *J. Path. Bact.* **64**, 715.

TAYLOR, J., MALTBY, M. P. & PAYNE, J. M. (1958). Factors influencing the response of ligated rabbit-gut segments to injected *Escherichia coli*. *J. Path. Bact.* **76**, 491.

TAYLOR, J., POWELL, B. W. & WRIGHT, J. (1949). Infantile diarrhoea and vomiting: a clinical and bacteriological investigation. *Brit. med. J.* ii, 117.

TAYLOR, J., WILKINS, M. P. & PAYNE, J. M. (1961). Relation of rabbit-gut reaction to enteropathogenic *Escherichia coli*. *Brit. J. exp. Path.* **42**, 43.

THOMSON, S. (1955*a*). The numbers of pathogenic bacilli in faeces in intestinal disease. *J. Hyg., Camb.* **53**, 217.

THOMSON, S. (1955*b*). The role of certain varieties of *Bacterium coli* in gastro-enteritis of babies. *J. Hyg., Camb.* **53**, 357.

VIOLLE, H. & CRENDIROPOULO, H. (1915). Note sur le choléra expérimental. *C.R. Soc. Biol., Paris*, **78**, 331.

WITHERS, F. W. (1952). Mortality rates and disease incidence in calves in relation to feeding, management and other environmental factors. *Brit. vet. J.* **108**, 315.

WRAMBY, G. (1948). *Investigations into the Antigenic Structure of* Bact. coli *Isolated from Calves*. Uppsala: Appelbergs Boktryckeriaktiebolag.

WOOD, P. C. (1955). The epidemiology of white scours among calves kept under experimental conditions. *J. Path. Bact.* **70**, 179.

YOUNG, V. M., GILLEM, H. C. & AKEROYD, J. H. (1962). Sensitization of infant red cells by bacterial polysaccharides of *Escherichia coli* during enteritis. *J. Pediat.* **60**, 172.

YOUNG, V. M., SOCHARD, M. R., GILLEM, H. C. & ROSS, S. (1960). Infectious agents in infantile diarrhoea: II. Serological reactions with *Escherichia coli* O1 through O25. *Proc. Soc. exp. Biol., N.Y.* **105**, 638.

SPECULATIONS ON PLANT
PATHOGEN–HOST RELATIONS

E. W. BUXTON

Rothamsted Experimental Station, Harpenden, Hertfordshire

INTRODUCTION

When invited to present this paper, I composed a title that rashly omitted the word 'speculations'. But it took only a little reading and still less reflexion to realize that our interpretations of the nature of the plant pathogen–host relationships are, to say the least, tentative: I quickly remedied the omission. Plants and their pathogens have evolved together and we have hardly begun to discover the full nature of the association or the specificity of the interaction. Optimism, backed by some experiments, suggests that characteristics of diseases are in some way related to the physiological characteristics that distinguish pathogens in culture. A great deal of fact finding, heuristic thinking, argument and speculation has already been expended towards connecting activities *in vitro* with those *in vivo*, so I contemplate the task of synthesis and judgement with considerable apprehension. Plant pathology grew out of the pressing need to protect crops from the depredations of fungi, bacteria and viruses, and many of its formative years were spent gathering information about the types of plant disease, the economic losses, and means of lessening these losses. Inevitably, epidemiology and control were studied at the expense of detailed work on the intimate host–parasite relationships or the biology of the causal organism. Even after 1807, when Prévost made it quite clear that micro-organisms incited plant disease, it took a further half century and the shock of the Irish potato famine before the idea was widely accepted. After a necessary period given over to describing and naming species of pathogens, the first inklings of investigation of the nature of the host–parasite relationship came, when De Bary (1886) extracted an enzyme from rotten carrot and simulated the rot by inoculating carrot tissue with the sterilized extract. This was also the starting-point for plant pathologists to investigate the biology of the organisms responsible for disease. As the techniques of biochemistry, genetics and physiology have been developed, plant pathologists have persistently tried to analyse the idiosyncrasies that pathogens reveal in the laboratory in the hope of finding mechanisms to explain their virulence *in vivo*.

However, we are a long way from finding the underlying nature of pathogenicity and of disease resistance, because we are not sure of the specific questions to ask. Host resistance may be based on biochemical or physical differences. Is there one type of resistance or are there many? Is the virulence of pathogens based on any particular metabolic activity that can be tagged *in vitro*? A discussion of these points fits well with the theme of this symposium.

THE PROBLEM

Two kinds of inquiry have been made. The first asks whether attributes of plant pathogens *in vitro*, such as production in cultures of specifically phytotoxic metabolites, give any clue to the underlying nature of pathogenicity. The second questions whether specific responses of pathogens *in vitro* to host-produced metabolites can tell us anything about differences in host resistance.

Metabolites produced in culture by plant pathogens fall into at least four classes:

(1) Toxic substances: usually implicated in wilts or lethal diseases in which tissues other than those infected may be affected; for example, those formed by *Fusarium* and *Verticillium*.

(2) Growth substances: associated with hyperplasia, gall formation and local lesions; for example, those formed by bacteria causing crown gall and wild fire of tobacco.

(3) Enzymes: possibly implicated in soft rots and wilts; for example, those formed by potato soft rot bacteria and *Botrytis*.

(4) Antibotics: these may affect interactions between microbes in soil, and especially on the surface of roots (the rhizosphere). Nearly all soil microbes produce some metabolite which is toxic *in vitro* to at least one other micro-organism.

The problem must be approached from the host as well as from the pathogen. Infection and pathogenicity can be influenced by plant products many of which have specific effects on pathogens that are also correlated with the host specificity of the pathogen. These products fall broadly into two groups:

(1) Host tissue extracts, some of which may be toxic, others growth-promoting to micro-organisms in culture.

(2) Exudates, from leaves or roots, that specifically control the growth of pathogens *in vitro*. Adaptation by pathogens towards tolerance of growth-inhibitory exudates is an important mechanism resulting in changes of pathogenicity.

In the following pages, the effects on pathogens *in vitro* of both types of products will be discussed, together with the potentialities of using genetically marked, nutritionally deficient, artificially produced mutants of pathogens to investigate the biochemical basis of their pathogenicity either *in vivo* or with the host products *in vitro*.

Before dealing with the evidence for the relevance of research in the laboratory to the underlying mechanism of pathogenicity, a few cautionary remarks must be made.

There is some hope of deriving logical conclusions by comparing the reactions to host metabolites of a pathogen and a non-pathogen of the same species, or by comparing the effects on a host of their products from culture *in vitro*. However, it is more rewarding to correlate a complex pattern of resistant and susceptible hosts with several pathogenic and non-pathogenic strains of one species of micro-organism. This can often be done by working with physiologic races of plant pathogens. Physiologic races are characterized by differences in their effects on a series of genetically different varieties of a host species. A relatively simple race pattern illustrates this point:

	Host varieties			
Pathogen races	A	B	C	D
1	+	−	−	−
2	+	+	−	−
3	+	+	+	−
4	+	+	+	+

+ = causes disease; − = no disease produced.

When host products of the different varieties have effects on the races *in vitro* that are in keeping with this pattern, or when race metabolites reproduce symptoms of the natural disease only on susceptible hosts, there is more reason to assume that they have a bearing on the underlying mechanisms of resistance or pathogenicity *in vivo*.

Specific differences in pathogenicity are very common between isolates of any one fungus. Usually these differences are unrelated to cultural type or morphology and sometimes other differences in behaviour *in vitro* have been sought. However, when some apparently important metabolite is produced by a pathogen *in vitro*, it is difficult to discover if it is also produced *in vivo*. The rules of the game are hard, and the criteria for implicating a metabolite of a pathogen in microbial behaviour *in vivo* should depend on satisfying, for each organism, at least the following requirements:

(1) The metabolite must be separated from crude filtrates and, if possible, identified.

(2) After inoculating the host with the metabolite, symptoms characteristic of the natural pathogen-incited disease must be reproduced.

(3) The metabolite must be re-isolated from diseased tissue and its origin from the pathogen confirmed.

These are no more than modified Koch's postulates, but they have often been ignored. Many workers have set out on the path of correlating activities *in vitro* with those *in vivo*, and found themselves in a blind alley, often after having merely demonstrated the presence of a toxin in a culture, or a fungitoxic substance in expressed sap of some disease-resistant variety. One major drawback in studying host–parasite relationship in plants is the lack of a serological procedure to identify either a toxin produced *in vitro* after it has been introduced into a host, or a toxin produced *in vivo* by a pathogen. Animal pathology does not suffer this burden. Furthermore, there is as yet no known way of specifically identifying immunity reactions in plants, although a great deal is known about serological techniques for differentiating pathogenic fungi, bacteria and viruses from each other *in vitro*.

THE EVIDENCE

Now that some of the pitfalls have been charted, the evidence for assigning activities *in vitro* to proposed mechanisms at work *in vivo*, can be examined.

Metabolites produced by plant pathogens in vitro

Production of toxins in vitro

Work in this subject presupposes, quite reasonably, that parasitism can be interpreted in terms of biochemical interactions between host and infecting organism, and that they are reflected by host responses to pathogen metabolites. The circumstantial evidence for this supposition rests on the fact that symptoms often occur in the host in advance of the site reached by the parasite, and that symptom-causing substances are released from infected host cells and become rapidly distributed throughout the host. Cell-free filtrates of many fungal cultures are phytotoxic and some cause symptoms characteristic of natural infection (Brian, 1955); this lends credence to the view that the toxins may also be produced by pathogens *in vivo*. On the other hand, some materials produced *in vitro* may not be important *in vivo*. Gäumann and his colleagues showed that a toxin, lycomarasmin (a polypeptide of mol. wt. 277), was produced in culture by the tomato wilt fungus *Fusarium oxysporum* f. *lycopersici* (Clauson-Kaas, Plattner & Gäumann, 1944).

Lycomarasmin, at concentrations as low as 5×10^{-3} M, wilted excised tomato leaves, and the inoculation of plants with the expressed sap of naturally diseased plants produced a similar wilt. However, although only 15 days were needed for wilt to develop in fungus-infected plants, Dimond & Waggoner (1953a) showed that lycomarasmin was produced *in vitro* only by senescent, 3-month-old cultures. The toxin's ability to produce wilt symptoms also depended on the presence of ferric ions, but infected iron-deficient plants wilted as quickly as normal infected plants (Dimond, 1955). Moreover, lycomarasmin-induced wilting was irreversible, whereas wilted foliage of naturally diseased plants became temporarily turgid when watered. Fusaric acid, another 'wilt toxin', also from *F. o. lycopersici* growing *in vitro* (Gäumann, 1957) has been detected on the root surface of infected plants (Kalyanasundaram, 1958). Dimond (1955), reviewing pathogenesis in wilt diseases, pointed out that symptoms of infected plants differed from those caused by either lycomarasmin or fusaric acid.

Dimond & Waggoner (1953b) found that ethylene, produced by *Fusarium o. lycopersici* in culture, caused epinasty, part of the wilt syndrome in tomato, and they also identified ethylene as a product of naturally wilted plants. This, apparently, was the first demonstration of an identified toxin *in vivo*, that is a substance produced in an infected host by the activities of the pathogen, which functions in the production of the disease and is not itself the initial inciting agent of the disease. So far ethylene and an unidentified substance centrifuged from diseased tomato stems (Gottlieb, 1943), and possibly helminthosporin isolated from oat plants infected with *Helminthosporium victoriae* (Pringle & Braun, 1957), qualify as important toxins produced *in vivo*. The tobacco wild fire organism *Pseudomonas tabaci* produces a toxin *in vitro*. Since both naturally diseased and toxin-treated leaves exhibit similar local lesions, it is reasonable to suppose that this toxin may be involved *in vivo* (Braun & Pringle, 1959).

Many toxins produced *in vitro*, although damaging to many tissues, are highly selective in that they affect some varieties of a plant species but not others: hence they may be to some extent responsible for host specificity of plant pathogens. However, without a specific relationship between toxin production and a physiologic race–host pattern, and without knowing the site of action in the host or the specific toxin-induced symptoms, so that this information can be correlated with the situation in the diseased plant, the fact that a pathogen produces a toxin *in vitro* may have no significance in explaining its pathogenesis. Moreover, not all toxins produced *in vitro* may be detectable in infected hosts, and there is equally no reason to assume that all toxins detectable in hosts will be contained in culture filtrates. Winstead & Walker (1954), working with strains of *Fusarium oxysporum* f. *conglutinans* that are pathogenic to radish, cabbage and cotton varieties, found a close correlation between toxin production by the different strains and the susceptibility to infection by these strains of the different varieties of hosts. This toxin, which has a much smaller molecule than lycomarasmin, seems to be the only one that so far provides a reasonable explanation for the behaviour of differently parasitic strains.

In future work in this field, it might be rewarding to look for specific toxic-inhibition mechanisms in disease-resistant plants and see whether they also specifically inhibit a pathogen that causes natural infections of closely related but susceptible hosts. This sort of approach, coupled with improvement in the methods of purifying toxins, of isolating substances from diseased tissue, and comparing the effects on different physiologic races, might clarify the hazy picture presented by the work to date.

Growth substances

The production *in vitro* of microbial metabolites that affect plant growth by causing hyperplasia, increased leaf area and plant height has been reviewed by Brian (1957) and Braun (1959). Galls resulting from microbial infection occur in at least 61 families and 142 genera. They are best illustrated by the crown gall disease of tomato caused by *Phytomonas tumefaciens*. Infected cells proliferate and form amorphous and often enormous tumours; secondary tumours, free from bacteria, can also occur, indicating a biochemical derangement or stimulation initiated by the bacteria in primary tumours. In culture, the bacteria produce indole-3-acetic acid, a much-studied growth substance. When applied to the cut ends of tobacco stems, indole-3-acetic acid from cultures *in vitro* causes tumours, depending on the strain of bacterium, the position of the cells to be treated and the inherent competence for regeneration of the treated cells. Many fungi and bacteria, especially in culture media containing tryptophan, produce indole-3-acetic acid which stimulates both cell extension and cell division in plants. Virulent strains of the bacterium causing tomato crown gall produce indole-3-acetic acid in culture and gall tissue contains more indole-3-acetic acid than other tissues, whereas attenuated strains produce galls only when indole-3-acetic acid is added to the inoculum. The same holds true for galls caused in maize by the smut fungus, *Ustilago maydis*. Moulton (1942) extracted an auxin both from the fungus and from diseased host tissue, and considered that the pathogenicity of different strains of the fungus was correlated with their ability to produce auxin in synthetic media. Wolf (1952), by colour tests and chromatography, identified the auxin from maize smut as indole-3-acetic acid. *U. nigra*, which does not produce tumours on its hosts, did not form indole-3-acetic acid *in vitro*, whereas all pathogenic strains of *U. zeae* did. *Nectria galligena*, the cause of apple canker, produces indole-3-acetic acid *in vitro*, whereas *N. corallina*, which is also parasitic on fruit trees but without causing canker, does not.

The work on crown gall and smut strongly indicates that growth substances produced by pathogens *in vitro* are also produced *in vivo*, and promote tumours in the host plants. Other products of plant pathogens which seem to have an implication *in vivo* are the important growth substances, the gibberellins. Tall thin rice seedlings, found to be infected with the pathogenic fungus *Fusarium moniliforme* are known as 'bakanae' (foolish in appearance). Kurosawa (1926) initiated research that has led to the gibberellins being isolated from cultures of *Gibberella fujikuroi*, the perfect state of *F. moniliforme*. Brian & Hemming (1955) found that gibberellic acid, produced by the fungus *in vitro*, greatly increased the internode length of Meteor peas, a dwarf variety, but had little effect on tall varieties.

Enzymes produced by pathogens in vitro

The first steps that plant pathogens take when invading plants is to break their way into cell walls or penetrate between cells by disrupting the middle lamellae. The ability to degrade cellulose and pectin seems therefore to play an important part in pathogenicity. Brown (1915), working with extracts from cultures of *Botrytis cinerea*, was the first to demonstrate that fungi contain something able to disintegrate cell walls. The extracts also killed some cells and their ability to kill depended on the nature of the walls of the cells. Brown's work led to a search for such enzymes, which have since been detected in culture filtrates of many fungi and bacteria. In the copious literature on this subject, many authors have come close to implicating the enzymes as basic components of the mechanism of pathogenicity *in vivo*, but several reviewers have pointed out that this is by no means established. No one has yet stated categorically that the ability to produce cell-degrading enzymes *in vitro* has explained specific differences in pathogenicity *in vivo*. Brown (1936, 1955) and Wood (1955, 1960) reviewed the evidence for production *in vitro* of enzymes, which are of two kinds, pectic and cellulolytic. Pectic enzymes include pectinesterase, which hydrolyses methyl ester groups in pectin to form pectic acid, poly-galacturonase, which uses pectic acid as substrate, and polymethyl-galacturase, which hydrolyses pectin to polygalacturonic acids of low molecular weight. The plant pathogens *Pythium debaryanum, Fusarium moniliforme, B. cinerea, Verticillium dahliae, Sclerotium rolfsii* and *Bacterium aroideae* all produce pectic enzymes in culture. *P. debaryanum* and *B. aroideae* do so whether or not the culture medium contains pectin, but the rest do so in quantity only when pectin is present.

Pectic enzymes produced *in vitro* usually reproduce symptoms of disease when introduced into plants, and pectic enzymes can certainly be extracted from diseased tissues. But if this seems to establish their role in pathogenesis, a number of detracting considerations must be remembered. First, different pathogens secrete numerous enzymes that act in different pH ranges, differ in thermolability, and produce diverse end-products: it is unlikely that each different enzyme can be correlated with the specificity of a pathogen for a particular host. Secondly, and detracting even further from their possible role *in vivo*, are the dis-concerting facts that healthy plant tissue contains pectic enzymes; and secondary invading organisms, which are usually non-pathogenic, also secrete pectic enzymes abundantly. Finally the evidence for correla-tion between the abilities of different micro-organisms to produce pectic enzymes and their specific differences in pathogenicity is not very strong. Lapwood (1957) compared the production *in vitro* of pectolytic enzyme by a *Flavobacterium* spp., *Pseudomonas* spp. and *P. syringae*, which was unable to rot potato tuber, with that of a strain of *Bacterium aroideae* which rots potatoes. On potato extract, all organisms produced equal amounts of enzyme. Parasitic ability seemed to depend not on enzyme production but on the fact that only *B. aroideae* grew fast enough to avoid the physical barriers of host cell reaction. Similarly, *Fusarium moniliforme* secreted protopectinase *in vitro*, but was not pathogenic to potato tubers, whereas *F. avena-ceum*, a common cause of rot in stored tubers, secreted little or no en-zyme *in vitro* (Singh & Wood, 1956). These facts make it difficult to attribute any part of the pathogenic process to pectic enzymes, especially as pectinesterase is often strongly absorbed on cell walls and there is no proof that enzyme properties are the same *in vivo* as *in vitro*.

The enzymes produced by wilt-inducing fungi have been investigated. The fact that the fungi are confined to the vascular system in the host and often cause symptoms in advance of their host penetration is compatible with the idea that they secrete active enzymes *in vivo*. Dimond (1955) reviewed the evidence for the production of pectinesterase and polymethylgalacturase by *Fusarium oxysporum* f. *lycopersici in vitro*, and considered the possibility that *in vivo* pectic enzymes hydrolyse pectin exposed at vessel pits to form calcium gels, which then plug the vascular system. Melanoid pigments are presumably trapped in the gels and form the characteristic vascular discoloration of wilted plants. Polymethylgalacturase can be detected in tracheal fluid, but whether it comes from the pathogen or from the host has never been made clear. Paquin & Coulombe (1962) found that a virulent strain of *F. o. lycopersici* produced more pectinesterase *in vitro* than did an avirulent strain, but a wider correlation between enzyme production by physiologic races is needed before such results can be used to implicate the enzyme *in vivo*. Wilting of cut tomato stems can be induced by enzymes occurring in culture filtrates of wilt-inducing pathogenic strains of *Pseudomonas solanacearum*, but non-pathogenic strains also produce plenty of pectolytic enzymes *in vitro* (Husain & Kelman, 1958). Similar, but even more telling evidence that the significance of tissue-degrading enzymes in parasitism is probably slight, was provided by McDonnell (1962), who found that ultra-violet mutants of *F. o. lycopersici* that lacked the capacity of the parent strains to produce cellulose nevertheless retained their pathogenicity.

It would be easy to dismiss the evidence that pectic and cellulolytic enzymes play a part *in vivo*, but it is only common sense to accept the fact that tissue degradation does involve some enzyme production on the part of the symptom-causing organism. Also Grossmann (1962) showed that tomato wilt could be prevented by applying to infected plants rufianic acid, a specific inhibitor of pectinase. There remains, however, the need to demonstrate more precisely the presence of enzymes produced by the pathogen *in vivo*. Refinement of techniques for purifying and identifying the different enzymes from sources *in vivo* is only one of the advances sorely needed.

Antibiotic metabolites

That bacteria, fungi and actinomycetes produce *in vitro* metabolites that adversely affect other microbes in their immediate vicinity has been known for 50 years or more. Their study was given a great impetus with the discovery of penicillin. In plant pathology, the practical motive for research has been to attempt the control of plant pathogens *in vivo* by adding to the host or its immediate surroundings either other microbes or their antibiotic products from culture *in vitro*. Soil-borne pathogens have been intensively studied in this respect (Brian, 1949, 1960), so in this context '*in vivo*' will be taken to imply 'in the soil at or near the site of infection of host roots'. There is intense competition between microbes in soil for space and nutrients, and one that produces a powerful antibiotic might have a considerable advantage over others. Weindling & Emerson (1936) demonstrated a fungitoxic substance in culture filtrate of the fungus *Trichoderma*, and Brian, Curtis, Hemming & McGowan (1946) identified viridin as a powerful antibiotic produced by *T. viride*. Since then Brian and his colleagues have studied metabolites produced by other organisms *in vitro*, such as gliotoxin, alternaric acid and griseofulvin, which are antibiotics. All such antibiotics are rapidly broken down in soil and, to assess the significance of antibiotics *in vivo*, a continuous source must be provided by adding the appropriate micro-organisms to soil. Their effects can be assessed on such things as changes in attack of host plants by pathogens, by changes in the rate at which partially sterilized soil is recolonized by other microbes, and by changes in the rhizosphere population (Brian, 1949).

There is nothing to indicate that the ability to produce an antibiotic affects the survival of the organism that produces it. In fact, many of the successful soil saprophytes do not produce antibiotics *in vitro*. One notable exception, however, is *Penicillium nigricans*, a common soil saprophyte, which produces griseofulvin *in vitro*. But can it do so near roots *in vivo* and so protect plants from soil-borne pathogens? Although griseofulvin disappeared rapidly when introduced into garden soil (Wright & Grove, 1957), extracts of coats of soil-sown seeds inoculated with *Trichoderma viride*, *P. frequentans* and *P. gladioli* contained the antibiotics gliotoxin, frequentin and gladiolic acid respectively; and uninoculated pea seeds sown in soil containing a gliotoxin-producing strain of *T. viride* retained gliotoxin in their seed coats (Wright, 1956). This strongly suggests that some antibiotics that are found in culture are also produced *in vivo*. Stevenson (1956a, b) investigated whether 8 actinomycetes that were antagonistic *in vitro* to the wheat root-rot fungus *Helminthosporium sativum* were also antagonistic in soil. Prevention of mycelial development, lysis, stunting, distortion and excessive branching of the fungal hyphae, all of which occur *in vitro* in the presence of the actinomycetes, also occurred when the fungus (on glass slides) was buried in soil to which the actinomycetes had been added. Thus, control of root rot by actinomycetes was correlated with the effects on the pathogenic fungus *in vitro*, but whether this control was due to the occurrence of actinomycin in soil was unknown because there was no effective means of detecting the antibiotic in soil. Fusarium blight of oat seedlings can similarly be controlled by adding strains of *Chaetomium* to soil, but any possibility that antibiotics are involved is ruled out by the fact that these strains are not strikingly antagonistic to the *Fusarium in vitro* (Tveit & Wood, 1955).

In general, therefore, although many soil fungi do produce antibiotics *in vitro*, evidence that they do so in soil is weak because it is technically difficult to gain. Antibiotics may well be produced, but act only in micro-environments. Germinating spores of many fungi produce antibiotic substances and they might maintain enough in the soil to keep nearby spores of plant pathogens from germinating or from growing towards host-plant roots. There is a need to investigate this possibility. Antibiotic-producing micro-organisms flourish in the rhizosphere and their ability to retard growth of plant pathogens could be profoundly affected by materials exuded from roots or produced from organic materials in soil. However, the fact that nearly all antibiotics produced *in vitro* rapidly disappear when added to soil tells against their biological significance *in vivo*, although this should not preclude the study of the few that are stable in soil. In their review of researches aimed at controlling disease by adding alien micro-organisms that produce antibiotics, Wood & Tveit (1955) list 190 references, but conclude that little or no success had been achieved up to that point. Perhaps understandably, interest in the possible production of antibiotics by pathogenic fungi *in vivo*, and their role in soil, has recently declined.

Specific reactions in vitro *of pathogens to plant products*

Resistance to disease by plants is the normal condition, susceptibility the exception. Both biochemical and physical differences between plant species are important in determining resistance. The biochemical nature of specific resistance lends itself to investigation in the laboratory through experiments that determine the reaction of pathogens to various host products. Emphasis is, of course, laid on materials exuded from leaves or roots at or near the point of infection. The problem of the nature of disease resistance is so vast that I shall have to make an arbitrary selection of the evidence for specific effects of host products on pathogens, that are correlated with compatible pathogenic reactions.

Reactions to host extracts

Walker (1921, 1923) observed that the dry outer scales of onions resistant to smudge disease caused by *Colletotrichum circinans* were red or yellow, whereas those of susceptible onions were white. He found that juice from the succulent scales of either resistant or susceptible varieties inhibited the germination of fungal spores; but whereas water extracts of the dry outer scales of the resistant onions also strongly inhibited germination, extracts of susceptible ones did not. The inhibiting substance in the coloured scales was found later to be protocatechuic acid. This early work still remains as one of the few clear-cut correlations between the reactions *in vitro* of a pathogen and a resistance mechanism *in vivo* reflecting a chemical difference between varieties. Kirkham (1959) showed that those apple varieties that contained most chlorogenic acid were also those in which apple scab fungus *Venturia inaequalis*, spread least readily. Tests *in vitro* showed that growth of the fungus was inhibited by chlorogenic acid and by some of its structural analogues such as *o*-coumaric acid and cinnamic acid. Similarly resistance in apple to the brown-rot fungus *Sclerotinia fructigena* seems to be related to the production of oxidized polyphenols which, *in vitro*, prevent the enzymes produced by the fungus from disintegrating the apple tissues (Byrde, 1957).

The occurrence of fungitoxic substances in higher plants has been studied by Wain and his colleagues at Wye College. Spencer (1962), reviewing the work, described the antifungal properties of extracts from stems of broad bean seedlings. *Botrytis cinerea*, which is pathogenic to broad bean, is relatively unaffected by the extracts, but *Alternaria brassicola*, which does not infect bean, is inhibited *in vitro* by concentrations of the antibiotic similar to those found in the plant.

Resistance of pods of *Pisum sativum* to infection by several fungi seems to depend on the formation in the pods of pisatin, one of the phytoalexin compounds (Cruickshank & Perrin, 1963). Fungi, both pathogenic and non-pathogenic, inoculated into pea pods stimulated pisatin production, but so did drops of cupric oxide and the chlorides of calcium, mercury and sodium in the absence of fungi. The amounts of pisatin produced *in vivo* after inoculating pods with the pathogenic fungi *Fusarium solani* var. *martii* and *Ascochyta pisi*, although large, were not enough to inhibit their growth when tested *in vitro*, whereas the non-pathogenic fungus *Sclerotinia fructicola* stimulated formation of less pisatin but was nevertheless inhibited by the smaller amount. Resistance may be based on the stimulation by infection of the production of sufficient phytoalexin to inhibit the infecting fungus.

There seems strength in the argument that substances contained in the extracts described here may be associated with disease resistance. But, as discussed in the other sections, translating ideas engendered from studies *in vitro* to the living plant is a dangerous exercise. A material produced *in vitro* might be enhanced by the environment *in vivo* and have different properties; for example, the benzoic ester of dinitro-ortho-cresol, which is a weak fungicide *in vitro*, when applied to the

surface of broad bean leaves becomes converted to the strongly fungicidal cresol (Bates, Spencer & Wain, 1962). By the same token, an
anti-fungal substance obtained from one plant, or from a pathogen,
may well be detoxicated within the tissues of a different plant to which
it is applied.

Specific responses to root exudates

Substances released from roots have profound effects on the soil
micro-flora. Clearly a pathogen must be able to compete successfully
with other microbes in the rhizosphere before it can invade host roots.
The stimulus provided by root exudates to pathogens has been well
documented, and techniques for obtaining root exudates and testing the
reactions of pathogens to them *in vitro*, have led towards an understanding of some of the factors that affect a pathogen's progress in the region
of the host root. Timonin (1941) tested the effects *in vitro* of root
exudates from wilt-resistant and wilt-susceptible flax on *Fusarium
oxysporum* f. *lini*, and found that exudates from wilt-resistant varieties
inhibited spore germination, whereas those from susceptible plants did
not. Hydrocyanic acid was detected in exudates from resistant but not
from susceptible plants. With three pathogenic races sharply defined on
three pea varieties, Buxton (1957) obtained a correlation between inhibition of *Fusarium* spores *in vitro* by exudates in resistant host + avirulent
race combinations and stimulation by exudates in susceptible host +
virulent race combinations. Race 1 of *F. oxysporum* f. *pisi* wilts pea
variety Onward but not Alaska or Delwiche Commando; race 2 wilts
the first two but not the last; and race 3 can wilt all three varieties. Both
root exudates from Alaska and Delwiche Commando inhibited spore
germination of race 1 but promoted growth of spores of race 3; only
the exudate from Alaska promoted the growth of race 2. Such a correlation between effects on spores and susceptibility or resistance to a
particular race, points to a significance *in vivo* of the interaction between
root exudates and the pathogen in the rhizosphere. The same kind of
relationship exists between banana root exudates and races of *F.
oxysporum* f. *cubense*, the cause of Panama wilt. The sugars and amino
acids in root exudates from resistant and susceptible banana varieties
differed in ways that suggested they may play a part in controlling the
success of pathogenic races *in vivo* (Buxton, 1962). Thus, banana variety
Lacatan is resistant to wilt caused by *F. oxysporum* f. *cubense* whereas
the widely grown high-yielding variety Gros Michel is very susceptible.
Root exudates from Lacatan inhibited spore germination of the fungus
in vitro, but root exudates from Gros Michel encouraged their germina-

tion. Eighteen amino acids were present in the root exudates; thirteen were common to both varieties but Gros Michel exudates contained no detectable cystine or threonine and Lacatan no leucine, serine or tyrosine. Gros Michel exudates contained one and a half times as much total carbohydrate as Lacatan exudates: eight sugars predominated, and Gros Michel root exudates contained more of each than did Lacatan at any stage of root growth. Materials that inhibited spore germination on agar media were demonstrated on chromatograms of Lacatan exudates but not on those from Gros Michel. That the specificity in responses of the pathogens to these exudates *in vitro* is the same as the specificity of race-variety relations *in vivo*, strongly indicates that these root exudates play some part in the mechanisms of wilt resistance.

There are other examples of stimulation and inhibition by root exudates. Some bacterial groups that had certain amino acid requirements *in vitro* were preferentially stimulated by exudates from wheat roots (Lochhead & Thexton, 1947; Lochhead, 1959). Rovira (1956a) showed that adding root exudates to media stimulated the growth of rhizosphere organisms more than the growth of non-rhizosphere ones, and further that Gram-negative bacteria were selectively increased in soil by adding such exudates (Rovira, 1956b).

Adaptation of pathogens to exudates can be important if associated with changes in pathogenicity. When race 1 of *Fusarium oxysporum* f. *pisi* was retained for 14 days in concentrated root exudates from a race 1-resistant plant, it changed to race 2-pathogenicity (Buxton, Perry, Doling & Reynolds, 1959). This again supports the suggestion that the effects of root exudates on the pathogen play a fundamental part in determining its ultimate effect on the host and that the phenotype as seen *in vitro* can be changed by a host metabolite produced *in vivo*.

Pellicularia filamentosa attacks tomato seedlings only when it can form infection cushions at the junction of epidermal cells on root surfaces. Non-pathogenic strains do not become organized in this way. When root exudates from susceptible plants are added to agar media and the fungus is allowed to grow over the media on cellophane, infection cushions occur as they do on root surfaces (Kerr & Flentje, 1957).

These examples of specific interactions between root exudates and pathogens derived from both observations *in vivo* and experiments *in vitro* need investigating further to obtain a better understanding of the nature of root resistance. The underlying differences between races of pathogens that enable them to overcome toxic exudates from resistant plants or to make the best use of available nutrients in exudates from susceptible ones also need to be sought.

The use of genetically marked pathogens in the search for a biochemical basis of pathogenicity

Irradiating plant pathogens can produce strains with artificially induced genetic markers which show as deficiencies in ability to synthesize specific amino acids, vitamins or nucleic acids. Such mutants are nearly always less pathogenic than their wild-type parents. Bacon, Burrows & Yates (1950), working with animal pathogens, found that some mutants had their pathogenicity restored when the required nutrient was added with the inoculum. Keitt & Boone (1956) showed that mutants of *Venturia inaequalis* (which causes apple scab), with requirements *in vitro* for a number of amino acids were less virulent than wild-types. Keitt and his co-workers found that virulence was restored when the required metabolite was added with the inoculum. Buxton (1956), with mutants of *Fusarium oxysporum* f. *pisi*, and Garber & Shaeffer (1957), with *Erwinia carotovorum*, similarly found that augmenting the environment of the host tissues with the required nutrient restored virulence. Non-pathogenic mutants of *Venturia* penetrate apple cuticle and become established in a subcuticular position, but go no further without the addition of the required metabolite. Does this mean that pathogenicity is determined by the ability to synthesize specific metabolites and that resistance may be based on the lack of that particular substance in the host? Garber (1956) proposed a nutrition–inhibition hypothesis that postulates a relationship between the availability of nutrients and pathogenic progess in the host. It is doubtful, however, that the difference between a pathogen and a non-pathogen will be entirely dependent on differences in single metabolite requirements. The artificial prevention of growth by imposing a mutant gene-controlled characteristic on a fungus, and subsequently allowing growth by supplying the specifically required substance whose metabolism is controlled at that locus, may have no more significance in pathogenicity than the fact that without the right food the fungus will not grow. We hardly need experimentation to reach that conclusion and virulence is not necessarily conditioned by growth rate of the pathogen. However, the use of mutants could yield much information, provided thorough physiological and biochemical studies were made in conjunction with analyses of available nutrients in resistant and susceptible hosts. In addition to the nutritionally deficient mutants, use could be made of genetic crosses between avirulent mutants which often lead to progeny with increased virulence, presumably from the complementary action of newly associated genes. Such experiments could provide more information on the physiological characters that are recognizable *in vitro* and that appear to be directly involved in pathogenicity. It should not be difficult to identify genes controlling specific pathogenicity with those determining the nutritional needs of the pathogen *in vitro*.

A relationship between the genotypes of a series of pathogenic races of flax rust (*Melampsora lini*) and those of flax varieties differing in resistance was uncovered by Flor (1955, 1959). Rust races that have a particular gene governing either virulence or avirulence to a particular host variety are met by that variety with its specific gene that controls resistance or susceptibility. Such a 'gene-for-gene' relationship was found by crossing the fungus strains and assessing the genotypes of the progeny by inoculating a range of flax varieties. Despite this satisfactory genetical relationship between host and parasite *in vivo*, the hopes of finding differences *in vitro* between races of rusts which might be important in pathogenicity, are exceedingly dim until someone finds a way of growing rusts in artificial culture. There is a possibility of pursuing the problem by comparing serological reactions of the rust races in the hopes of relating them to differences in metabolism *in vivo*, that may in turn be related to specific pathogenicity. Perhaps a more simple route is to look for similar systems between different pathogens and hosts in which the pathogens can be grown in culture.

CONCLUSION

Although attempts to define specific pathogenicity in terms of events that occur *in vitro* have not proved that such attributes are implicated in the natural process of disease, there are many correlations between the peculiarities found *in vitro* and *in vivo*; at least some of these correlations should provide leads towards understanding the mechanisms that determine the pathogenicity of individual strains of a species.

Pragmatically it might be argued that plant diseases can be satisfactorily controlled without knowing in detail how parasites operate, but this would be intellectually frustrating. Not enough is known either of the important metabolites formed by a pathogen *in vivo*, or even of the fate of enzymes, toxins or growth substances produced by the pathogen *in vitro* and experimentally introduced into hosts. Are they changed into unrecognizable forms? How far do they permeate in plant tissues? Does an enzyme, that macerates plant tissue *in vitro*, do so in the host? These may seem simple questions, but there are no cogent answers.

Improved biochemical techniques and a wider use of specific inhibitors of toxins or enzymes in experiments on host chemotherapy should help to uncover possible ways in which pathogens differ intrinsically from each other. A worthwhile line of inquiry may consider the possibilities of using a gene-for-gene relationship between host resistance and specific pathogen virulence. If, for example, a gene conditioning pathogenicity were found to be linked with a gene controlling specific enzyme production *in vitro*, useful information on pathogenicity could be gained by investigating the effects of associating, through out-breeding, such a set of genes with other sets in pathogens which infect the same host but produce different disease symptoms or have different degrees of virulence. Similarly the artificially induced mutations that deprive pathogens of specific abilities *in vitro* should be studied further for their effects on pathogenicity; such studies may provide, albeit indirectly, further information on what is, and equally important what is not, a part of the mechanism of virulence.

REFERENCES

BACON, G. A., BURROWS, T. W. & YATES, M. (1950). The effects of biochemical mutation on the virulence of *Bacterium typhosum*. *Brit. J. exp. Path.* **31**, 714.

BATES, A. N., SPENCER, D. M. & WAIN, R. L. (1962). Investigations on fungicides. V. The fungicidal properties of 2-methyl-4:6-dinitrophenol (DNC) and some of its esters. *Ann. appl. Biol.* **50**, 21.

BRAUN, A. C. (1959). Growth is affected. In *Plant Pathology—An Advanced Treatise*, **1**, 189. New York: Academic Press.

BRAUN, A. C. & PRINGLE, R. B. (1959). Pathogen factors in the physiology of disease—toxins and other metabolites. In *Plant Pathology—Problems and Progress 1908–1958*, p. 88. Wisconsin: University of Wisconsin Press.

BRIAN, P. W. (1949). The production of antibiotics by microorganisms in relation to biological equilibria in soil. *Symp. Soc. exp. Biol.* **3**, 357.

BRIAN, P. W. (1955). The role of toxins in the etiology of plant diseases caused by fungi and bacteria. In *Mechanisms of Microbial Pathogenicity. Symp. Soc. gen. Microbiol.* **5**, 294.

BRIAN, P. W. (1957). The effects of some microbial metabolic products on plant growth. *Symp. Soc. exp. Biol.* **11**, 166.

BRIAN, P. W. (1960). Antagonistic and competitive mechanisms limiting survival and activity of fungi in soil. In *The Ecology of Soil Fungi*, p. 115. Liverpool: University of Liverpool Press.

BRIAN, P. W., CURTIS, P. J., HEMMING, H. G. & McGOWAN, J. C. (1946). The production of viridin by pigment-forming strains of *Trichoderma viride*. *Ann. appl. Biol.* **33**, 190.

BRIAN, P. W. & HEMMING, H. G. (1955). The effect of gibberellic acid on shoot growth of pea seedlings. *Physiol. Plant.* **8**, 669.

BROWN, W. (1915). Studies in the physiology of parasitism. I. The action of *Botrytis cinerea*. *Ann. Bot., Lond.* **29**, 318.

BROWN, W. (1936). The physiology of host–parasite relations. *Bot. Rev.* **2**, 236.

BROWN, W. (1955). On the physiology of parasitism in plants. *Ann. appl. Biol.* **43**, 325.

BUXTON, E. W. (1956). Heterokaryosis and parasexual recombination in pathogenic strains of *Fusarium oxysporum. J. gen. Microbiol.* **15**, 133.

BUXTON, E. W. (1957). Some effects of pea root exudates on physiologic races of *Fusarium oxysporum* f. *pisi* (Linf.). *Trans. Brit. mycol. Soc.* **40**, 145.

BUXTON, E. W. (1962). Root exudates from banana and their relationship to strains of the Fusarium causing Panama wilt. *Ann. appl. Biol.* **50**, 269.

BUXTON, E. W., PERRY, D. A., DOLING, D. A. & REYNOLDS, J. D. (1959). The resistance of pea varieties to Fusarium wilt. *Plant Pathology*, **8**, 39.

BYRDE, R. J. W. (1957). The varietal resistance of fruits to brown rot. II. The nature of resistance in some varieties of cider apple. *J. hort. Sci.* **32**, 227.

CLAUSON-KAAS, N., PLATTNER, P. A. & GÄUMANN, E. (1944). Über ein welkeer- zeugendes Stoffwechselprodukt von *Fusarium lycopersici* Sacc. *Ber. schweiz. bot. Ges.* **54**, 523.

CRUICKSHANK, I. A. M. & PERRIN, DAWN M. (1963). Studies on Phytoalexins. VI. Pisatin: the effect of some factors on its formation in *Pisum sativum* L. and the significance of pisatin in disease resistance. *Aust. J. biol. Sci.* **16**, 111.

DE BARY, A. (1886). Über einige Sclerotinien und Sclerotienkrankheiten. *Bot. Ztg*, **44**, 377.

DIMOND, A. E. (1955). Pathogenesis in the wilt diseases. *Annu. Rev. Pl. Physiol.* **6**, 329.

DIMOND, A. E. & WAGGONER, P. E. (1953*a*). The physiology of lycomarasmin production by *Fusarium oxysporum* f. *lycopersici. Phytopathology*, **43**, 195.

DIMOND, A. E. & WAGGONER, P. E. (1953*b*). The cause of epinastic symptoms in *Fusarium* wilt of tomatoes. *Phytopathology*, **43**, 663.

FLOR, H. H. (1955). Host–parasite interaction in flax rust—its genetic and other implications. *Phytopathology*, **45**, 680.

FLOR, H. H. (1959). Genetic controls of host–parasite interactions in rust diseases. In *Plant Pathology—Problems and Progress, 1908–1958*, p. 137. Wisconsin: University of Wisconsin Press.

GARBER, E. D. (1956). A nutrition-inhibition hypothesis of pathogenicity. *Amer. Nat.* **90**, 183.

GARBER, E. D. & SCHAEFFER, S. G. (1957). Free histidine content of turnip varieties and their resistance to histidine-requiring mutants of *Erwinia aroideae. J. Bact.* **74**, 392.

GÄUMANN, E. (1957). Fusaric acid as a wilt toxin. *Phytopathology*, **47**, 342.

GOTTLIEB, D. (1943). The presence of a toxin in tomato wilt. *Phytopathology*, **33**, 126.

GROSSMANN, F. (1962). Untersuchungen über die Hemmung pektolytischer Enzyme von *Fusarium oxysporum* f. *lycopersici. Phytopath. Z.* **45**, 139.

HUSAIN, A. & KELMAN, A. (1958). The role of pectic and cellulolytic enzymes in pathogenesis by *Pseudomonas solanacearum. Phytopathology*, **48**, 377.

KALYANASUNDARAM, R. (1958). Production of fusaric acid by *Fusarium lycopersici* Sacc. in the rhizosphere of tomato plants. *Phytopath. Z.* **32**, 25.

KEITT, E. W. & BOONE, D. M. (1956). Use of induced mutations in the study of host–pathogen relationships. In *Genetics in Plant Breeding. Brookhaven Symposia in Biology*, **9**, 209.

KERR, A. & FLENTJE, N. T. (1957). Host infection in *Pellicularia filamentosa* controlled by chemical stimuli. *Nature, Lond.* **179**, 204.

KIRKHAM, D. S. (1959). Host factors in the physiology of disease. In *Plant Pathology—Problems and Progress, 1908–1958*, p. 110. Wisconsin: University of Wisconsin Press.

KUROSAWA, E. (1926). Experimental studies on the reaction of *Fusarium hetero-sporum* in rice plants. *J. nat. Hist. Soc., Formosa*, **16**, 213.

LAPWOOD, D. H. (1957). Studies in the physiology of parasitism. XXIII. On the parasitic vigour of certain bacteria in relation to their capacity to secrete pectolytic enzymes. *Ann. Bot., Lond.* **21**, 167.

LOCHHEAD, A. G. (1959). Rhizosphere microorganisms in relation to root disease fungi. In *Plant Pathology—Problems and Progress, 1908–1958*, p. 327. Wisconsin: University of Wisconsin Press.

LOCHHEAD, A. G. & THEXTON, R. H. (1947). Qualitative studies of soil micro-organisms. VII. The 'rhizosphere effect' in relation to the amino acid nutrition of bacteria. *Canad. J. Res. C*, **25**, 20.

MCDONNELL, K. (1962). Relationship of pectic enzymes and pathogenicity in the Fusarium wilt of tomatoes. *Trans. Brit. mycol. Soc.* **45**, 55.

MOULTON, J. E. (1942). Extraction of auxin from maize, from smut tumours of maize, and from *Ustilago zeae*. *Bot. Gaz.* **103**, 725.

PAQUIN, R. & COULOMBE, L. J. (1962). Pectic enzyme synthesis in relation to virulence in *Fusarium oxysporum* f. *lycopersici* (Sacc.) Snyder & Hansen. *Canad. J. Bot.* **40**, 533.

PRINGLE, R. B. & BRAUN, A. C. (1957). The isolation of the toxin of *Helminthosporium victoriae*. *Phytopathology*, **47**, 369.

ROVIRA, A. D. (1956a). Plant root excretions in relation to the rhizosphere effect. II. A study of the properties of root exudate and its effect on the growth of micro-organisms isolated from the rhizosphere and control soil. *Plant & Soil*, **7**, 195.

ROVIRA, A. D. (1956b). Plant root excretions in relation to the rhizosphere effect. III. The effect of root exudate on the numbers and activity of micro-organisms in soil. *Plant & Soil*, **7**, 209.

SINGH, R. K. & WOOD, R. K. S. (1956). Studies in the physiology of parasitism. XXI. The production and properties of pectic enzymes secreted by *Fusarium moniliforme* Sheldon. *Ann. Bot., Lond.* **77**, 89.

SPENCER, D. M. (1962). Antibiotics in seeds and seedling plants. In *Antibiotics in Agriculture*, p. 125. London: Butterworths.

STEVENSON, I. L. (1956a). Antibiotic activity of actionomycetes in soil and their controlling effects on root-rot of wheat. *J. gen. Microbiol.* **14**, 440.

STEVENSON, I. L. (1956b). Antibiotic activity of actinomycetes in soil as demonstrated by direct observation techniques. *J. gen. Microbiol.* **15**, 372.

TIMONIN, M. I. (1941). The interaction of higher plants and soil microorganisms. III. Effect of by-products of plant growth on the activity of fungi and actinomycetes. *Soil Sci.* **52**, 395.

TVEIT, M. & WOOD, R. K. S. (1955). The control of *Fusarium* blight in oat seedlings with antagonistic species of *Chaetomium*. *Ann. appl. Biol.* **43**, 538.

WALKER, J. C. (1921). Onion smudge. *J. agric. Res.* **20**, 685.

WALKER, J. C. (1923). Disease resistance to onion smudge. *J. agric. Res.* **24**, 1019.

WEINDLING, R. & EMERSON, O. H. (1936). The isolation of a toxic substance from the culture filtrate of *Trichoderma*. *Phytopathology*, **26**, 1068.

WINSTEAD, N. N. & WALKER, J. C. (1954). Toxic metabolites of the pathogen in relation to *Fusarium* resistance. *Phytopathology*, **44**, 159.

WOLF, F. T. (1952). The production of indole acetic acid by *Ustilago zeae*, and its possible significance in tumor formation. *Proc. nat. Acad. Sci., Wash.* **38**, 106.

WOOD, R. K. S. (1955). Pectic enzymes secreted by pathogens and their role in plant infection. In *Mechanisms of Microbial Pathogenicity. Symp. Soc. gen. Microbiol.* **5**, 263.

Wood, R. K. S. (1960). Pectic and cellulolytic enzymes in plant disease. *Annu. Rev. Pl. Physiol.* **11**, 299.

Wood, R. K. S. & Tveit, M. (1955). Control of plant diseases by use of antagonistic organisms. *Bot. Rev.* **21**, 441.

Wright, J. M. (1956). The production of antibiotics in soil. IV. Production of antibiotics in coats of seeds sown in soil. *Ann. appl. Biol.* **44**, 561.

Wright, J. M. & Grove, J. F. (1957). The production of antibiotics in soil. V. Breakdown of griseofulvin in soil. *Ann. appl. Biol.* **45**, 36.

SUBSTANCES PRODUCED BY PATHOGENIC ORGANISMS THAT INDUCE SYMPTOMS OF DISEASE IN HIGHER PLANTS

B. J. DEVERALL

Department of Botany, Imperial College, London

INTRODUCTION

This review will be confined to diseases of higher plants caused by fungi and bacteria. In these diseases, visible symptoms are rather restricted, comprising degrees of chlorosis and necrosis, wilting and deformation of growth. Physiological symptoms may include rises in temperature and respiration rates, and changes involving growth-regulating substances. Some pathogenic organisms kill the invaded tissues quickly but others, particularly obligate parasites, live in a degree of harmony with the host cells for some time, perhaps because they require to be associated with living cells in order to grow and reproduce.

One can conceive of three ways in which parasitic fungi or bacteria may damage their host. If fungi penetrate cell structures or tissues mechanically, damage would be caused, but although this damage may be an important factor in the initiation of disease, it probably has little effect on the economy of the plant as a whole. Then there is the damage which results from the use by pathogenic organisms of some metabolites produced by the host; this has been little studied probably because of the technical difficulties involved in distinguishing it from the third type of damage. This third type is undoubtedly the most important in all diseases and is the only one to be considered in this paper. It depends upon the fact that when growing in or on a plant, the parasite produces a multitude of substances which pass into the environment; this is composed largely of the cells of the host plant. Some of these substances may disrupt the metabolism of the host with no obvious microscopic symptoms at first, but later symptoms may be visible at cellular, tissue, organ or whole plant level. One of the main objects of physiological and biochemical plant pathology is the identification of substances produced by parasites that initiate changes in host metabolism and express themselves as disease symptoms. A favoured approach has been to seek in culture filtrates of the organism substances which, when applied to the host tissue, can induce some or all of the symptoms of the disease. This has proved to be relatively simple, but the difficulty has arisen when attempts have been made to determine whether these same substances are produced in the plant and if so whether this occurs at the correct time and in appropriate concentration to cause the symptoms.

The purposes of this paper are twofold. First, evidence will be presented to show that in some plant diseases there is knowledge of the way in which the pathogenic organism induces disease symptoms. Secondly, examples are provided which indicate the need for caution in applying results from systems *in vitro* to the interpretation of events in the diseased plant. The word 'toxin' will be used frequently so, before proceeding further, its use in plant pathology will be discussed.

TOXINS AND VIVOTOXINS

The term 'toxin' has been used by plant pathologists for poisonous substances found in culture filtrates of fungi and bacteria. The *Concise Oxford Dictionary* defines a toxin as 'a poison, especially one secreted by a microbe and causing some particular disease', and a poison as 'a substance that when introduced into or absorbed by a living organism destroys life or injures health, especially (popularly) one that destroys life by rapid action and when taken in small quantity'. The term as used by plant pathologists does not necessarily imply rapid action in response to small doses. Ludwig (1960) mentioned many toxins produced by plant pathogens, and Braun & Pringle (1959) have discussed some of the most interesting examples. Dimond & Waggoner (1953a) distinguished between toxins produced in culture and 'vivotoxins', because a role for toxic components of culture filtrates had been too readily ascribed in disease. A vivotoxin was defined as a substance produced in the infected host by the pathogenic organism or its host, which functioned in the production of disease, but was not itself the initial inciting agent. On the other hand, a toxin was any compound produced by a micro-organism in any environment which was toxic to plants. Before a substance can be regarded as a vivotoxin it must be regularly separated from the diseased plant, purified, and reproduce at least a portion of the disease syndrome in a healthy plant. Probably Dimond & Waggoner introduced the term vivotoxin as a result of their studies of the vascular wilt of tomato plants caused by the fungus *Fusarium oxysporum* f. *lycopersici*, and the role of the toxins produced by this fungus in culture. The need to use the concept of vivotoxins is clear when one attempts to reconcile the apparently conflicting reports on the role of fungal metabolites in the production of symptoms of wilt.

THE ROLE OF VIVOTOXINS IN VASCULAR WILT DISEASES

Most attention has been paid to the wilt of tomatoes caused by *Fusarium oxysporum* f. *lycopersici*, but there are a number of studies of the similar disease of the same host caused by *Verticillium albo-atrum*. Infected plants show symptoms which may be secondary to the loss of turgor of the upper parts of the stem and of leaves; these are browning of the walls of xylem vessels in the stem, epinasty of petioles, adventitious root development and premature defoliation. Some of these symptoms may be induced by substances different from those which cause wilting.

Gottlieb (1943, 1944), in a study of *Fusarium* wilt, reviewed the detection in filtrates from cultures of vascular wilt fungi, of toxic substances which would cause wilt of healthy cuttings of the host plant. For example, Rosen (1926) found that filtrates from cultures of *F. vasinfectum* (*F. oxysporum* f. *vasinfectum*) grown in a nutrient solution containing inorganic nitrogen wilted cotton stems, but that those obtained from solutions containing organic nitrogen were non-toxic. These results showed that substances produced *in vitro* may not be produced *in vivo*, because the nutrients available might be different. Gottlieb (1943) collected tracheal fluids from tomato stems by centrifugation under nitrogen, and found that fluids from infected plants showing chronic or severe wilt symptoms caused wilt when applied to cut shoots; no effect was caused by fluids from healthy plants or from plants wilted because of lack of water. Anaerobic conditions were important in this extraction because sap obtained by pressing either healthy or diseased plants in air was toxic. Gottlieb (1943, 1944) studied the properties of toxic tracheal fluids, but the toxins were not characterized so it is not known if they are related to the numerous substances which have been found recently in culture filtrates of pathogenic organisms.

Gottlieb's pioneer work on vivotoxins was followed by an extensive series of investigations on the same disease by Gäumann, Kern, and their associates in Zürich. They detected and identified toxins produced by pathogenic organisms *in vitro*, and studied ways in which these substances affected the water relations of plant cells with a view to understanding their role in causing symptoms in the infected plant. The first wilt toxin to be studied was lycomarasmin, which was isolated in pure form from culture filtrates by Clausson-Kaas, Plattner & Gäumann, (1944). A review by Gäumann (1951) dealt in detail with its mode of action when introduced into cuttings of tomato plants. More recently, a large number of unrelated compounds have been shown to induce wilting in healthy cuttings of tomato. Without considering for the moment whether they are vivotoxins, the following substances have been shown to cause wilt: high molecular weight compounds (Hodgson, Peterson & Riker, 1949; Scheffer & Walker, 1953; Wood, 1961); pectic enzyme preparations (Gothoskar, Scheffer, Walker & Stahmann, 1953; Wood, 1961; Blackhurst & Wood, 1963); cellulases (Husain & Dimond, 1960); lycomarasmin (Gäumann, 1951) and fusaric acid (Gäumann, 1957, 1958). Furthermore, Gäumann & Obrist (1960) compared the effects on the permeability of protoplasts of *Rhoeo discolor* of eleven 'wilt toxins', lycomarasmin, fusaric acid, skyrin, enniatin B, baccatin, diaporthin, lycomarasmic acid, fusarubin, enniatin, substance 'J' and dehydrofusaric acid. These substances isolated from culture filtrates of fungi were not necessarily vivotoxins but were called 'wilt toxins' simply because they caused wilting of tomato shoots under test conditions. These wilt toxins, hydrolytic enzymes and high molecular weight substances may or may not be vivotoxins and they almost certainly act by different mechanisms. This will be evident in the following sections.

Wilt toxins

The two metabolites produced *in vitro* by *Fusarium oxysporum* f. *lycopersici*, namely lycomarasmin (Gäumann, 1951) and fusaric acid (Gäumann, 1957, 1958) cause wilt of tomato cuttings at concentrations as low as 10^{-4}M. Both affect a wide range of plant tissues, and fusaric acid is produced by six members of the *Hypocreales*; this lack of specificity is in contrast to the fungus, strains of which are highly specific in their effects on different tomato varieties.

Of the wilt toxins, only fusaric acid qualifies for consideration as a vivotoxin because it appears to be produced *in vivo*. Following the demonstration by Kern & Sanwal (1954) that mycelium labelled by feeding methyl-^{14}C glycine produced radioactive metabolites which accumulated preferentially in severely affected leaves of tomato plants, Kern & Kluepfel (1956) detected radioactive fusaric acid by chromatographic examination of extracts of plants inoculated with non-radioactive mycelium but kept in an atmosphere of $^{14}CO_2$. This does not prove that the fusaric acid was of fungal origin and there are difficulties in detecting fusaric acid by paper chromatography. However, Lakshminarayanan & Subramanian (1955) detected fusaric acid in cotton plants infected with *Fusarium oxysporum* f. *vasinfectum*; the extracted fusaric acid and authentic fusaric acid migrated in a similar way on chromatograms in three different solvent systems. Page (1959) also detected fusaric acid in banana plants infected with *F. oxysporum* f. *cubense*, which causes Panama wilt disease.

However, an important point in considering whether a toxic substance is significant *in vivo* is its concentration at the critical stage of the disease. Kluepfel (1957) used fusaric acid isotopically labelled in the carboxyl carbon atom to investigate the fate of the molecule after its entry into tomato plants. He recovered 83 % of the original radioactivity in fusaric acid or derivatives after 48 hr. Ten per cent of the original fusaric acid was decarboxylated, but much of it was still free. This suggests that decomposition rates in the tomato plant are low. However, a reliable figure for the production rate of fusaric acid *in vivo* has still to be calculated because the quantitative estimation of fusaric acid is difficult (Page, 1959). The present method depends upon the chromatography of free fusaric acid, whereas in plants the acid might be present as chelates with iron.

Thus although fusaric acid or derivatives are present in diseased plants, it is not known whether the concentrations are sufficiently high to affect the cells of the host.

Discrepancies arise when symptoms caused by fusaric acid alone and by infection are compared. Gäumann, Kern, Schüepp & Obrist (1958) found that fusaric acid at 20 mg./kg. fresh weight increased the rate of transpiration and water uptake at first, although water uptake declined later. Dimond & Waggoner (1953c) observed that infection depressed the rate of transpiration to one-third of the rate of healthy leaves. Scheffer & Walker (1953) recorded a progressive decline in water loss after infection of tomato plants, and observed no damage to leaf blades. High concentrations of fusaric acid (e.g. 150 mg./kg. fresh weight) caused the edges of tomato leaves to become necrotic and to transpire at high rates (Sivadjian & Kern, 1958). This comparison suggests that fusaric acid does not play a role in the development of symptoms of wilt disease in tomato. It is possible that fusaric acid, at concentrations too low to show an effect when used alone, may act in conjunction with other metabolites of the fungus.

Lycomarasmin does not appear to have been sought in diseased plants. Dimond & Waggoner (1953b) estimated, from rates of production *in vitro*, the amount of lycomarasmin that might be present in a diseased plant 2 weeks after inoculation; making the most lenient assumptions in terms of rates of decomposition *in vitro*, they concluded that the maximum quantity would be one-eighth of the minimum toxic dose. On these grounds, they decided that lycomarasmin could not be responsible for wilting of infected plants.

Scheffer & Walker (1953) confirmed that after increasing the water loss from plants, lycomarasmin caused wilt of cut stems of tomato.

Furthermore, lycomarasmin failed to cause plugging and browning of the vessels, symptoms prominent in pathological wilting. Iron-deficient and normal plants responded to infection with equally severe symptoms of wilt, but Kern (1959) showed that cuttings of iron-deficient plants were not affected by lycomarasmin. The toxicity of lycomarasmin to cuttings is much increased if supplied together with iron with which it chelates. Waggoner & Dimond (1953) showed that the toxicity of lycomarasmin plus iron could be abolished by adding 8-quinolinol, which preferentially chelates the iron, or by adding copper, which is preferentially chelated by lycomarasmin. However, treatment of infected plants with 8-quinolinol or with copper ions had no effect on the severity of symptoms. Therefore, although lycomarasmin will cause wilt, details of symptoms caused by the toxin are different from those caused by infection. The evidence suggests that lycomarasmin does not cause symptoms *in vivo*.

There is little evidence that other toxins produced *in vitro* cause *Fusarium* wilt of tomatoes. Little is known about the toxin isolated by Gottlieb (1943), and the only other metabolites which need to be considered are enzymes which hydrolyse components of cell walls.

Enzymes that hydrolyse cell walls

Scheffer & Walker (1953) found that crude filtrates from cultures of different ages of *Fusarium oxysporum* f. *lycopersici* caused leaf blade injury in tomato cuttings within 24 hr., but no vascular plugging or browning. Diluted filtrates caused plugging and/or browning in 3–4 days. It was suspected that leaf-blade injury was caused by toxic substances accumulating in cultures which were of no importance in disease, so mycelial mats were collected, washed and transferred to new medium for a short period. Filtrates from these replacement cultures caused wilting, vascular browning and plugging but little leaf blade injury. The active component of the replacement filtrates was non-dialysable and heat labile. Gothoskar *et al.* (1953) found that the active filtrates contained pectic enzyme activity. When commercial preparations of hydrolases were tested only those containing pectic enzymes produced typical disease symptoms, including vascular browning. The replacement filtrates contained considerable pectin-methylesterase activity; of the commercial pectinases, those with high pectinmethylesterase activity caused the most wilting and vascular browning. These findings suggested that pectinmethylesterase played an important role, which Gothoskar, Scheffer, Walker & Stahmann (1955) confirmed. Winstead & Walker (1954) found that culture filtrates of many fungi caused wilting of tomato cuttings, and confirmed the fact that high pectinmethylesterase activity was often correlated with ability to cause vascular browning. However, in cotton wilt, Kamal & Wood (1956) found that culture filtrates of *Verticillium dahliae* caused wilting only when they contained tissue macerating and polygalacturonase activities, and preparations with high pectinmethylesterase activity caused the least browning.

Husain & Kelman (1958) found that purified cellulase from *Myrothecium verrucaria* would cause wilt of leaves on tomato cuttings in 12 hr. Husain & Dimond (1960) showed that cellulase produced by *Fusarium oxysporum* f. *lycopersici*, partially purified by repeated $(NH_4)_2SO_4$ treatment, caused a complete wilt of tomato cuttings in 20 hr. They suggested that cellulases could have been present in the preparations having pectinmethylesterase activity used by Gothoskar et al. (1955).

Wood (1960) reviewed these and other reports, which suggested an absence of correlation between ability to produce certain hydrolytic enzymes in culture and ability to cause disease. He drew attention to the fact that the detection of the enzymes in diseased plants and the concentrations required to cause symptoms should be investigated; he also mentioned the possible difficulty of isolating enzymes from diseased tissue in the light of experience with the brown rots of apple.

Some attempts have been made to extract the hydrolytic enzymes from diseased plants but the activities in the extracts have been too low for reliable measurement. Host tissues can inactivate enzymes (Wood, 1961), and recently Blackhurst & Wood (1963) showed that the low levels of enzyme activity were of the order to be expected on the basis of the known rate of enzyme secretion *in vitro*.

The role of hydrolytic enzymes in vascular wilt diseases depends on whether sufficient activity is present in infected plants. Even low activity over a period of 1–2 weeks in vascular tissue could cause considerable hydrolysis of cell-wall components. A major point in favour of an enzyme theory is the close agreement between symptoms caused by active filtrates and by infection. These important symptoms are wilting, vascular plugging and browning without leaf-blade injury. Some of the evidence concerning the relationship of plugging to wilting and the implication of high molecular weight substances will now be considered.

High molecular weight substances

Scheffer & Walker (1953) permitted infected stems to take up dyes and also forced dyes under pressure through infected stems; many of the vessels failed to conduct and solutions of dyes could not be forced through the infected vessels. Similarly, Dimond & Waggoner (1953c) measured the rate of movement of radioactive phosphorus through stems and found that the rate in healthy stems was 0·25 cm./sec., in moderately diseased stems 0·0091 cm./sec. and in severely diseased stems 0·0045 cm./sec. Also a wilted leaf with a diseased petiole recovered its turgor if placed in water after removal of the discoloured portions of the petiole. Scheffer & Walker (1953) could induce plugging of vessels and wilting of healthy stems by placing cut ends of shoots in dilute solutions of polysaccharides of large molecular weight. The plugs were near the cut end, and if this part was removed the stem recovered its turgor when put into water. Wood (1961) and Blackhurst & Wood (1963) caused wilting by immersing the cut ends of stems in very dilute solutions of pectic polymers.

Thus wilting in diseased stems is associated with impaired water flow. Moreover wilted shoots recover when cut and placed in water if damaged vessels are first removed, irrespective of whether the damage is caused by infection or by high molecular weight substances.

One must conclude that, in the absence of any evidence for highly active toxins such as those described later for other diseases, wilting after infection appears to be caused by a physical blocking of the vascular system by substances probably released as a result of the action of hydrolytic enzymes secreted by the fungus. The wilt toxins produced *in vitro* cause wilting of cut shoots by mechanisms involving the impairment of the permeability of cells, but there is no evidence that they act in this way in diseased plants.

The introduction of the term vivotoxin has great value, particularly in emphasizing the wisdom of trying to show that a toxin is present in the diseased plant in adequate concentration to cause disease. However, recent studies have shown that it may not be possible to satisfy the criteria for establishing a vivotoxin, and in certain diseases other criteria may have to be used. A few examples of vivotoxins and of the alternative criteria will now be given.

BACTERIAL TOXINS WHICH CAUSE CHLOROSIS

One of the most likely candidates for acceptance as a vivotoxin is the wildfire toxin produced in culture by *Pseudomonas tabaci*; this toxin has been purified, characterized and introduced into tobacco to produce typical disease symptoms (Johnson & Murwin, 1925; Braun, 1950; Woolley, Schaffner & Braun, 1955). The purified toxin caused a highly characteristic yellow halo when placed in punctures on the leaves of tobacco, at a concentration as low as 0·1 μg./ml.; this was considered to be strong evidence that it was the vivotoxin causing this disease (Woolley, Pringle & Braun, 1952). There are some indications of its mode of action *in vivo*. Braun (1950) found that the inhibitory action of the toxin on the growth of *Chlorella* was reversed by L-methionine. Also methionine sulphoxime, an antagonist of methionine, produced the same symptoms as the toxin. The suggestion that the toxin was an anti-metabolite of methionine in tobacco was not supported by the failure of methionine to prevent the formation of chlorotic lesions in tobacco leaves. Despite this, there are considerable structural similarities between tabtoxinine, a component amino acid of the toxin, and methionine (Woolley, Schaffner & Braun, 1952). The possibility of wildfire toxin acting *in vivo* by competing with methionine has yet to be proved con-clusively. Some indication of a fruitful approach to this problem is suggested by the recent study by Patel & Walker (1963) on the halo-blight of beans (*Phaseolus vulgaris*) caused by *Pseudomonas phaseolicola*. There are profound changes in the quantities of some amino acids, including ornithine, histidine and methionine, in leaves inoculated with the bacteria and in subsequently emerging leaves, which do not contain the organism but show chlorosis: Skoog (1952) reported toxin production by *P. phaseolicola*. The techniques of amino acid analysis used by Patel and Walker might be used with profit to study the re-sponse of tobacco leaves to the wildfire toxin, and it would be of interest to know how closely the halo-blight toxin resembles the wildfire toxin.

THF TOXIN OF *HELMINTHOSPORIUM VICTORIAE*

The wildfire toxin is non-specific and will damage cells which are not parasitized by the bacterium. In contrast the Victoria blight toxin is toxic only to those varieties of oats which are susceptible to attack by *Helminthosporium victoriae*. Meehan & Murphy (1946, 1947) found that a toxin was produced both in culture and in the host plant, and that it only affected oat varieties with Victoria parentage. Although the fungus remains at the base of infected plants, symptoms consisting of bronzing followed by chlorosis and necrosis appear generally on the shoot system; this was taken as evidence for the action of a toxin produced locally but spreading systemically to cause generalized symptoms. Pringle & Braun (1958) showed that a toxin produced *in vitro* prevented the growth of roots of susceptible oats at a con-centration of 0·01 μg./ml., but that a thousand times this concentration was required to affect roots of resistant oats. After hydrolysis of the toxin with $NaHCO_3$, two ninhydrin positive components were obtained, and one of these, called victoxinine, was inhibitory at 75 μg./ml. to growth of roots of both resistant and susceptible varieties. At this time victoxinine was thought to be the main toxic factor, and the other component, a peptide, was responsible for specificity. However Scheffer & Pringle (1963 a, b) showed that the purified toxin stimulates the respiration of susceptible oats but has no effect on oxidation of succinate by mitochondria, whereas victoxinine inhibits tissue respiration and mitochondrial succinoxidase. It seems, therefore, that toxicity depends upon the activity of the entire molecule and that victoxinine alone has no role in the induction of disease symptoms. This interpretation is supported by Grimm & Wheeler (1963) who showed that infection caused an increase in tissue respiration on the 3rd and 4th day. Pringle & Braun (1960) found that many isolates of *H. victoriae* produced little or no toxin *in vitro* but that some filtrates contained victoxinine in high concentration. Passage of isolates of *H. victoriae*, which produce victoxinine but no toxin in culture, through susceptible host plants restored the ability of these isolates to produce the toxin in culture (Braun & Pringle, 1959).

The work on the Victoria blight toxin illustrates a number of important points. First, a close correlation between the pathogenicity of the fungus and toxicity of a substance produced by it *in vitro* to different strains of a host, can be of value in assessing the importance of a substance *in vivo*. Secondly, it might be profitable to use parasites freshly isolated from diseased plants when attempting to establish toxin production *in vitro*. Thirdly, assay of culture filtrates might reveal one substance of relatively low toxicity and specificity but produced in large quantities, yet miss another, which might be present in lower concentration but may be more significant because of its specificity and high toxicity. This last point is demonstrated by victoxinine which is a component of the more active Victoria blight toxin.

Scheffer & Pringle (1961) detected another specific toxin, produced by *Periconia circinata*, which caused the 'milo' disease of *Sorghum vulgare* var. *subglabrescens*. Only pathogenic strains of the fungus produced the toxin in culture. The toxin, which may be a polypeptide, was toxic only to the roots of susceptible seedlings. Although it remains to be shown that the toxin is a vivotoxin, the evidence based on specificity of action is highly suggestive.

FUMARIC ACID IN THE HULL ROT OF ALMONDS

This is one of the most thorough demonstrations of the role of a fungal metabolite in causing disease symptoms, and is also an example of a disease in which the toxin has not been isolated from tissues damaged by the parasite. Mirocha & Wilson (1961) showed that *Rhizopus* spp. were the fungi most frequently associated with the rotting of the fruit mesocarp of almonds and the subsequent blighting of the adjoining twigs and leaves. They found that the organism was confined to the fruit, and that vascular discoloration in the twig and necrosis of the leaf occurred on the side of the stem with vascular connexions to the peduncle of the rotted fruit. They postulated that a translocatable toxin was formed in the mesocarp tissue. Mirocha, DeVay & Wilson (1961) showed that ^{32}P applied to the mesocarp tissue of healthy fruit accumulated in the leaves in a pattern similar to the distribution of symptoms. Mycelium of the parasite grown on a medium supplying glucose-^{14}C was placed in fruit on orchard trees, and the subsequent distribution of radioactivity in the twigs and leaves was found to correspond to the distribution of the blight symptoms. If the radioactive mycelium was killed before inoculation there was no spread of radioactivity. *Rhizopus* spp. grown on water extracts of mesocarp tissue produced a number of metabolites, which included fumaric acid, in considerable quantities. Fumaric acid was the only metabolite which induced blight when tested on shoots. When fumarate-^{14}C was applied to fruit tissue, it was shown to be translocated to the shoot and to accumulate in the leaves before symptoms appeared. During the development of symptoms, fumarate disappeared but the radioactivity appeared in citric, malic and tartaric acids. It was concluded that either fumaric acid or a derivative of it was the important toxin in diseased plants. Using gas chromatography Mirocha & DeVay (1961) estimated the concentration of fumaric acid in leaves and fruit of healthy and diseased almond trees. They detected trace quantities of fumaric acid in healthy leaves and fruit, and in blighted leaves. The concentration of fumaric acid in rotted mesocarp tissue was over 200 μg./g. dry weight of tissue. Thus fumaric acid is produced by the fungus *in vitro* and *in vivo* in large quantities, and it is translocated to and accumulates in leaves near the site of infection. Damage to these leaves occurs, but, because fumaric acid is converted to other acids, examination of the diseased leaves does not reveal more than trace quantities of this, the metabolite which initiates the development of symptoms.

The amount of fumaric acid produced in almond fruit is considerably higher than that of other more specialized toxins, such as that of *Helminthosporium victoriae* which is active at concentrations of 0·01 μg./ml. Hence the labelled mycelium of *Rhizopus* spp. was more likely to yield labelled toxin detectable in other parts of the plants.

The use of ^{14}C and ^{3}H labelled mycelium in the manner described had been suggested by Wheeler (1953) to detect microbial toxins and to distinguish them from host reaction products. He recommended that initial radioactivity supplied in the inoculum must be 10^3 to 10^4 times greater than the minimum required for detection in isolated compounds, because of the small fraction of isotope likely to be found in any one metabolite. A difficulty in this method would be possible damage to organisms or host by the emission of β-particles. However, he was able to grow *Helminthosporium victoriae in vitro* for 5 days on a medium containing 100 μC./ml. of ^{14}C, and then to obtain radioactive metabolites from the mycelium derived from harvested spores. The toxicity of the filtrates of the radioactive mycelium was the same as that of filtrates of non-radioactive mycelium. Thus viability of spores, and toxin production, were not adversely affected under these conditions.

PECTIC ENZYMES IN ROTS OF APPLE FRUIT

Cole & Wood (1961 *a*, *b*) examined some of the rots of apple fruit caused by a number of fungi. Each fungus produced *in vitro* enzymes capable of macerating pieces of plant tissue and of hydrolysing pectic substances. In the rot caused by *Penicillium expansum*, the tissues became soft but remained white, water-insoluble pectic substances were reduced by about 70 % and macerating and polygalacturonase activity could be detected in extracts from rotted tissues. In the rots caused by *Sclerotinia fructigena* and *Botrytis cinerea*, the tissues were firm and brown, water-insoluble pectic substances were reduced by 10–20 % but no macerating or polygalacturonase activity could be detected in extracts. In each rot monogalacturonic acid, presumably derived from cell walls by hydrolysis of pectic substances, was detected. Despite the evidence that hydrolysis of pectic substances, had occurred, it was not possible to detect hydrolysing enzymes in brown rots possibly because of their inhibition by polyphenols. Thus, Cole & Wood (1961 *b*) showed that some oxidized polyphenols were capable *in vitro* of reducing the activity of fungal polygalacturonases. Browning in rots is presumably caused by the oxidation of phenols, and it was shown that there was a considerable reduction in amounts of phenols in rotted tissues. Byrde (1957), and other workers demonstrated the inactivation of pectic enzymes by phenolic substances and their oxidation products in apples. Failure to detect the enzyme which was probably responsible for causing much of the damage in apple rots has particular relevance to attempts to isolate proteinaceous vivotoxins.

In similar work with the rot of pods of cocoa by *Phytophthora palmivora*, Spence (1961) showed that extracts from tissues of infected pods had no pectolytic activity, although active extracts were obtained from potato tubers parasitized by this fungus. The activity of potato tuber extracts was greatly reduced when mixed with oxidized extracts from pods, but there was much less reduction in activity when oxidation of pod extracts was prevented, or when oxidized extracts were treated with gelatin to remove the oxidation products before they could act upon the pectolytic and macerating enzymes.

Grossman (1962 *a*, *b*, *c*) showed another way in which pectic enzymes may be inactivated so that their detection is difficult. Some commonly used inhibitors had no effect on the activity of pectolytic enzymes produced by *Fusarium oxysporum* f. *lycopersici*, but effective inhibitors were found in tannins, quinones and certain acridine derivatives. Both polygalacturonase and pectinmethylesterase were inhibited by oxidized extracts of potato tubers, broad bean plants and apple and pear fruits, and by non-oxidized acetone extracts of some varieties of apples and pears. The inhibitors obtained from fruits were thought to belong to a group of condensed tannins or to be quinones derived by oxidation of low molecular-weight phenols. These inhibitory substances were of the type which might accumulate when cells were killed, and as such could interfere with the detection of these hydrolysing enzymes in the diseased plant.

It is not clearly known to what extent the inactivation of proteinaceous enzymes is a general response to their exposure to oxidizing or oxidized phenol. There may be some degree of specificity between enzymes and oxidation products of phenols in this reaction.

While considering the rotting of tissues by plant pathogens, it should be mentioned that Byrde & Fielding (1962) claimed to have separated a macerating factor from the polygalacturonases produced *in vitro* by *Sclerotinia fructigena*. The factor was active in low concentrations and caused separation of plant cells. Naef-Roth, Gäumann & Albersheim (1961) found that *Dothidea ribesia* produced a macerating factor, termed phytolysine, *in vitro*. They distinguished this substance from pectic enzyme and claimed that it attacked the cell membrane as distinct from the cell wall. These macerating factors may prove difficult to isolate from diseased plants in the light of experiences with the causative factors of brown rots of apple fruit and of other parts of the plant.

ALTERNARIC ACID IN THE EARLY BLIGHT
DISEASE OF TOMATO

A substance called alternaric acid was found to be produced *in vitro* by *Alternaria solani*, the cause of the early blight disease of tomatoes. It produced a severe wilt when applied to seedlings of a number of plant species (Brian *et al.* 1949). Pound & Stahmann (1951) and Brian, Elson, Hemming & Wright (1952) found that alternaric acid caused necrotic lesions on the aerial parts of tomato plants, and these symptoms were similar to those of the natural disease. Although a similar substance was found in infected tomato fruit, the importance of alternaric acid as a vivotoxin has not been stressed because of lack of correlation between the ability of different isolates of *A. solani* to produce the toxin and to cause disease. Experience with the Victoria blight toxin suggests that re-examination of the role of alternaric acid might yield encouraging results.

CHOCOLATE SPOT DISEASE OF BEANS

The chocolate spot disease of beans, *Vicia faba*, caused by species of *Botrytis*, will be considered because recent studies suggest that characteristic lesions are caused by the bean phenolase enzyme which may be activated by the infecting organism in different ways. Kenten (1957, 1958) found that browning of extracts of leaves occurred immediately the activity of latent phenolase was released by a number of means, including exposure to extreme pH conditions, to ammonium sulphate or to anionic wetting agents at pH 6. Kenten suggested that activity was released by the physical separation of oxidase protein from a masking protein. Deverall & Wood (1961 *a*, *b*) observed that fungal germ tubes caused brown leaf spots within a few hours of germinating on the leaf surface. They found that solutions of pectic substances also released the oxidase activity in extracts of leaves provided they were at pH 4·5. They suggested that the germ tubes of the fungi may secrete hydrolytic enzymes, as they do *in vitro*, which damage cells and affect the pectic substances of the cell wall so releasing phenolase. However, although small drops of an anionic wetting agent, placed on leaf surfaces, caused brown lesions to appear within an hour, solutions of pectic substances had no such effect. In these tests the high molecular weight substances may fail to penetrate leaves or may fail to act at the pH of leaves, which is about pH 6. After infection, however, the action of pectinmethylesterase, which is produced by the fungus *in vitro* and also present in healthy leaves, may depress the pH in the lesion to values at which pectic substances might release phenolase activity. Hence the involvement of pectic enzymes in lesion formation has not been proved although the evidence is suggestive; the relevant point is that there may be several ways of initiating the browning response. Metabolites of the fungus may be able to affect host substances other than pectins which can release phenolase activity, or the fungus may produce substances capable of acting like anionic wetting agents. For example, Kenten (1958) showed that some fatty acids were able to activate the phenolase of beans.

Although leaf spot of beans may be one of the simplest responses to infection, necrosis is a common part of disease syndromes and it would be interesting to know if browning is caused by the release of phenolase activity in a similar way in other plants.

ROLE OF TOXINS IN DEFORMATION OF GROWTH

Two general types of alteration of growth induced by parasites can be conceived. One is a permanent shift in metabolism probably based upon changes in the nucleic acid control systems of cells, which causes uncontrolled growth and produces galls. The other is a disturbance of hormone control systems in diseases involving elongation, stunting and excessive branching of plants. In only a few instances is much known about the way in which the pathogenic organisms induce alterations of growth.

Braun (1962) reviewed recent progress in the study of the induction of galls in higher plants by *Agrobacterium tumefaciens,* but this interesting problem will not be discussed here. Braun suggested that resting cells of the host plant were induced to grow and divide permanently through the action of a tumour-inducing principle, the nature of which was not established, but current thought favours the possibility that it is a nucleic acid derived from the bacterium or a virus carried and transmitted by it.

Most of the work on disturbances of hormone control in plants concerns two growth-regulating substances, gibberellic acid and indole-acetic acid, both of which may be produced by pathogenic organisms or hosts.

A good example of the way in which a pathogenic organism affects the growth of the host is provided by the action of *Gibberella fujikuroi* which causes increase in height of rice plants in bakanae disease. About thirty years ago in Japan, it was found that sterile culture filtrates of the fungus caused a similar elongation of rice seedlings and that the active substances were gibberellic acid and close relatives. In recent years, gibberellins have been found as natural components of higher plants (Stowe & Yamaki, 1957). The claim that gibberellins are the vivotoxins causing bakanae disease rests upon the similarity between the symptoms characteristic of the infected plants and those produced by the gibberellins. It is possible that pathogenic organisms may induce growth responses in higher plants by affecting the host systems which control the concentration of gibberellins instead of producing gibberellins themselves.

Changes in the metabolism of indole-3-acetic acid after infection

The most studied system of growth regulation in higher plants involves indole-3-acetic acid, and it is not surprising that, in many diseases characterized by alterations of growth, attention has been focused on changes in concentration and rates of metabolism of this acid. Rarely, though, has the way in which the pathogenic organism causes the changes been satisfactorily established.

A mechanism involving the destruction of indole-3-acetic acid was proposed by Sequeira & Steeves (1954) in relation to the destructive disease of coffee caused by the fungus *Omphalia flavida*. The disease is characterized by severe premature defoliation of the trees. Leaves fell prematurely when the fungus caused a lesion near the junction of blade and petiole. This was not caused by mechanical damage, and Sequeira & Steeves considered that defoliation might result from interference with the supply of indole-3-acetic acid to the petioles, because work with other plants had shown that defoliation could be delayed by supplying the petiole with indole-3-acetic acid. The fact that the fungus produced *in vitro* an enzymic system which destroyed the acid rapidly, suggested that the shedding of diseased leaves resulted from the reduction of the amount of indole-3-acetic acid in the petioles by this fungal enzyme. This attractive hypothesis requires evidence showing that indole-3-acetic acid is at a lower concentration in diseased petioles and that its destruction increases in rate after infection.

Increases in amounts of growth-promoting substances in diseased plants have been demonstrated. For example, Pegg & Selman (1959) found abnormally high levels of indole-3-acetic acid or related substances in tomato plants infected by *Verticillium albo-atrum*, and concluded that many of the symptoms of vascular wilt disease, including petiolar epinasty, tylosis, pith hyperplasia and excessive adventitious rooting might result from the increases. They found that the fungus would produce indole-3-acetic acid *in vitro*, and that some of the disease symptoms could be caused by treating healthy plants with solutions of the acid (1–5 μg./l.). Similarly, Sequeira & Kelman (1962) showed that tobacco plants infected with *Pseudomonas solanacearum*, which causes bacterial wilt of solanaceous plants, contained between 0·03 and 3·3 μg./100 g. tissue more growth-promoting substances than corresponding healthy plants. The bacterium produced indole-3-acetic acid *in vitro* in both semisynthetic media and media based on extracts of plant tissue, and they suggested that, together with other growth-regulating substances, accumulation of the acid *in vivo* played a part in the development of disease symptoms.

Increases in amount of indole-3-acetic acid in diseased plants may be caused by its synthesis by pathogenic organisms, by increased synthesis by the host, or by reduced destruction of the acid by infected tissue. Demonstration of changed rates of synthesis of a metabolite *in vivo* depends upon a comparison of the rates of conversion of radioactive precursors into the metabolite in healthy and diseased plants. A satisfactory method for measuring rates of synthesis of indole-3-acetic acid from labelled precursors has not been applied to diseased plants, so there is no proof, as yet, of the production of the acid by pathogenic organisms *in vivo*. However, in studies of rust diseases, several attempts have been made to determine the rates of destruction of indole-3-acetic acid *in vivo*, and these have yielded results which suggested that increased synthesis of the acid was not the only way in which higher levels were achieved.

Increased quantities of indole-3-acetic acid were found in plant organs infected by rust fungi, both when growth was disturbed (Pilet, 1957, 1960; Daly & Inman, 1958) and when no morphological change was observed as in wheat rust (Shaw & Hawkins, 1958). Using a colorimetric method of estimating the acid, Pilet (1957, 1960) found that extracts of rusted leaves and stems of *Euphorbia cyparissias* had lower activities of indole-3-acetic acid oxidase than similar extracts of healthy leaves. He postulated that the destroying enzyme might be inhibited in diseased tissues by fungal toxins. Shaw & Hawkins (1958) measured the rate at which indole-3-acetic acid isotopically labelled in the carboxyl carbon atom was decarboxylated in pieces of leaf of wheat, and found that heavily rusted leaves were less active in this respect than healthy leaves. They also suggested that its oxidase was inhibited in rusted leaves. Daly & Deverall (1963) confirmed that the rate of decarboxylation by rusted leaves was low, but observed that the uptake of the acid by rusted leaf sections from solutions was also depressed. Associated with the low uptake and decarboxylation of indole-3-acetic acid was the near absence in diseased tissues of two radioactive derivatives, and Deverall & Daly (1963) suggested that there might be an impaired transport of indole-3-acetic acid caused by the low rate of formation of these derivatives in rusted leaves. Thus further analysis of the situation in rusted tissues is necessary before reaching any conclusion on the role of fungal toxins as inhibitors of indole-3-acetic acid metabolism.

The use of labelled indole-3-acetic acid in the above experiments led to the discovery by Shaw & Hawkins (1958) of stimulation of decarboxylation immediately after inoculation of wheat leaves with uredospores of *Puccinia graminis tritici.* Daly & Deverall (1963) obtained a similar result using an inoculation technique which eliminated mechanical injury as the cause of the stimulation. The significance of the reaction in the infection process is unknown; it might be caused by the indole acetic acid oxidase of the fungus, shown by Oaks & Shaw (1960) to be present in mycelium derived from a form of tissue culture of flax rust, *Melampsora lini.* Alternatively fungal metabolites may affect the control system of the host which was shown by Daly & Deverall (1963) to be sensitive to environmental influences.

In summary, changes in the metabolism of indole-3-acetic acid have been found in a number of diseases but, although it is known that fungi and bacteria may synthesize the acid and produce its oxidase *in vitro*, very little is known about the reactions which occur *in vivo*.

CONCLUSION

In most of the examples discussed above, one or more substances active in inducing symptoms similar to those caused by infection have readily been detected in culture filtrates of pathogenic organisms. It is possible that absence of correct nutritional balance or of precursors will prevent the production of a particular substance *in vitro*, but this has not been a difficulty in the past. The success of biochemists in isolating and identifying these metabolites produced *in vitro* is clearly illustrated by the work in Zürich on wilt toxins and in New York on wildfire and Victoria blight toxins.

The weakness in many studies on the role of toxins is the failure to show the toxin to be present in adequate concentration *in vivo* to cause the particular disease symptoms. There are many reasons why some vivotoxins cannot be detected in diseased tissue. In these cases, a demonstration *in vivo* or during extraction of the reactions responsible for the inactivation of the toxin would be valuable. On the other hand there may be cases in which sufficient refinement of a biochemical technique would permit estimation of the amount of a substance *in vivo* and possibly its isolation from this source.

It might be considered pedantic to insist on this demonstration if the toxin will cause highly characteristic symptoms in healthy plants or if the toxin is confined in its action to those varieties of host plant which are susceptible to the organism. However, if the nature of symptoms is used to support a claim that a substance is the vivotoxin, it is obviously necessary to expect a favourable comparison of the detailed morphological symptoms and possibly the biochemical changes caused by the toxin and by infection. The different details of the symptoms associated with wilting caused by fusaric acid and by infection emphasize this point.

Situations occur in which the active substance is a metabolite of both host and pathogen, and in order to show that *in vivo* the substance has a pathogenic origin it is necessary to use a radioactive inoculum, of the pathogenic organism, as in the study of the role of fumaric acid in the hull rot of almonds. This technique might be the only way of showing that a toxic metabolite had fungal origin in those diseases caused by obligate parasites.

I wish to express my gratitude to Dr R. K. S. Wood for his advice and helpful criticism during the preparation of this article.

REFERENCES

BLACKHURST, F. M. & WOOD, R. K. S. (1963). *Verticillium* wilt of tomatoes: further experiments on the role of pectic and cellulolytic enzymes. *Ann. appl. Biol.* (in the Press).

BRAUN, A. C. (1950). The mechanism of action of a bacterial toxin on plant cells. *Proc. nat. Acad. Sci., Wash.* **36**, 423.

BRAUN, A. C. (1962). Tumor inception and development in the crown gall disease. *Annu. Rev. Pl. Physiol.* **13**, 533.

BRAUN, A. C. & PRINGLE, R. B. (1959). Pathogen factors in the physiology of disease—toxins and other metabolites. In *Plant Pathology, Problems and Progress, 1908–1958*, p. 88. Madison: University of Wisconsin Press.

BRIAN, P. W., CURTIS, P. J., HEMMING, H. G., UNWIN, C. H. & WRIGHT, J. M. (1949). Alternaric acid, a biologically active metabolic product of the fungus, *Alternaria solani. Nature, Lond.* **164**, 534.

BRIAN, P. W., ELSON, G. W., HEMMING, H. G. & WRIGHT, J. M. (1952). The phytotoxic properties of alternaric acid in relation to the etiology of plant diseases caused by *Alternaria solani* (Ell. & Mert.) Jones & Grout. *Ann. appl. Biol.* **39**, 308.

BYRDE, R. J. W. (1957). The varietal resistance of fruits to brown rot. II. The nature of resistance in some varieties of cider apples. *J. hort. Sci.* **32**, 227.

BYRDE, R. J. W. & FIELDING, A. H. (1962). Resolution of endopolygalacturonase and a macerating factor in a fungal culture filtrate. *Nature, Lond.* **196**, 1227.

CLAUSSON-KAAS, N., PLATTNER, P. L. A. & GÄUMANN, E. (1944). Über ein Welkeerzeugendes Stoffwechselprodukt von *Fusarium lycopersici* Sacc. *Ber. schweiz. bot. Ges.* **54**, 523.

COLE, M. & WOOD, R. K. S. (1961a). Types of rot, rate of rotting, and analysis of pectic substances in apples rotted by fungi. *Ann. Bot., Lond.* **25**, 417.

COLE, M. & WOOD, R. K. S. (1961b). Pectic enzymes and phenolic substances in apples rotted by fungi. *Ann. Bot., Lond.* **25**, 435.

DALY, J. M. & DEVERALL, B. J. (1963). Metabolism of IAA in rust diseases. I. Factors influencing rates of decarboxylation. *Plant Physiol.* **38** (in the Press).

DALY, J. M. & INMAN, R. E. (1958). Changes in auxin levels in safflower hypocotyls infected with *Puccinia carthami. Phytopathology*, **48**, 91.

DEVERALL, B. J. & DALY, J. M. (1963). Metabolism of IAA in rust diseases. II. Metabolites of IAA carboxyl ^{14}C in tissues. *Plant Physiol.* (in the Press).

DEVERALL, B. J. & WOOD, R. K. S. (1961a). Infection of bean plants (*Vicia faba* L.) with *Botrytis cinerea* and *Botrytis fabae. Ann. appl. Biol.* **49**, 461.

DEVERALL, B. J. & WOOD, R. K. S. (1961b). Chocolate spot of beans (*Vicia faba* L.)—interactions between phenolase of host and pectic enzymes of the pathogen. *Ann. appl. Biol.* **49**, 473.

DIMOND, A. E. & WAGGONER, P. E. (1953a). On the nature and role of vivotoxins in plant disease. *Phytopathology*, **43**, 229.

DIMOND, A. E. & WAGGONER, P. E. (1953b). Effect of lycomarasmin decomposition upon estimates of its production. *Phytopathology*, **43**, 319.

DIMOND, A. E. & WAGGONER, P. E. (1953c). The water economy of *Fusarium* wilted tomato plants. *Phytopathology*, **43**, 619.

GÄUMANN, E. (1951). Some problems of pathological wilting in plants. *Advanc. Enzymol.* **11**, 401.

GÄUMANN, E. (1957). Fusaric acid as a wilt toxin. *Phytopathology*, **47**, 342.

GÄUMANN, E. (1958). The mechanisms of fusaric acid injury. *Phytopathology*, **48**, 670.

GÄUMANN, E., KERN, H., SCHÜEPP, H. & OBRIST, W. (1958). Der Einfluss der Fusarinsäure auf den Wasserhaushalt abgeschnittener Tomatensprosse. *Phytopath. Z.* **32**, 225.

GÄUMANN, E. & OBRIST, W. (1960). Über die Schädigung der Wasserpermeabilität pflanzlicher Protoplasten durcheinige Welketoxine. *Phytopath. Z.* **37**, 145.

GOTHOSKAR, S. S., SCHEFFER, R. P., WALKER, J. C. & STAHMANN, M. A. (1953). The role of pectic enzymes in *Fusarium* wilt of tomato. *Phytopathology*, **43**, 535.

GOTHOSKAR, S. S., SCHEFFER, R. P., WALKER, J. C. & STAHMANN, M. A. (1955). The role of enzymes in the development of *Fusarium* wilt of tomato. *Phytopathology*, **45**, 381.

GOTTLIEB, D. (1943). The presence of a toxin in tomato wilt. *Phytopathology*, **33**, 126.

GOTTLIEB, D. (1944). The mechanism of wilting caused by *Fusarium bulbigenum* var. *lycopersici. Phytopathology*, **34**, 41.

GRIMM, R. B. & WHEELER, H. E. (1963). Respiratory and enzymatic changes in Victoria blight of oats. *Phytopathology*, **53**, 436.

GROSSMAN, F. (1962a). Untersuchungen über die Hemmung pektolytischer Enzyme von *Fusarium oxysporum* f. *lycopersici*. I. Hemmung durch definierte Substanzen *in vitro. Phytopath. Z.* **44**, 361.

GROSSMAN, F. (1962b). Untersuchungen über die Hemmung pektolytischer Enzyme von *Fusarium oxysporum* f. *lycopersici*. II. Hemmung durch Pflanzenextrakte *in vitro. Phytopath. Z.* **45**, 1.

GROSSMAN, F. (1962c). Untersuchungen über die Hemmung pektolytischer Enzyme von *Fusarium oxysporum* f. *lycopersici*. III. Wirkung einiger Hemmstoffe *in vivo. Phytopath. Z.* **45**, 139.

HODGSON, R., PETERSON, W. H. & RIKER, A. J. (1949). The toxicity of polysaccharides and other large molecules in tomato cuttings. *Phytopathology*, **39**, 47.

HUSAIN, A. & DIMOND, A. E. (1960). Role of cellulolytic enzymes in pathogenesis by *Fusarium oxysporum* f. *lycopersici*. *Phytopathology*, **50**, 329.

HUSAIN, A. & KELMAN, A. (1958). The role of pectic and cellulolytic enzymes in pathogenesis by *Pseudomonas solanacearum*. *Phytopathology*, **48**, 377.

JOHNSON, J. & MURWIN, H. F. (1925). Experiments on the control of wildfire of tobacco. *Res. Bull. Wis. agric. Exp. Sta.* no. 62.

KAMAL, M. & WOOD, R. K. S. (1956). Pectic enzymes secreted by *Verticillium dahliae* and their role in the development of the wilt disease of cotton. *Ann. appl. Biol.* **44**, 322.

KENTEN, R. H. (1957). Latent phenolase in extracts of broad bean (*Vicia faba* L.) leaves. I. Activation by acid and alkali. *Biochem. J.* **67**, 300.

KENTEN, R. H. (1958). Latent phenolase in extracts of broad bean (*Vicia faba* L.) leaves. II. Activation by anionic wetting agents. *Biochem. J.* **68**, 244.

KERN, H. (1959). Der Einfluss des Lycomarasmins auf den Wasserhaushalt von unterschiedlich mit Eisen ernährten Tomatenpflanzen. *Phytopath. Z.* **35**, 232.

KERN, H. & KLUEPFEL, D. (1956). Die Bildung von Fusarinsäure durch *Fusarium lycopersici in vivo*. *Experientia*, **12**, 181.

KERN, H. & SANWAL, B. D. (1954). Untersuchungen über den Stoffwechsel von *Fusarium lycopersici* mit Hilfe von radioaktiven Kohlenstoff. *Phytopath. Z.* **22**, 449.

KLUEPFEL, D. (1957). Über die Biosynthese und die Umwandlungen der Fusarinsäure in Tomatenpflanzen. *Phytopath. Z.* **29**, 349.

LAKSHMINARAYANAN, K. & SUBRAMANIAN, D. (1955). Is fusaric acid a vivotoxin? *Nature, Lond.* **176**, 697.

LUDWIG, R. A. (1960). Toxins. In *Plant Pathology*, **2**, 315. Ed. by J. G. Horsfall & A. E. Dimond. New York: Academic Press.

MEEHAN, F. & MURPHY, H. C. (1946). A new *Helminthosporium* blight of oats. *Science*, **104**, 413.

MEEHAN, F. & MURPHY, H. C. (1947). Differential phytotoxicity of metabolic by-products of *Helminthosporium victoriae*. *Science*, **106**, 270.

MIROCHA, C. J. & DEVAY, J. E. (1961). A rapid gas chromatographic method for determining fumaric acid in fungus cultures and diseased plant tissues. *Phytopathology*, **51**, 274.

MIROCHA, C. J., DEVAY, J. E. & WILSON, E. E. (1961). Role of fumaric acid in the hull rot disease of almonds. *Phytopathology*, **51**, 851.

MIROCHA, C. J. & WILSON, E. E. (1961). Hull rot disease of almonds. *Phytopathology*, **51**, 843.

NAEF-ROTH, ST, GÄUMANN, E. & ALBERSHEIM, P. (1961). Zur Bildung eines mazerierenden Fermentes durch *Dothidea ribesia* Fr. *Phytopath. Z.* **40**, 283.

OAKS, A. & SHAW, M. (1960). An indole-acetic acid oxidase system in the mycelium of *Melampsora lini* (Pers.) Lev. *Canad. J. Bot.* **38**, 761.

PAGE, O. T. (1959). Fusaric acid in banana plants infected with *Fusarium oxysporum* f. *cubense*. *Phytopathology*, **49**, 230.

PATEL, P. N. & WALKER, J. C. (1963). Changes in free amino acid and amide content of resistant and susceptible beans after infection with the halo-blight organism. *Phytopathology*, **53**, 522.

PEGG, G. F. & SELMAN, I. W. (1959). An analysis of the growth response of young tomato plants to infection by *Verticillium albo-atrum*. II. The production of growth substances. *Ann. appl. Biol.* **47**, 222.

PILET, P. E. (1957). Activité anti-auxines—oxydasique de l'*Uromyces pisi* (Pers.) de parasite d'*Euphorbia cyparissias* L. *Phytopath. Z.* **31**, 162.

PILET, P. E. (1960). Auxin content and auxin catabolism of the stems of *Euphorbia cyparissias* L. infected by *Uromyces pisi* (Pers.). *Phytopath. Z.* **40**, 75.

POUND, G. S. & STAHMANN, M. A. (1951). The production of a toxic material by *Alternaria solani* and its relation to the early blight disease of tomato. *Phytopathology*, **41**, 1104.

PRINGLE, R. B. & BRAUN, A. C. (1958). Constitution of the toxin of *Helminthosporium victoriae*. *Nature, Lond.* **181**, 1205.

PRINGLE, R. B. & BRAUN, A. C. (1960). Isolation of victoxinine from cultures of *Helminthosporium victoriae*. *Phytopathology*, **50**, 324.

ROSEN, H. R. (1926). Efforts to determine the means by which the cotton wilt fungus, *Fusarium vasinfectum*, induces wilting. *J. agric. Res.* **33**, 1143.

SCHEFFER, R. P. & PRINGLE, R. B. (1961). A selective toxin produced by *Periconia circinata*. *Nature, Lond.* **191**, 912.

SCHEFFER, R. P. & PRINGLE, R. B. (1963a). Respiratory effects of the selective toxin of *Helminthosporium victoriae*. *Phytopathology*, **53**, 465.

SCHEFFER, R. P. & PRINGLE, R. B. (1963b). Toxicity of victoxinine. *Phytopathology*, **53**, 558.

SCHEFFER, R. P. & WALKER, J. C. (1953). The physiology of *Fusarium* wilt of tomato. *Phytopathology*, **43**, 116.

SEQUEIRA, L. & KELMAN, A. (1962). The accumulation of growth substances in plants infected by *Pseudomonas solanacearum*. *Phytopathology*, **52**, 439.

SEQUEIRA, L. & STEEVES, T. A. (1954). Auxin inactivation and its relation to leaf drop caused by the fungus *Omphalia flavida*. *Plant Physiol.* **29**, 11.

SHAW, M. & HAWKINS, A. R. (1958). Physiology of host–parasite relations. V. A preliminary examination of the level of free endogenous indole-acetic acid in rusted and mildewed cereal leaves and their ability to decarboxylate exogenously supplied radioactive indole-acetic acid. *Canad. J. Bot.* **36**, 1.

SIVADJIAN, J. & KERN, H. (1958). Über den Einfluss von Welketoxinin auf die Transpiration von Tomatenblättern. *Phytopath. Z.* **33**, 241.

SKOOG, H. A. (1952). Studies on host–parasite relations of bean varieties resistant and susceptible to *Pseudomonas phaseolicola* and toxin production by the parasite. *Phytopathology*, **42**, 475.

SPENCE, J. A. (1961). Black-pod disease of cocoa. II. A study of host–parasite relations. *Ann. appl. Biol.* **49**, 723.

STOWE, B. B. & YAMAKI, T. (1957). The history and physiological action of the gibberellins. *Annu. Rev. Pl. Physiol.* **8**, 181.

WAGGONER, P. E. & DIMOND, A. E. (1953). Role of chelation in causing and inhibiting the toxicity of lycomarasmin. *Phytopathology*, **43**, 281.

WHEELER, H. E. (1953). Detection of microbial toxins by the use of radioisotopes. *Phytopathology*, **43**, 236.

WINSTEAD, N. N. & WALKER, J. C. (1954). Production of vascular browning by metabolites from several pathogens. *Phytopathology*, **44**, 153.

WOOD, R. K. S. (1960). Pectic and cellulolytic enzymes in plant disease. *Annu. Rev. Pl. Physiol.* **11**, 299.

WOOD, R. K. S. (1961). *Verticillium* wilt of tomatoes—the role of pectic and cellulolytic enzymes. *Ann. appl. Biol.* **49**, 120.

WOOLLEY, D. W., PRINGLE, R. B. & BRAUN, A. C. (1952). Isolation of the phytopathogenic toxin of *Pseudomonas tabaci*, an antagonist of methionine. *J. biol. Chem.* **197**, 409.

WOOLLEY, D. W., SCHAFFNER, G. & BRAUN, A. C. (1952). Isolation and determination of structure of a new amino acid contained within the toxin of *Pseudomonas tabaci*. *J. biol. Chem.* **198**, 807.

WOOLLEY, D. W., SCHAFFNER, G. & BRAUN, A. C. (1955). Studies on the structure of the phytopathogenic toxin of *Pseudomonas tabaci*. *J. biol. Chem.* **215**, 485.

HOST–PARASITE RELATIONS AND ENVIRON-
MENTAL INFLUENCES IN SEED-BORNE
DISEASES

V. R. WALLEN

Plant Research Institute, Ottawa, Canada

One of the principal means of dissemination of many plant diseases is by seed. On the surface and internally, seed may carry fungi, bacteria, viruses and other organisms which can introduce diseases into previously uninfected areas and countries. A high proportion of the research on seed-borne diseases has been mycological in character. As a result, hundreds of seed fungi both pathogenic and saprophytic have been isolated and identified. Less work has been done with bacterial and virus diseases. Investigations of seed-borne disease have usually been a part of a more complex study of the parasitization of plants. Certain aspects of the problem have been reviewed by Porter (1949), Noble (1951) and Dykstra (1961). A list of seed-borne diseases was prepared by Noble, de Tempe & Neergaard (1958).

In this paper I have omitted many aspects of the problem that have been previously reviewed. I have attempted to group into sections research that embraces the general theme of the symposium. The subject will be discussed under the following headings:

The infection cycle. This section includes examples of studies on fungi and bacteria that have led to the development of important techniques for their identification in seed. The infrequency of virus infection in seed is also discussed.

Effects of the pathogen on the host. The physical and chemical changes that take place when the seed is infected by the pathogen are discussed.

Differential sensitivity of host and pathogen. The sensitivity of the pathogen and host to changing environmental conditions is reviewed.

Longevity of the pathogen and the host. This section brings together for the first time studies on the survival *in vitro* of seed-borne fungi, bacteria and viruses and their seed hosts.

The bulk of the work to be described comes from practical studies on the prevention of seed-borne disease in the field or during storage.

In the main the studies are *in vivo* as the presence of the pathogenic organism is determined either by the production of symptoms on a plant typical of a specific seed-borne disease, or by isolation of the pathogenic organism from the seed sometime after harvest. Only a few relevant studies have been made *in vitro* on the relative properties of the isolated pathogenic organisms and the uninfected seed. Perhaps this paper will encourage the establishment of techniques for the isolation of seed-borne pathogenic organisms and for their further study *in vitro* and *in vivo*.

THE INFECTION CYCLE

Seed infection by fungi, bacteria and viruses may arise in many ways such as by means of floral infection, systemic invasion through the vascular system of the plant to the seed, direct penetration through pods or other fruits and invasion of the seed in storage after harvest.

Fungi

Research has been done with loose smut in barley and wheat, caused by *Ustilago nuda* and *U. tritici* respectively, in relation to floral infection in the field and the pathway of infection to the seed. An excellent technique for the recognition of the fungus in the seed embryo has resulted from this research (Simmonds, 1946; Popp, 1958). In Popp's technique embryos are separated from the seed by boiling in a mixture of 3 % (w/v) sodium hydroxide, 12 % (w/v) sodium silicate and 0·4 % (w/v) detergent. The whole embryos are then boiled in 12 % (v/v) ethanol containing 15 % (w/v) sodium hydroxide and heated in a 3:1 mixture of ethanol and glacial acetic acid to clarify the tissue and remove oil droplets. The embryos are then heated in 45 % (v/v) acetic acid containing 0·1 % trypan blue to stain the fungus hyphae. The embryos are finally heated and mounted in 45 % (v/v) lactic acid. Infected embryos can be quickly located by microscopic examination.

Loose smut can cause losses of up to 100 % of the crop (Persons, 1954). The normal kernels are replaced by black sori of chlamydospores. Wind-blown chlamydospores are released from infected wheat and barley heads and infect young florets. According to Sreeramulu (1962), the maximum infection of barley takes place during a 3- to 4-week period corresponding with flowering. Ohms & Bever (1956) found that the optimum time for infection was 1 day before, and at anthesis. Campbell & Tyner (1959) found that plants were most susceptible to *Ustilago nuda* within the first 2 days following heading of the crop. Atkins *et al.* (1963) found that humidity was the most important environmental factor in re-infection of winter wheat in Texas. Of four locations under test over a 5-year period, the one that had the highest incidence of loose smut had also the highest mean daily relative humidity (77·5). High wind velocity also aided the dissemination of spores. Rainfall had an adverse effect on infection as it washed the spores from the smutted heads. In England, however, Marshall (1959) noted that, despite a fluctuating environment during the infection period, there was little change in the percentage of loose smut in barley over a 3-year period. Tapke (1948) made an extensive review of the environmental factors necessary for the development of loose smut in barley and wheat.

All workers do not agree on the location of primary infection. Lang (1917) stated that chlamydospores of *Ustilago nuda* either germinated on the surface of styles and followed the path of the pollen tubes to the integuments and the embryo; or they germinated on the surface of the ovary, penetrated the ovary wall and the integuments, and continued between the integuments and the nucellar epidermis until they reached the embryo. Batts (1955) found that *U. tritici* entered the ovary wall through any part of the epidermis. At the point of entry an appressorium was formed, and beneath the appressorium a bulbous swelling occurred through which the penetrating hypha passed in a narrowed condition; a funnel-like structure resulted. After penetration of the ovary, the fungus reached the testa in approximately 2 days, and after entering the aleurone layer it finally entered the scutellum in 26 days. Profuse development of mycelium occurred in the embryo. The mycelium was mainly intracellular in the pericarp and testa and intercellular in the aleurone, scutellum and other tissues. Lang (1917) and Vanderwalle (1942) reported all mycelium to be intercellular. Pedersen (1956) could not confirm Lang's work regarding infection through the styles, and proposed that during its early growth the embryo secretes substances which attract the fungal hyphae. These substances diffuse through the epithelial layers of the scutellum and the integuments and enters the pericarp. Pedersen agreed with Batts that primary infection took place only on the surface of the ovary despite the high germination of spores on styles and ovaries. Once infection has been established both fungus and seed lie dormant in an almost symbiotic state.

When the germination process begins, the fungus becomes active if mycelium is present in the growing point of the embryo, and it is carried upward by the elongation of the epicotyl. The fungus ramifies through the tissues of the crown and penetrates the rudimentary inflorescences as they form. As the young ear is later carried upward by elongation of the lowest internodes, the fungus is taken with it and the mycelium increases as the ear develops, until on emergence the ear is replaced by spores. Contrary to early reports that the growth of the fungus keeps pace with that of the plant, Batts & Jeater (1958b) found that the mycelium was present in the nodes, but not in the internodes. The growth of the mycelium could not keep pace with that of the cells of the internodes.

Using the method for detection of infected embryos (Simmonds, 1946), several workers found a good correlation between the percentage of infected barley embryos and the number of infected plants that were produced in the field (Russell, 1950; Russell & Popp, 1951). However Batts & Jeater (1958a) noted that some varieties of field resistant wheat were grown from seed with infected embryos. The embryo became infected normally, but the mycelium did not spread from the scutellum to the growing point as in the case of field susceptible varieties. They concluded that detection of loose smut by the embryo method was of value only in the examination of susceptible varieties of wheat. In Canada, the embryo method of detection is no longer used for the determination of *Ustilago nuda* in Keystone barley because this variety shows extreme embryo susceptibility and mature plant resistance. Popp (1959) found that resistance or susceptibility in wheat was indicated by the following reactions of specific tissues of the embryo to the fungus: resistance of the entire embryo; susceptibility of all the embryo except the plumular bud; susceptibility of all tissues including the plumular bud. Only infection of the plumular bud tissue was correlated with the percentage of adult plants which later became infected with smut.

Recently another type of reaction between fungus and host has been observed. Mantle (1961) noted an abnormal reaction of Kota wheat to loose smut in which the fungus and the host were not compatible. Infections of the apical growing point resulted in dead seedlings and dwarfed plants, or plants which were normal as a result of the death of infected shoots and recovery by the production of healthy tillers. Mumford & Rasmusson (1963) studied a similar phenomenon during the infection of barley variety Jet with *Ustilago nuda*. This fungus sometimes reacts similarly with Keystone barley. Although Oort (1944) used the term hypersensitivity to describe this reaction between host and parasite, other workers have objected to its use; they regard the reaction as simply an incompatible host–parasite reaction.

Bacteria

Seed-borne blights of beans, caused by *Xanthomonas phaseoli* (common blight) and *Pseudomonas phaseolicola* (halo blight), are present wherever beans are grown. Although the bacteria may overwinter in the soil of certain areas, the principal means of dissemination has been by seed. Burkholder (1930) isolated a bacterium causing symptoms on bean plants similar to common blight, but this organism differed from *X. phaseoli* in producing a brown pigment on proteinaceous media. The third blight disease was called 'fuscous' blight (Burkholder & Bullard, 1946).

Zaumeyer (1929) observed that infection of seed by *Xanthomonas phaseoli* may take place through the vascular system of the plant without the production of any visible symptoms. Infection may also occur from dorsal suture infections of the pod followed by invasion of the funiculus, the raphe and the seed coat. Bacteria may enter the seed through the micropyle after they have penetrated the pod cavity either through stomata or by breaking through the vascular system of pod sutures. Once in the seed coat, the bacteria invade the intercellular spaces and surround the cotyledons. At germination the seed swells through imbibing water and epidermal cells of the cotyledons are pulled apart, allowing the bacteria to enter these openings, to penetrate the cotyledons and thus to systemically invade the young seedling.

In the past decade, the use of specific bacteriophages *in vitro* has provided an excellent method for the rapid identification of phyto-pathogenic bacteria in seed, and given an insight into their complex inter- and intraspecific relationships. Studies on *Xanthomonas phaseoli* and *Pseudomonas phaseolicola* are described which illustrate the potentiality of this method for other studies.

Katznelson & Sutton (1951) first developed the phage method to detect and identify the seed-borne pathogens *Xanthomonas phaseoli* and *Pseudomonas phaseolicola* in bean seed. When a single susceptible bacterial cell is infected by a specific phage particle the cell is lysed and many phage particles are liberated. This multiplication of specific phage forms the basis of the method which is as follows: 250 surface sterilized bean seeds (2%, w/v, sodium hypochlorite solution for 10 min.) are ground for 5–7 min. in sterile nutrient broth (1 l.) in a blendor. The macerated sample is incubated for 24 hr. to permit the multiplication of phage sensitive cells of the suspected bacterial pathogen. Portions (10 ml.) are removed aseptically to sterile flasks and a phage suspension containing 4000 to 5000 phage particles is added. Immediately samples of the mixture (0·1 ml.) from each flask are spread on plates seeded with suitable indicator bacteria; each phage particle forms zones of lysis (plaques) which can be counted. Six to 12 hr. later another sample is plated. A significant increase in the number of phage particles indicates the presence of the homologous bacteria, either *X. phaseoli* or *P. phaseolicola*, in the seed sample.

Recently Wallen, Sutton & Grainger (1963) have used this method to understand the complex situation that exists among phytopathogenic bacteria in bean seed. They showed that although *Xanthomonas phaseoli* and *Pseudomonas phaseolicola* are present to some extent in samples from the province of Ontario, the principal pathogen causing blight is *X. phaseoli* var. *fuscans*. Inspections of registered bean fields in 1961 and 1962 revealed a high incidence of bacterial blight, but the phage method detected *X. phaseoli* and *P. phaseolicola* in only a small percentage of the samples tested. One or more of three possibilities were considered: the organism had not established itself in the seed, despite heavy pod infections; the phage method had failed to detect the organisms in the seed; or another strain of the organism heretofore unrecognized and insensitive to the phages was present in the crop. The last possibility was shown to be correct by repeated studies *in vitro*, involving the isolation of a yellow bacterium that produced a dark brown pigment in nutrient agar, and the determination of its sensitivity to a bacteriophage for *X. phaseoli* var. *fuscans* which had been isolated from soil of the field where the beans were grown. On the basis of its cultural characteristics, pigment production, phage sensitivity and pathogenicity this bacterium was *X. phaseoli* var. *fuscans*; it was present in over 50 % of the bean seed samples. The fact that *X. phaseoli* var. *fuscans* has become the principal pathogen in this area, where *X. phaseoli* and *P. phaseolicola* were formerly predominant, has probably been due to the use of bean varieties with less resistance to the fuscans organism. In this work the phage method has given a hint of an evolutionary trend. Pigmentation of the various cultures of the fuscans organism varied from light to dark brown when grown on nutrient agar. The fuscans phage lysed these cultures and also a small group of yellow pigmented isolates which did not produce the brown pigment, and which were insensitive to the common blight phage. However, these cultures did produce symptoms typical of 'fuscous' blight. At present, we can only theorize as to the nature of these organisms in bean seed. It is possible that the yellow pigmented xanthomonads are in the process of transition to a true fuscans type of organism, or we have isolated an intermediate strain between a true *X. phaseoli* organism and a fuscans type. Further, Canadian strains of *fuscans* differ from a European strain (E 1299) in that they adsorb the homologous phage of Klements & Lovas (1960) for strain E 1299, but on plates no plaques are produced at the critical test dilution.

Viruses

Of the many virus diseases of plants, only a small percentage are thought to be seed transmitted. The low incidence of seed transmission cannot be attributed to any one cause. Allard (1915) introduced the theory of induced sterility when he observed malformations of certain floral parts as a result of infection by tobacco mosaic virus. He concluded that the normal functions of pistils and stamens were disturbed to such an extent as to cause sterility. Caldwell (1952) demonstrated that in the aspermy disease of tomato, virus in the microspore mother cell produced a complex interference with meiosis which resulted in abnormal pollen formation. These theories are adequate for these particular diseases, but do not account for other diseases that are not seed-borne in crops with abundant seed set. Bennett & Esau (1936), in a study of the curly top virus in sugar beet and tobacco, believed that the lack of a vascular connexion between the embryo and the adjoining tissues would prevent the passage of the virus into the embryo. This theory satisfactorily explains the lack of transmission for virus diseases that are restricted to the vascular tissues of the host. Duggar (1930) postulated that seed transmission of diseases might be prevented by the activity of some specific protein or other inhibitor; and Kausche (1940) showed that tobacco mosaic virus was not transmissible by seed because of an inactivating substance formed in the seed during the process of ripening and germination. Bennett & Esau (1936) postulated that the lack of a plasmodesmatal connexion between parent and embryo may prevent seed transmission of highly infectious virus diseases.

Some viruses are transmissible by seed. Gold, Suneson, Houston & Oswald (1954) found that barley mosaic virus would infect 100 % of developing barley embryos and subsequently 50–90 % of the seedlings. Rod-shaped particles were found in leaves, embryos, endosperm, pollen and unfertilized pistils. Pollen transmission was suggested also by presence of virus-like rods in seed produced from healthy pistils pollinated by pollen from diseased plants. Crowley (1957) worked with five viruses; bean mosaic virus, common bean mosaic virus, tomato spotted virus, cucumber mosaic virus and tobacco mosaic virus. Common bean mosaic virus was present in 83 % of the embryos of bean seeds as well as 21 % of the testas. Of the other viruses examined, none was present in the embryos and most were confined to the testas of their hosts. Crowley concluded that in order to be seed transmitted a virus must invade its host systemically, be able to infect and survive in the haploid gametophytic cells and to survive in the embryo throughout its development, maturation, storage and germination. Athow & Bancroft (1959) found that seed transmission of tobacco ringspot virus, the cause of bud blight of soybean, depends on the time of infection of seed-bearing plants. In this study, the virus was associated with the embryonic tissue but not with the seed coat. Athow & Laviolette (1962) obtained 93% transmission of the virus in soybean seeds grown from plants infected with tobacco ringspot virus and noted that the same percentage of infected seeds occurred in one-, two- and three-seeded pods. No multiple infected pods contained more than one non-infected seed. Infection was not associated with apical, central or basal position in the seed pod. The authors postulated that the virus concentration may not be sufficient in the nucellar tissue and other tissues over the entire plant to achieve 100% infection, or the virus may not survive in all the embryos throughout the life of the seed.

Most authors now agree that in the majority of virus diseases that are seed transmitted, the virus must be able to invade the host systemically and establish a compatible relationship with the embryo of the seed so that both host and virus can perpetuate themselves.

The effect of environment on seed infection

Excessive rainfall and high humidity have been shown to influence the degree of infection in many seed-borne diseases. The fungus *Colletotrichum gossypii* caused heavy losses to cotton-seed production under certain conditions (Arndt, 1956). The degree of infestation was directly related to the amount and frequency of rainfall at the time the boll opened. Rainfall also affected seed viability. Mead & Cormack (1961) conducted a survey of alfalfa seed grown in Canada and infected with *Ascochyta imperfecta*; the heaviest infestations occurred on seed from Manitoba, where heavy rains fell at harvest time and the crop remained in the field over winter.

The increasing appreciation of the factors producing diseased seed, especially high humidity and high rainfall, has resulted in a shift of seed production to dry areas such as Oregon, California, and Idaho. Formerly, much of the bean and pea seed used for planting purposes was grown in New York and the New England states, where anthracnose and bacterial blights on beans, and *Ascochyta* species on peas, caused extensive losses. A similar situation was present in Ontario where pea seed crops suffered heavy losses from *Ascochyta* species. Seed for the pea-canning industry is now grown in the arid regions of western United States, but growers of field peas for the pea soup industry in Ontario are still plagued with a high disease incidence, because of using locally grown seed for planting purposes.

The time of storage of grain has increased to five years in the United States and Canada, and the control of environmental conditions plays an important role in maintaining wheat, barley, corn and other seeds free from infection by moulds. A group at Minnesota has studied the influence of various factors on the deterioration of grains in storage (Christensen, 1957). Moisture content, temperature and time of storage affect the growth of moulds in stored grain. In general, the higher the moisture and temperature (within the limits of the growth of the fungi involved) the shorter was the permissible storage time. The oxygen–carbon dioxide ratio was also critical. As the oxygen concentration was lowered, mould growth, germ damage, fat acidity and respiration rate decreased. At high levels of carbon dioxide, seed viability remained high and fungal activity was reduced.

EFFECTS OF THE PATHOGEN ON THE HOST

Noble (1957), in her review of the transmission of plant pathogens by seed, stated that the relationship of seed-borne pathogen and host may extend from near symbiosis to rapid killing of the seed. She further divided seed-borne pathogens into three groups: pathogens which do not injure the seed; pathogens which injure the seed to a certain extent; and pathogens which kill the seed.

A symbiotic relationship during early infection and seed dormancy exists between the fungus *Diaporthe phaseolorum* and soybean seed (Wallen & Seaman, 1963). At germination the fungus becomes virulent and in most cases kills the seed. Wallen & Cuddy (1960) showed that 32 % of the 1959 soybean crop was infected with this fungus and that germination was reduced to 64 %. The high proportion of infected seeds was due to the fact that weather conditions in August and early September were unusual in 1959; a higher than normal relative humidity, mean dew-point and temperature prevailed. As the fungus is primarily a facultative saprophyte, it attacks maturing and senescent plants. The fungus may attack only stems and petioles if the crop matures late in the growing season; however, if the crop matures early under favourable environmental conditions, as it did in 1959, the pods and seeds may also be attacked. In 1960, the soybean crop, which was grown under more normal environmental conditions contained only slightly more than 1 % infected seeds whose germination was over 92 %, yet it originated to a great extent from the highly infected seed stocks of the year before.

Obviously, not many infected seeds had germinated and further infection of the subsequent crop had been reduced by normal weather conditions. Although this fungus would establish a near symbiotic relationship with the seed, the fungus died out with age and germination was restored to the level of a healthy seed sample. The germinability of many infected seeds can be restored after a period of storage as a result of loss of viability of the fungus; this indicates an inability of the fungus, which flourishes during germination, to establish itself firmly during early infection and seed dormancy. It is possible that during early infection of the seed, certain specific nutrients are not available to the fungus but later, if the fungus does not die during seed dormancy, it flourishes with the onset of germination when these nutrients become available. Tiffany (1951) found that seed-borne inoculum of *Colletotrichum truncatum* was responsible for three types of infection on soybean: pre-emergence killing, seedling blight and symptomless establishment of internal mycelium throughout the whole plant and seed. Mycelium from infected cotyledons became established in the cortical cells of the stem without apparently affecting them; it remained localized in this stem area until flowering. The mycelium resumed growth and penetrated the lower stem, petioles, leaves, the developing seeds and pods without the development of symptoms. Under favourable environmental conditions, the fungus advanced and sporulated abundantly on stems and pods.

There are many other examples of fungi remaining quiescent during seed dormancy and growing during germination to infect and possibly kill the seedling. Deutschmann (1953) found that the fungus *Cercospora kikuchii* became localized and dormant in the seed coat of soybean. When germination began the fungus became virulent and migrated to other tissues of the young seedlings. Garofalo (1955) observed that a species of *Fusarium* infected the seed coats of egg-plant seeds and at germination the fungus became active and killed the seed. Sato & Shoja (1954) found that a species of *Colletotrichum* and a species of *Fusarium* was present on the surface and in the dormant tissues of black locust seed at germination. These fungi became active, killed many seeds and produced diseased seedlings. Sackston & Martens (1959) observed that microsclerotia of *Verticillium albo-atrum* were confined to the hull and testa of sunflower seeds but caused a 50 % loss to germinating seedlings. Dekker (1957) observed the same phenomenon with *Ascochyta pisi*, but rapid germination of the seed at high temperatures allowed a high proportion of the seedlings to escape infection. In 80 % of these seeds the fungus had penetrated deeper than the seed coat and

40 % of the embryos were also infected. In studies with this organism, we found that approximately 25 % of infected seeds produced diseased seedlings, a high proportion of the embryo infected seeds were killed and healthy seedlings were produced by infected seeds only when the infection was on, or just beneath, the seed coat.

Some fungi kill the seed before germination. Wallace (1959) isolated the fungus *Podosporiella verticillata* in trace amounts from common and durum wheats. Seeds infected with this organism did not germinate. Wallace suspected that a toxin was produced that killed the seed. Mathur (1961) noted a loss in germination and a slight retardation in emergence of winter wheat infected with *Ustilago tritici*. Wells & Platt (1949) showed that loose smut infection in barley influenced germination and the establishment of barley plants but varieties differed in their susceptibility. Blattny & Osvald (1954) found that a seed-transmitted hop virus could cause extensive reduction of germination. A mild form of the virus produced a 20 % reduction in germination while a virulent strain reduced the germination by as much as 90 %.

Seed-borne pathogens may cause seeds to develop abnormally. They may even replace the principal tissues of the seed as in bunt and ergot infections of cereals, but these malformations are then not true seeds and will not be discussed here. Kernkamp & Hemerick (1952) found that the fungus *Ascochyta imperfecta* caused alfalfa seeds to be light, dark, and shrivelled with the production of spotted, distorted and necrotic seedlings. Seedling losses were proportional to the severity of foliar infection. *Alternaria brassicae* on rape caused a high percentage of shrunken seeds (McDonald, 1959). Sackston (1950) observed that flax seeds infected with *Septoria linicola*, the cause of pasmo disease, were smaller and lighter in weight than healthy seeds. Losses in yield were attributed to reduction in size of the individual seeds. McFadden, Kaufmann, Russell & Tyner (1960) found that a high percentage of small barley seeds were infected with *Ustilago nuda*. A virus that causes a ringspot of tobacco does not cause any floral modifications or capsule malformation, but the seeds are smaller and lighter (Marcelli, 1955).

Certain species of fungi produce physiological changes in seed. Species of *Penicillium* and *Mucor* may influence the quality of rape seed oil. Low temperature and humidity in seed storage are important in keeping these organisms at a low level (Czyzewska, 1958). Nova-kovoskaya (1959) found that the growth of the fungus *Fusarium sporotrichiella* on rye and wheat decreased their content of alcohol soluble proteins and certain amino acids. When farm animals were fed grain parasitized by this fungus, symptoms of endemic osteoarthritis deformans or Kaschin-Beck's disease were produced. Goodman (1951) inoculated ground corn with *P. solitum*, *Aspergillus flavus* and *A. amstelo-dami*, which can use for growth the constituents of corn oil as a sole carbon source. The fat acidity of corn containing oil rose initially and then decreased, whereas the fat acidity remained low in corn having no oil.

Seed discoloration influences seed quality to a great extent. Bean seed infected in the field with *Colletotrichum lindemuthianum* turns brown. When infected seed is soaked in preparation for canning, the seeds turn black: electric-eye machines are used now to separate anthracnose-infected seed from healthy seed. The yellowed varnished appearance of bean seed infected with bacterial blight make them unacceptable by the canning industry. Darkening of the pericarps of wheat seeds may be caused by various fungi and bacteria in the field before or during ripening (Hagborg, 1936; Hanson & Christensen, 1953; Machacek & Greaney, 1938). Blackening of the germ of wheat caused by storage fungi and termed 'sick' wheat has influenced greatly the grading of wheat and other cereals (Christensen, 1957).

In this section, the effects on germinability, size, weight and colour of seed, and certain physiological disorders caused by the growth of many fungi under various environmental conditions have been described. These factors affect the future performance of any seed sample with regard to quality and yield.

DIFFERENTIAL SENSITIVITY OF HOST AND PATHOGEN

Heat therapy has been used to inactivate a number of seed-borne diseases; the fundamental principle is that the infecting organism must be killed by the heat treatment without significant injury to the host seed. Injury usually means loss of germinability. Early workers concentrated on eradicating the pathogen rather than preserving the seed properties, for example the disinfestation of crucifer seeds infected with bacteria causing black rot (Walker, 1923), and cereals containing fungi causing smuts (Jensen, 1888).

Dry heat treatment is effective for eradicating certain pathogens (Lehman, 1925; Borisenko, 1955) but not widely used. Di-electric heating by passing an electric current through the seed to be treated has been employed experimentally; however, serious losses in germin-ability accompanied any satisfactory degree of control (Grainger & Simpson, 1950).

The standard treatment for the control of many seed-borne diseases is treatment with hot water. This was introduced by Jensen (1888) and in the past 15 years there has been continued interest in devising the optimum combinations of temperature and time of treatments; also treatment under anaerobic conditions and with water containing various chemicals have been used. Most studies have involved the pathogen in association with the host rather than studies of the pathogen *in vitro*. Few studies *in vitro* have been made because this work, particularly with fungi and bacteria, has dealt with control of seed-borne organisms *in vivo* without attempting to ascertain information about the organisms *in vitro*. Studies such as thermal death-points, although giving an indication of the thermal stability of pathogen and host proteins, show little relationship to those appertaining to the complex conditions of the pathogen–host combinations. For example, Kassanis (1954) found that cucumber mosaic virus, which has a thermal inactivation point of 70° *in vitro*, was inactivated at 36° *in vivo* when infected cucumber cuttings were held at this temperature for 32 days. Tyner (1953) first demonstrated that barley loose smut could be controlled at room temperature (72–77° F.) if the seed were allowed to soak for 56–64 hr.; or if the seed was pre-soaked for 6, 8 or 10 hr. followed by immersion for 40, 44 or 48 hr. in a suspension of 0·2 % Spergon (tetrachloropara-benzoquinone); or by soaking the seed in a suspension of 0·2 % Spergon for 48 hr. with no pre-soak period. He considered (Tyner, 1957) that the efficacy of the simple soaking in water was possibly due to quinones being formed in the germ of barley during fermentation and these inactivating the fungus mycelium. This work attracted much attention and soon other methods of controlling infection by soaking followed, with other explanations for their efficacy. Wallen & Skolko (1953) showed that a fungistatic principle that controlled *Ascochyta pisi* in naturally infected pea seed and present after seeds were soaked 18 hr. in water and dried, could be inactivated by chlorine. Similarly, soaked seed, but without the chlorine after-treatment, exhibited low amounts of *A. pisi*. Known antibiotics were similarly inactivated by chlorine, and Wallen (1954) showed that bacteria isolated from the soaking process were fungistatic to *A. pisi*. Hebert (1955) inactivated *Ustilago nuda* by anaerobic conditions after first soaking the infected barley seed for periods up to 6 hr. in water. The soaked seed was removed, drained, placed in rubber-stoppered test tubes and stored at various tempera-tures for periods up to 60 hr. He found that ability to produce smut was controlled by a pre-soak for 2 hr. followed by anaerobic storage for 22 hr. at 32°; at lower temperatures a longer period of anaerobic storage was needed. Hebert (1956) later showed that agar slants of *U. nuda*, grown at various temperatures and kept in an O_2 deficient atmosphere, were non-viable in a short time. He found a close correlation between the length of time required for control of the fungus *in vivo* and the duration of viability of the pathogen grown *in vitro* in the absence of O_2. To control *Helminthosporium sativum* and *H. victoriae* in barley and oats, Jacquet, Leben & Arny (1957) soaked the seeds in a synthetic mixture of the approximate amounts of organic acids found in the filtrates from water-soaked seed. The authors concluded that organic acids were important factors in the control of the diseases. Russell & Chinn (1958) controlled *U. nuda* in barley seed by steeping the infected barley in a 1 % NaCl solution and the advantage of this method was that germination was not seriously impaired. Ivanoff (1958) investigated the use of water-soak methods for the control of seed rots of unshelled peanuts due to several species of fungi, and of oat seedling blight caused by *H. sativum*. During the soaking of peanuts diffusion resulted in a significant loss of sugars and other soluble substances. Ivanoff concluded that starvation and limited oxygen supply were the main factors contributing to inactivation of the fungi.

Many theories have been advanced to explain the inactivation of fungi in seed during the water soak treatment. Unfortunately, no theory has been verified because much of the work has been devoted to practical considerations of control with fewer studies aimed directly at determining the principles responsible for such control. There is therefore a fertile field of research *in vivo* and *in vitro*, to see whether the inhibition is by microbial products, by toxic agents in the steep water, by depletion of oxygen, by nutrient improverishment or by a combination of these factors.

LONGEVITY OF THE PATHOGEN AND HOST

Longevity studies on pathogen and host have been prompted by the need to find suitable means of eliminating pathogenic organisms from seed. Successful control is dependent upon the ability of the seed to retain its germinability until the pathogen has died completely or has been reduced to a low level. An indication of the longevity of the host can be obtained from a review by Crocker (1945) who listed the life span of many seeds. Most tree seeds live only a few days or months and must therefore quickly find a suitable environment for germination. Certain vegetable crop seeds may survive 10 years and many of the cereal grains 2 to 4 years. Hard coated seeds live the longest; these seeds are mostly legumes and are long lived because of the impermeability of the seed coat which prevents any exchange of gases or water. Three factors increase the longevity of seeds in storage: low and constant moisture content; low temperature; and the absence of oxygen.

Longevity studies of pathogenic organisms in seeds have dealt with maintenance of viability and not, unfortunately in most cases, with ascertaining if the surviving organisms were still in a virulent state. Studies on fungal pathogens predominated in the early work in this field because of the relative ease of their identification in seed and because most of the early research in plant pathology involved mycological studies. Studies on seed-borne bacteria and viruses involve more complicated methods of detection; in addition to isolating the pathogen from seed, it must be shown to cause characteristic symptoms in a host plant.

Early research was conducted on celery seed infected with the blight organisms *Septoria apii* and *S. apii-graveolentis*. The pathogens occur in the seed and initiate infection of seedlings in the seed bed. Krout (1921) found that it was relatively easy to kill the spores of the fungus on the surface of the seed by seed disinfection. Many seeds, however, harbour the pathogen in pycnidia embedded in the pericarp. Any attempt to kill the fungus in this state by chemical means affected the germinability of the seed. Krout studied the effect of ageing on the viability of the spores and mycelium. The mycelium of the fungus died before the spores. He found that spores isolated from the surface of seed stored in the laboratory were dead after 11 months, but that spores isolated from the pericarp lived as long as 2 years. He also noted that the germ tubes from spores 2 years old were slow in developing, and that spores in the pericarp region of the seed for 1 to 2 years lacked the vigour to produce germ tubes capable of producing an infection. Celery seed germinated well up to 4 years, but after 5 years germination was so reduced that it was advisable not to use the seed. This early work on longevity still serves as a basis for the control of these organisms in celery seed.

The fungus *Ascochyta pisi* occurs in seed of certain *Vicia* species. Sprague (1929) isolated *A. pisi* from 9-year-old seed of *Vicia faba* L., and Crosier (1939) studied the longevity of the fungus in seeds of the hairy vetch, *V. villosa* Roth. Crosier found that most fungal infections died out in 3 years and only rarely did any infections survive 7 years; many seed samples retained their germinability at a high level after the fungus had died out completely, and most samples showed no decline during the first 4 years. Wallen (1955) confirmed Crosier's work using *Pisum sativum* L. as the host plant. Seed infections by *A. pisi* were eliminated after storage for 7 years without any significant loss in

germination or lack of emergence in pea seedlings. Most other saprophytic fungi associated with the seed had also died out during 7 years of
storage in a cool basement. *In vitro, A. pisi* can survive long periods; a
culture isolated by the author in 1949 and maintained on pea agar at
10° with occasional transference, is still virulent and in use in studies on
plant resistance.

Longevity and virulence studies, *in vitro* and *in vivo*, have been
conducted on a number of cereal seed-borne pathogens. Porter (1955)
found that barley seed infected with the fungus *Ustilago nuda* and first
planted and harvested in 1942 was still viable in 1953 and produced a
number of smutted heads. Although the fungus was still pathogenic
after 11 years in the seed, and spores could form a promycelium, no
sporidia were found. Hence the new generation of spores could not
perpetuate the fungus. Russell (1961) found that certain samples of
barley maintained *U. nuda* at the same level of infection in a virulent
state for 9 years before some of the infections died out; other samples
harboured the pathogen for only 6 years and in each succeeding year
the percentage of loose smut declined until all infections were eliminated. The average number of seeds, which on germination formed plants
with loose smut at the start of the study, was 14%; 11 years later 3·2 %
of the seeds were still infected. One reason for the decline in the apparent
percentage of infected seed was that successively larger proportions of
the seeds failed to germinate. It would appear that the pathogen,
although in a dormant state for an extended period, was still highly
virulent and killed the host as the latter was weakened by ageing.
Although the pathogen successfully killed the host, the normal procedure whereby the pathogen reproduces itself in the form of loose smut
in the maturing crop was not possible, and both pathogen and host had
failed to survive.

Tapke (1953) compared the longevity and virulence of *Ustilago nuda*
in vitro and *in vivo* by exposing chlamydospores of the fungus and
infected seed to different temperatures. Chlamydospores were non-
viable after a few months when stored *in vitro* at room temperature but
storage at 28–32° F. prolonged viability: 14-year-old spores stored at
28–32° F. were able to infect flowers and produce smutted heads.
However, *in vivo* the fungus is less durable; infected seed stored at these
temperatures produced plants with smutted heads only up to 7 years.
Grasso (1957) maintained *in vitro* monosporidial cultures of *U. avenae*
and *U. levis* on potato dextrose agar. Viability and pathogenicity were
retained for 36 months, although the cultures were maintained at room
temperature without transfer to new media.

Machacek & Wallace (1952) found that *Helminthosporium avenae* in oats and *H. teres* in barley lost their viability slowly and could still be recovered from the seed after 10 years. *Alternaria tenuis, H. sativum* and *Septoria nodorum* died out rapidly in wheat seed. After 10 years' storage germination of the seed varied in the different samples from 18 to 85 % for wheat, 88 to 96 % for oats, and 62 to 98 % for barley. Hence only certain pathogens are eliminated before the death of the host. Although Machacek & Wallace found that *H. sativum* was short-lived in wheat seed, Christensen (1922) showed it to survive in barley seed for 11 years. Leukel, Dickson & Johnson (1933) and Shands (1937) found that *H. gramineum* could live in barley seed for 5 and 10 years, respectively. Cappellini & Haenseler (1958) studied 32 species of fungi pathogenic for cereals after storage *in vitro* under sterile mineral oil. The most significant change observed after 7 years was a loss in sporulation in 3 species. None of the storage cultures were tested for survival of pathogenicity.

Dungan & Koehler (1944) investigated the decline in viability of 5 samples of maize seed infected with different fungi. An appreciable percentage of the seed was still viable after all 5 fungi had died out. *Gibberella zeae* died out completely in 2 years and most seeds were free of the pathogen after 15 months. A small number of seeds still harboured the pathogen *Fusarium moniliforme* after 8 years and *Cephalosporium acremonium* died out after 7 years. Infection by *Diplodia zeae* and *Nigrospora oryzae* was reduced about 50 % when seed was held for 1 year. Other workers have stated that maize seed will store for long periods at a low relative humidity. Robertson, Lute & Kroeger (1943) stated that seed corn stored in an unheated room at a low relative humidity retained its viability for long periods with many seeds still viable after 21 years. Kiesselbach (1937) found that properly stored corn gave satisfactory yields in the field after storage for 4 years. They stated that humidity of the atmosphere in which the seed is stored seems to play an important role in the retention of viability (Dungan & Koehler, 1944).

Cormack (1961) studied the longevity of the organism *Corynebacterium insidiosum*, the cause of bacterial wilt of alfalfa. Infected seed was stored in the laboratory at 70–80° F. and under these conditions the pathogen was viable for about 3 years. Bacteria isolated from 3-year-old seed were still pathogenic to alfalfa roots. Alfalfa straw was a better medium for survival of the bacteria than seed; the organism was still viable after infected hay had been stored for 10 years. Studies with *C. insidiosum* were carried out *in vitro* by Bordewick (1960). He maintained two cultures of differing virulence, on various media, under oil and without oil, and at 24°, 5° and $-30°$. The virulence of the cultures was measured by the severity of the symptoms that they produced on inoculated alfalfa plants. The more virulent organisms remained pathogenic 3–6 months longer than the less virulent but all virulence disappeared after 20 months storage under optimum conditions. Five degrees was the best temperature for the maintenance of virulence and the organisms remained virulent longer under oil.

The longevity of various pathogens on soybean has been studied. The fungus *Diaporthe phaseolorum* (Wallen & Seaman, 1963) is present in Canadian soybean seed to some extent in each year. Although longevity studies have only been carried on for 3 years, soybean seed, originally 50 % infected with internal *D. phaseolorum*, now contain less than 10 % of infected seeds. Kent (1945) and Welch (1946) found *Pseudomonas glycinea*, which causes bacterial blight of soybeans, could be isolated from seed after 16 months; this would allow the pathogen to survive the winter. The bacterial pustule disease of soybean is caused by *Xanthomonas phaseoli* var. *sojensis* which could be isolated from seed at least 17 months old. Graham (1953) studied the longevity and virulence of these two pathogens and *P. tabaci* in soybean. Soybean seed was harvested from severely diseased plants and sown in soil after storage of from 6–30 months. As indicated by the appearance of primary infections on young plants, *X. phaseoli* var. *sojensis* was alive in or on seed after 30 months of storage, whereas *P. tabaci* and *P. glycinea* was evident only on seed that had been stored for 6 months.

Miscellaneous studies on the longevity of various fungi *in vivo* and *in vitro* are as follows. Wenzl (1959) studied the fungus *Cercospora beticola* in beet seed; the fungus gradually lost its ability to produce conidia but infectiousness was not correlated with the production of conidia. The fungus could survive for about $2\frac{1}{2}$ years in the beet seed. Lloyd (1959) found that *Phoma lingam* was present in small amounts in New Zealand swede, turnip, rape and certain other field crops. It was transmitted primarily in the small seeds and about 50 % of the infected seeds were killed by the fungus. Although the organism could survive for 14 months in the seed, some infections regressed after 8 months of storage. Neergaard (1961) found that during storage of rice under unspecified laboratory conditions, the number of seeds infected with a species of *Epicoccum* decreased over a 6-month period from 24 to less than 4 %, while those containing *Alternaria tenuis* remained at the same level during this time and those having *Septoria nodorum* decreased from 26·5 to 10 % of the seed. Barnes (1962) noted that of 7 fungi tested on agar and under mineral oil for their longevity *in vitro*, *Fusicladium carpophilum*, *F. effusum*, *Monilinia fructicola*, *Trichothecium roseum* and *Venturia inaequalis* survived a 5-year period. Phillips (1956) observed that pycnospores of the fungus *Didymella lycopersici* did not survive more than 9 months on the surface of artificially inoculated tomato seed. Essentially the same result was obtained in the field; the fungus was found on 30 % of samples of fresh seed but was not detected on the same seed a year later.

In most of the studies described above, the pathogen died before the host. Colhoun & Muskett (1948) described the reverse phenomenon in the case of flax seed and certain pathogenic seed-borne fungi. They studied 4 organisms, *Colletotrichum linicola*, *Polyspora lini*, *Botrytis cinerea* and a species of *Phoma*. Flax seed stored in a basement where the temperature was not constant retained its germinability for 18–24 months. The 4 organisms outlived the host under these normal storage conditions and could survive more than 4 years in the seed, well beyond the longevity of the seed; other organisms were non-viable after 16 months.

Records of virus longevity in seed appear in the more recent literature. Bennett & Costa (1961) found that the virus causing sowbane mosaic could be transmitted in seeds of *Chenopodium murale* that had been stored for 6 months and $1\frac{1}{2}$ years. McNeal, Milts & Berg (1961) found that the percentage of infected seedlings produced from wheat seed infected with barley stripe mosaic virus did not vary in some samples whether the seed was 1 or 3 years old, indicating no loss of viability or virulence during this time. Germination of the seed was not markedly affected by storage for 3 years. Alexander (1960) found that although tobacco mosaic virus declined rapidly on tomato seeds after storage for 1 year, some virus could be recovered after 3 years. Blattny & Osvald (1954) found that a hop virus could survive for 2 years in the seed.

In the future, tests to determine longevity and maintenance of virulence of a specific organism in seed should be done under the optimum conditions for seed storage. Such factors as temperature and humidity should be regulated throughout the experiment and should be well defined in any publication of results. In the past, contradictory results have been obtained, and, without question, the differing environmental conditions were responsible for the varying results. Another factor which must be considered is the possibility of selecting mutants of the pathogen during seed storage. Burkholder (1948), in his review of bacteria as plant pathogens, wrote: 'many bacterial species retain their virulence in culture for years while others lose it gradually or rapidly.

Some species mutate into strongly virulent or weakly virulent types.'
The selection of mutants introduces the possibility that the initial
fungus, bacterium or virus in culture or in seed may not be the same
strain that is isolated later after differential survival has occurred.

CONCLUSIONS AND GENERAL OBSERVATIONS

Certain factors influencing seed infection and host–parasite relationships in seed-borne diseases have been reviewed in this paper. Much of
the research discussed in the four sections is a by-product of applied
research on control of seed-borne diseases. It is obvious that fundamental research in this field has been lacking and that our understanding
is fragmentary both of the normal interaction of host and pathogenic
organisms and of how it is influenced by conditions used to control
seed-borne disease. The theme of this Symposium indicates the lines
along which future basic research could proceed. There appears a need
for the increased study of more seed-borne pathogenic organisms
in vitro, for comparison with their general behaviour *in vivo* which is
known from the practical studies. The morphology of many pathogenic
organisms *in vitro* may be different from that *in vivo* (cf. Dr Mariat's
paper in this Symposium) and some pathogenic organisms may be more
easily identified *in vitro* than *in vivo*. The effect of extracts from seed
tissues on the growth of appropriate pathogenic organisms could be
examined *in vitro* and conversely the effect on uninfected seeds of
products from growth *in vitro* could be studied (cf. the papers by
Dr E. W. Buxton and Dr B. J. Deverall in this Symposium). With a
view to designing better control measures, the effects of storage time,
temperature and fungicides on the isolated pathogenic organisms, the
uninfected seed and the infected seed could be compared. Areas of
study which might benefit from research using techniques *in vivo* and
in vitro are as follows.

An interesting field of research lies in the interaction of the pathogenic
organisms with the germinating seed. As indicated earlier, many cases
have been observed where the organism and the seed are in a near
symbiotic state during seed dormancy. With the breaking of seed dormancy and initiation of germination, the organism becomes active and
in many cases kills the young seed or causes serious infection of the
developing seedling. The processes involved in this relationship have
not been explored to any extent. The physiologist and biochemist could
add considerably to a more complete understanding of this complex
problem. It is possible that, in the process of germination, nutrients are

set free which stimulate the growth of the pathogenic organisms; this could be investigated by experiments *in vivo* and *in vitro*. Also, tissue culture of plant cells may have some use in these studies.

Another problem corresponding to tissue specificity of animal pathogenic organisms is why some pathogenic fungi such as *Ascochyta pisi* or *Ustilago nuda* attack the seed, whereas others such as *Pythium* or *Phytophthora* are primarily root and foliage parasites. Similarly, why do some seed-borne organisms prefer the vascular route to the seed, whereas others invade through the pod, flowers or other parts of the fruit to become deep-seated in the seed? Again, why are certain smut fungi carried simply on the surface of the seed and attack primarily the young seedling?

The nature of the microbial products which are formed *in vivo* and which either kill the seed or produce other undesirable effects is an investigation for future study. Also, the mechanisms of the various examples of seed disinfection, by soaking them in water under various conditions, need elucidating. Perhaps the most useful research where rapid advances can be made is that dealing with methods for identifying seed-borne organisms *in vitro*. The mycologists have developed many techniques for identifying seed-borne fungi, but little has been done for the causative organisms of bacterial and virus disease in seed. At the moment, the only safeguard against the propagation of seed-borne viruses is the inspection of the crop in the field before harvesting. Much could be done to extend to seed-borne pathogenic organisms the methods of identification already used for organisms infecting animals, for example culture of bacteria on selective media followed by serological or phage typing, and plant or tissue culture of viruses followed by serological typing.

Finally methods are needed to assess the effect in the field, either on growing crops or on seed storage, of different levels of seed infection. At the moment almost arbitrary rules are used to make specifications on maximum disease tolerance. Research programmes should be designed for the development of methods for the detection, identification and evaluation of the potential destructiveness of seed-borne disease. The host, the pathogenic organisms, the environment and the interplay of these factors must therefore receive proper consideration.

In all, there is plenty of scope in elucidating the mechanisms of seed-borne infection, for fundamental and complementary experiments *in vivo* and *in vitro*.

Contribution No. 324 from the Plant Research Institute, Canada Department of Agriculture, Ottawa, Canada.

REFERENCES

ALEXANDER, L. J. (1960). Inactivation of tobacco mosaic virus from tomato seed. *Phytopathology*, **50**, 627.

ALLARD, H. A. (1915). Distribution of the virus of the mosaic disease in capsules, filaments, anthers, pistils of tobacco plants. *J. agric. Res.* **5**, 151.

ARNDT, C. H. (1956). Cotton seed produced in South Carolina in 1954 and 1955, its viability and infestation by fungi. *Plant Dis. Reptr*, **40**, 1001.

ATHOW, K. L. & BANCROFT, J. B. (1959). Development and transmission of tobacco ringspot virus in soybean. *Phytopathology*, **49**, 697.

ATHOW, K. L. & LAVIOLETTE, F. A. (1962). Relation of seed position and pod location to tobacco ringspot virus seed transmission in soybean. *Phytopathology*, **52**, 714.

ATKINS, I. M., MERKLE, O. G., PORTER, K. B., LAHR, K. A. & WEIBEL, D. E. (1963). The influence of environment on loose smut percentages, re-infection and grain yields of winter wheat at four locations in Texas. *Plant Dis. Reptr*, **47**, 192.

BARNES, G. L. (1962). A new survival record for certain fungi on agar under a mineral oil seal. *Plant Dis. Reptr*, **46**, 192.

BATTS, C. C. V. (1955). Observations on the infection of wheat by loose smut (*Ustilago tritici* (Pers.) Rostr.). *Trans. Brit. mycol. Soc.* **38**, 465.

BATTS, C. C. V. & JEATER, A. (1958a). The reaction of wheat varieties to loose smut as determined by embryo, seedling, and adult plant tests. *Ann. appl. Biol.* **46**, 23.

BATTS, C. C. V. & JEATER, A. (1958b). The development of loose smut (*Ustilago tritici*) in susceptible varieties of wheat, and some observations on field infection. *Trans. Brit. mycol. Soc.* **41**, 115.

BENNETT, C. W. & COSTA, A. S. (1961). Sowbane mosaic caused by a seed-transmitted virus. *Phytopathology*, **51**, 546.

BENNETT, C. W. & ESAU, K. (1936). Further studies on the relation of the curly top virus to plant tissues. *J. agric. Res.* **53**, 595.

BLATTNY, C. & OSVALD, V. (1954). Prenos viros chmele (*Humulus lupulus* L.) na potomstvo semenem. *Preslia*, **26**, 1.

BORDEWICK, B. E. (1960). Studies on maintenance of virulence of *Corynebacterium insidiosum* (McCull.) H. L. Jens. in culture and the inheritance of resistance to *C. insidiosum* in diploid *Medicago falcata* L. *Dissertation Abstr.* **21**, 1016.

BORISENKO, S. I. (1955). Borba a pylnoy golovney na semenovodcheskikh posevakh. *Zemledyelie Sofiya*, 1955, p. 114.

BURKHOLDER, W. H. (1930). The bacterial diseases of the bean, a comparative study. *Mem. Cornell agric. Exp. Sta.* No. 127.

BURKHOLDER, W. H. (1948). Bacteria as plant pathogens. *Annu. Rev. Microbiol.* **2**, 389.

BURKHOLDER, W. H. & BULLARD, F. T. (1946). Varietal susceptibility of beans to *Xanthomonas phaseoli* var. *fuscans*. *Plant Dis. Reptr*, **30**, 446.

CALDWELL, J. (1952). Some effects of plant viruses on nuclear divisions. *Ann. appl. Biol.* **39**, 98.

CAMPBELL, W. P. & TYNER, L. E. (1959). Comparison of degree and duration of susceptibility of barley to ergot and true loose smut. *Phytopathology*, **49**, 348.

CAPPELLINI, R. A. & HAENSELER, C. M. (1958). Longevity of some graminicolous species of *Helminthosporium* under mineral oil. *Phytopathology*, **48**, 695.

CHRISTENSEN, C. M. (1957). Deterioration of stored grains by fungi. *Bot. Rev.* **23**, 108.

CHRISTENSEN, J. J. (1922). Studies on the parasitism of *Helminthosporium sativum*. *Tech. Bull. Minn. agric. Exp. Sta.* No. 11.

COLHOUN, J. & MUSKETT, A. E. (1948). A study of the longevity of the seed-borne parasites of flax in relation to the storage of the seed. *Ann. appl. Biol.* **35**, 429.

CORMACK, M. W. (1961). Longevity of the bacterial wilt organism in alfalfa hay, pod debris, and seed. *Phytopathology*, **51**, 260.

CROCKER, W. (1945). Longevity of seeds. *J. N.Y. bot. Gdn*, **46**, 25.

CROSIER, W. (1939). Occurrence and longevity of *Ascochyta pisi* in seeds of hairy vetch. *J. agric. Res.* **59**, 683.

CROWLEY, N. C. (1957). Studies on the seed transmission of plant virus diseases. *Aust. J. biol. Sci.* **10**, 449.

CZYZEWSKA, S. (1958). Badania fitopatologiczno-mykologicsne nasion rzepaku (*Brassica napus* L. var. *oleifera* D.C.). *Boczn Nauk rol. Ser. A*, **78**, 283.

DEKKER, J. (1957). Inwendige ontsmetting van door *Ascochyta pisi* aangetaste erwtezaden met de antibiotica rimocidine en pimaricine, benevens enkele aspecten van het parasitisme van deze schimmel. *Tijdschr. PlZiekt.* **63**, 65.

DEUTSCHMANN, F. (1953). Über die 'Purple Stain' Krankheit der Sojabohne und die Farbstoffbildung ihres Erregers *Cercospora kikuchii* Mats. et Tom. *Phytopath. Z.* **20**, 297.

DUGGAR, B. M. (1930). The problem of seed transmission of the typical mosaic of tobacco. *Phytopathology*, **20**, 133.

DUNGAN, G. H. & KOEHLER, B. (1944). Age of seed corn in relation to seed infection and yielding capacity. *J. Amer. Soc. Agron.* **36**, 436.

DYKSTRA, T. F. (1961). Production of disease-free seed. *Bot. Rev.* **27**, 445.

GAROFALO, F. (1955). Osservazioni scie semi di melanzona infetti de Fusarium. *Nuovo G. bot. ital.* **62**, 545.

GOLD, A. H., SUNESON, C. A., HOUSTON, B. R. & OSWALD, J. W. (1954). Electron microscopy and seed and pollen transmission of rod shaped particles associated with the false stripe disease of barley. *Phytopathology*, **44**, 115.

GOODMAN, J. J. (1951). Lipolytic activity of certain fungi important in grain deterioration. *Phytopathology*, **41**, 14.

GRAHAM, J. H. (1953). Overwintering of three bacterial pathogens of soybean. *Phytopathology*, **43**, 189.

GRAINGER, J. & SIMPSON, D. E. (1950). Electronic heating and control of seed-borne diseases. *Nature, Lond.* **165**, 532.

GRASSO, V. (1957). Un metodo per la conservazione della colture dei carboni dell'avena. *Ric. sci.* **27**, 88.

HAGBORG, W. A. F. (1936). Black chaff, a composite disease. *Canad. J. Res.* C, **14**, 347.

HANSON, E. W. & CHRISTENSEN, J. J. (1953). The black point disease of wheat in the U.S.A. *Tech. Bull. Minn. agric. Exp. Sta.* No. 206.

HEBERT, T. T. (1955). A new method of controlling loose smut of barley. *Plant. Dis. Reptr*, **39**, 20.

HEBERT, T. T. (1956). Mode of action of the wet anaerobic storage treatment for the control of loose smut in barley. *Phytopathology*, **46**, 14.

IVANOFF, S. S. (1958). The water-soak method of plant disease control in relation to microbial activities, oxygen supply, and food availability. *Phytopathology*, **48**, 502.

JACQUET, P., LEBEN, C. & ARNY, D. C. (1957). Organic acids and the control of two seedling blights by the water-soak method. *Phytopathology*, **47**, 377.

JENSEN, J. L. (1888). The propagation and prevention of smut in oats and barley. *J. R. agric. Soc.* (Ser. 2), **24**, 397.

KASSANIS, B. (1954). Heat therapy of virus infected plants. *Ann. appl. Biol.* **41**, 470.

KATZNELSON, H. & SUTTON, M. D. (1951). A rapid phage plaque count method for the detection of bacteria as applied to the demonstration of internal bacterial infection of seed. *J. Bact.* **61**, 689.

KAUSCHE, G. K. (1940). Über eine das Virusprotein inaktivierende Substanz im Samen von *Nicotiana tabacum* var. *Samsun. Biol. Zbl.* **60**, 423.

KENT, G. C. (1945). A study of soybean disease and their control. *Rep. Ia. agric. Exp. Sta.* p. 221.

KERNKAMP, M. F. & HEMERICK, G. A. (1952). Alfalfa seed losses due to *Aschochyta imperfecta* Pk. (Blackstem). *Phytopathology*, **42**, 468.

KIESSELBACH, T. A. (1937). Effects of age, size and source of seed on the corn crop. *Bull. Neb. agric. Exp. Sta.* No. 305, 1.

KLEMENTS, Z. & LOVAS, B. (1960). Biological and morphological characterization of the phage for *Xanthomonas phaseoli* var. *fuscans. Phytopath. Z.* **37**, 321.

KROUT, W. S. (1921). Treatment of celery seed for the control of *Septoria* blight. *J. agric. Res.* **21**, 369.

LANG, W. (1917). Zur Ansteckung der Gerste durch *Ustilago nuda. Ber. dtsch. bot. Ges.* **35**, 4.

LEHMAN, S. G. (1925). Studies on treatment of cotton seed. *Tech. Bull. N.C. agric. Exp. Sta.* No. 26, 1.

LEUKEL, R. W., DICKSON, J. G. & JOHNSON, A. G. (1933). Effect of certain environmental factors on stripe disease of barley, and on the control of the disease by seed treatment. *Tech. Bull. U.S. Dep. Agric.* No. 341.

LLOYD, A. B. (1959). The transmission of *Phoma lingam* (Tode) Desm. in the seeds of swede, turnip, chou, moellier, rape, and kale. *N.Z. J. agric. Res.* **2**, 649.

MACHACEK, J. E. & GREANEY, F. J. (1938). The 'black point' or 'kernel smudge' disease of cereals. *Canad. J. Res.* C, **16**, 84.

MACHACEK, J. E. & WALLACE, H. A. H. (1952). Longevity of some common fungi in cereal seed. *Canad. J. Bot.* **30**, 164.

MANTLE, P. G. (1961). Further observation on an abnormal reaction of wheat to loose smut. *Trans. Brit. mycol. Soc.* **44**, 529.

MARCELLI, E. (1955). Osservazioni su di una nuova virosi del tabacco transmissibile per seme. *Tabacco*, **59**, 404.

MARSHALL, G. M. (1959). The incidence of certain seed-borne diseases in commercial seed samples. 1. Loose smut of barley, *Ustilago nuda* (Jens.) Rostr. *Ann. appl. Biol.* **47**, 232.

MATHUR, S. C. (1961). The effect of *Ustilago tritici* (Pers.) Rostr. on the growth and morphological characters of winter wheat. *Dissertation Abstr.* **22**, 702.

McDONALD, W. C. (1959). Gray leaf spot of rape in Manitoba. *Canad. J. Plant Sci.* **39**, 409.

McFADDEN, A. D., KAUFMANN, M. L., RUSSELL, R. C. & TYNER, L. E. (1960). Association between seed size and the incidence of loose smut in barley. *Canad. J. Plant Sci.* **40**, 611.

McNEAL, F. H., MILTS, I. K. & BERG, M. A. (1961). Variation in barley mosaic virus incidence in wheat seed due to storage and continuous propagation and the effect of the disease on yield and test weight. *Agron. J.* **53**, 128.

MEAD, H. W. & CORMACK, M. W. (1961). Studies on *Ascochyta imperfecta* Peck. Parasitic strains among fifty isolates from Canadian alfalfa seed. *Canad. J. Bot.* **39**, 793.

MUMFORD, D. L. E. & RASMUSSON, D. C. (1963). Resistance of barley to *Ustilago nuda* after embryo infection. *Phytopathology*, **53**, 125.

NEERGAARD, P. (1961). Report of the Committee on Plant Diseases, 1959–62. *I.S.T.A. Proc. Compt. Rend. Mitteil.* **27**, 104.

NOBLE, M. (1951). Seed pathology. *Nature, Lond.* **168**, 534.

NOBLE, M. (1957). The transmission of plant pathogens by seed. In *Biological Aspects of the Transmission of Disease*. Ed. by C. Horton-Smith, p. 81. Edinburgh: Oliver and Boyd.

NOBLE, M., TEMPE, J. DE & NEERGAARD, P. (1958). *An Annotated List of Seed-borne Diseases*. Commonwealth Mycological Institute, Kew.

NOVAKOVOSKAYA, E. S. (1959). Izmeneniya belkov rzhi i pshenitsy pod vliyaniem rosta nekotorykh shtammov griba *Fusarium sporotrichiella* na zerne (k izuchenitv etiologii i patogeneza urovskoi Kashina-Beka bolezni). *Vop. Pitan.* **18**, 54.

OHMS, R. E. & BEVER, W. M. (1956). Effect of time of inoculation of winter wheat with *Ustilago tritici* on the percentage of embryos infected and the abundance of hyphae. *Phytopathology*, **46**, 157.

OORT, A. J. F. (1944). Hypersensitivity of wheat to loose smut. *Tijdschr. PlZiekt.* **50**, 73.

PEDERSEN, P. N. (1956). Infection of barley by loose smut, *Ustilago nuda* (Jens.) Rostr. *Friesia*, **5**, 341.

PERSONS, T. D. (1954). Destructive outbreak of loose smut in a Georgia wheat field. *Plant Dis. Reptr*, **38**, 422.

PHILLIPS, D. H. (1956). Tomato seed transmission of *Didymella lycopersici* Kleb. *Trans. Brit. mycol. Soc.* **39**, 319.

POPP, W. (1958). An improved method of detecting loose smut mycelium in whole embryos of wheat and barley. *Phytopathology*, **48**, 641.

POPP, W. (1959). A new approach to the embryo test for predicting loose smut of wheat in adult plants. *Phytopathology*, **49**, 75.

PORTER, R. H. (1949). Recent developments in seed technology. *Bot. Rev.* **15**, 221.

PORTER, R. H. (1955). Longevity of loose smut *Ustilago nuda* (Jens.) Rostr. in barley seed. *Phytopathology*, **45**, 637.

ROBERTSON, D. W., LUTE, A. M. & KROEGER, H. (1943). Germination of 20-year old wheat, oats, barley, rye, sorghum, and soybeans. *J. Amer. Soc. Agron.* **35**, 786.

RUSSELL, R. C. (1950). The whole-embryo method of testing barley for loose smut as a routine test. *Sci. Agric.* **30**, 361.

RUSSELL, R. C. (1961). The influence of aging of seed on the development of loose smut in barley. *Canad. J. Bot.* **39**, 1741.

RUSSELL, R. C. & CHINN, S. H. F. (1958). The salt-water-soak treatment for the control of loose smut of barley. *Plant Dis. Reptr*, **42**, 618.

RUSSELL, R. C. & POPP, W. (1951). The embryo test as a method of forecasting loose smut infection in barley. *Sci. Agric.* **31**, 559.

SACKSTON, W. E. (1950). Effect of pasmo disease on seed yield and thousand kernel weight of flax. *Canad. J. Bot.* **28**, 493.

SACKSTON, W. E. & MARTENS, J. W. (1959). Dissemination of *Verticillium alboatrum* on seed of sunflower (*Helianthus annuus*). *Canad. J. Bot.* **37**, 759.

SATO, K. & SHOJA, T. (1954). On disinfection and hastening of germination for black locust seeds. *J. Jap. For. Soc.* **34**, 244.

SHANDS, R. G. (1937). Longevity of *Gibberella saubinetii* and other fungi in barley kernels and its relation to the emetic effect. *Phytopathology*, **27**, 749.

SIMMONDS, P. M. (1946). Detection of loose smut fungi in embryos of barley and wheat. *Sci. Agric.* **26**, 51.

SPRAGUE, R. (1929). Host range and life history studies of some leguminous Ascochytae. *Phytopathology*, **19**, 917.

SREERAMULU, T. (1962). Aerial dissemination of barley loose smut (*Ustilago nuda*). *Trans. Brit. mycol. Soc.* **45**, 373.

TAPKE, V. F. (1948). Environment and the cereal smuts. *Bot. Rev.* **14**, 359.

TAPKE, V. F. (1953). Longevity in *Ustilago nuda*. *Phytopathology*, **43**, 407.

Tiffany, L. H. (1951). Delayed sporulation of *Colletotrichum* on soybean. *Phytopathology*, **41**, 975.

Tyner, L. E. (1953). The control of loose smut of barley and wheat by Spergon and by soaking in water at room temperature. *Phytopathology*, **43**, 313.

Tyner, L. E. (1957). Factors influencing the elimination of loose smut from barley by water-soak treatments. *Phytopathology*, **47**, 420.

Vanderwalle, R. (1942). Note sur la biologie d'*Ustilago nuda tritici* Schaf. *Bull. Inst. agron. Gembloux*, **11**, 103.

Walker, J. C. (1923). The hot water treatment of cabbage seed. *Phytopathology*, **13**, 251.

Wallace, H. A. H. (1959). A rare seed-borne disease of wheat cause by *Podosporiella verticillata*. *Canad. J. Bot.* **37**, 509.

Wallen, V. R. (1954). Antibiosis and some internally seed-borne pathogens. Thesis for Ph.D. degree of McGill University, Montreal.

Wallen, V. R. (1955). The effect of storage for several years on the viability of *Ascochyta pisi* in pea seed and on the germination of the seed and emergence. *Plant Dis. Reptr*, **39**, 674.

Wallen, V. R. & Cuddy, T. F. (1960). Relation of seed-borne *Diaporthe phaseolorum* to the germination of soybeans. *Proc. Ass. off. Seed Anal. N. Amer.* **50**, 137.

Wallen, V. R. & Seaman, L. (1963). Seed infection of soybean by *Diaporthe phaseolorum* and its influence on host development. *Canad. J. Bot.* **41**, 13.

Wallen, V. R. & Skolko, A. J. (1953). The inactivation of antifungal antibiotics by chlorine. *Plant Dis. Reptr*, **37**, 421.

Wallen, V. R., Sutton, M. D. & Grainger, P. N. (1963). A high incidence of fuscous blight in Sanilac beans from southwestern Ontario. *Plant Dis. Reptr*, **47**, 652.

Welch, A. W. (1946). A study of soybean diseases and their control. *Rep. Ia. agric. Exp. Sta.* p. 191.

Wells, S. A. & Platt, A. W. (1949). The effect of loose smut on the viability of artificially inoculated barley seeds. *Sci. Agric.* **29**, 45.

Wenzl, H. (1959). Bedeutung und Bekämpfung der Infektion von Rübensaatgut durch *Cercospora beticola* Sacc. *PflSchBer.* **23**, 33.

Zaumeyer, W. J. (1929). Seed infection by *Bacterium phaseoli*. *Phytopathology*, **19**, 96.

THE BEHAVIOUR OF MICROBIAL PARASITES IN RELATION TO PHAGOCYTIC CELLS *IN VITRO* AND *IN VIVO*

G. B. MACKANESS

Department of Microbiology, University of Adelaide, South Australia

INTRODUCTION

For almost a century students of infectious disease have inquired into the origins, physiology and behavioural characteristics of phagocytic cells. Much has been learnt of their role, their potential and their limitations in so far as they affect the relationship between a host and its parasite. Their pre-eminent role in determining both the fate of microbial parasites and the efficiency of the host's reaction against them was recognized. Even viruses with a specialized mechanism that enables them to parasitize host cells other than phagocytes are not free to pursue a parasitic career unhindered by these ubiquitous cells (Kantoch, 1961).

It has been usual to think that experimental conditions *in vivo* are too complex for the proper analysis of events; and much effort has been expended in creating experimental models outside the body in the hope of achieving simplicity. We have to admit, however, that the experimental situation of cells in tissue culture is both static and unphysiological (Rowley, 1963). It has shortcomings that must be clearly recognized. Nevertheless, much of what is already known has been learnt from experiments conducted outside the body. For in a sense the experimenter feels intellectually obliged to demonstrate *in vitro* the events that are thought to occur under natural conditions in the host; and events observed for the first time *in vitro* have no real significance until they have also been shown to take place *in vivo*, in the absence of possible artifact. Surface phagocytosis is a case in point. Although it had been convincingly demonstrated *in vitro* (Wood, Smith & Watson, 1946), its significance in infectious disease was a matter for conjecture until Cohn (1962*a*) observed that encapsulated staphylococci were ingested *in vivo* more rapidly than they were under similar conditions of opsonization *in vitro* (Cohn & Morse, 1959). The mechanical advantage enjoyed by phagocytes that are able to make use of surface phagocytosis *in vivo* was thought to provide the most likely explanation for the observed difference.

The studies of the last-mentioned workers illustrate the complementary nature of observations made under natural and contrived experimental conditions (Cohn & Morse, 1959, 1960; Cohn, 1962a, b, c). They also serve to remind us, that though we can often make valid deductions from observations in vitro, we must carefully assess the influence of such variables as the density of phagocyte populations (which are continuously augmented by diapedesis) the availability of opsonins and the concentration of bacterial toxins (which are influenced by the available blood supply), and many other factors which are absent from the experimental situation in vitro. Of the many variables involved, perhaps the most important with which the experimenter must contend is the dynamic nature of the immune response that develops in the course of all infectious diseases. The only immune responses that have been induced in tissues maintained in vitro are those studied by Fishman (1961), Fishman & Adler (1963) and Stevens & McKenna (1958). Although important these studies fall short of realistic representations of the dynamics of equivalent immunological responses in vivo. In a way it is perhaps fortunate that the immunological accompaniments of infectious processes do not occur in vitro, otherwise we would lose much of the simplicity which we aim to achieve under artificial conditions.

The infectious process is a problem in ecology. It concerns the livelihood of a relatively homogeneous population of micro-organisms amidst an inimical, highly diversified and changing population of host cells. Some of the latter are equipped to ingest and to destroy the parasite, others to elaborate substances that facilitate this process. These two categories of cells, phagocytes and immunologically competent cells, are provoked to added functional activity in the course of an infection. Together they bring about ecological changes that can be revealed by observations made either in vivo or in vitro. Our objective is to examine whether or not the more crucial observations that have been made in these two situations are consistent with each other. For this purpose microbial parasites may be divided into two classes; those which are killed following ingestion by phagocytes, and those which survive ingestion. In dealing with organisms of the first class the host must provide adequate numbers of phagocytic cells and the opsonic factors needed for their prompt ingestion. Organisms of the second class pose a different problem because some are readily ingested in the absence of immune opsonins. In both cases, however, there is the same underlying problem of restricting the growth of pathogenic organisms in host cells.

THE REACTIONS OF PHAGOCYTES WITH
EXTRACELLULAR MICROBIAL PARASITES

From an ecological viewpoint, the host's essential problem in dealing with organisms of this class is to transfer rapidly the bacterial population from the extracellular to the intracellular phase. The manoeuvre can only succeed if the rate of transfer exceeds the rate of bacterial multiplication. The ingestion of a population of bacteria is neither instantaneous nor complete; it is a process the rate of which is influenced by many factors. The earliest quantitative studies on phagocytosis established that the two most important factors were the frequency of contact and the efficiency of the union between the phagocyte and the bacterial cell (Mudd, McCutcheon & Lucké, 1934). The former is influenced by the number of phagocytes available, and the latter by the physical properties of the bacterial surface. Although this information has been available for many years, only recently have the kinetics of the interaction of cells and micro-organisms been studied and compared *in vivo* and *in vitro*.

Quantitative measurement of phagocytosis and
bacterial inactivation in vitro

If we regard the rate of phagocytosis as crucial to the outcome of infections caused by extracellular parasites, it is important that we have an accurate means of measuring it, and of determining the fate of organisms following phagocytosis. They must be measured simultaneously because phagocytosis and bacterial killing are sequential events that cannot be meaningfully dissociated. Failure to recognize the latter has led to many misleading reports.

Two methods have been successfully employed to measure the rate of phagocytosis and bacterial killing. They involve the physical separation of intracellular from extracellular bacterial populations so that each can be estimated independently. Cohn & Morse (1959) adapted a technique described by Maaløe (1946) for studying the interaction between rabbit polymorphonuclear leucocytes and staphylococci. Mixtures of cells and bacteria were kept in suspension by constant shaking. Samples of the mixture were taken at intervals and separated by centrifugation into a supernatant phase containing extracellular organisms, and a deposit containing cell-associated bacteria. The rate of disappearance of organisms from the fluid phase was used to express the rate of phagocytosis, while the difference between the total population and the extracellular population was used to estimate the rate of intracellular killing.

Rowley & Whitby (1959) devised a method applicable to macrophages in monolayers; when these were exposed to a bacterial suspension the rate of disappearance of organisms from the fluid phase was a measure of the rate of phagocytosis, while viable counts on the monolayer after washing provided information on intracellular survival. These two methods are useful only for brief observations because of uncertainty concerning the rate of extracellular multiplication. For this reason Cohn & Morse (1959) used lag phase organisms. As the immediate fate of exponentially growing and lag phase staphylococci in rabbit macrophages is identical (Mackaness, 1960), it may be permissible to use this means of minimizing the disturbing effect of extracellular growth. However, the initial advantages soon disappear, and may be largely offset by unsuspected deficiencies in the physiological state of lag phase organisms.

In spite of their imperfections these procedures have provided important information. Cohn & Morse (1959) showed that an apparent difference in the survival of coagulase-positive and coagulase-negative staphylococci in the presence of rabbit granulocytes was due mainly to differences in their relative rates of phagocytosis, and not to a difference in their capacity to survive ingestion. Adequate opsonization of virulent organisms rendered them virtually indistinguishable from avirulent strains.

Rowley & Whitby (1959) obtained similar results in studying the interaction between mouse macrophages and some Gram-negative organisms. The half-life of bacteria within mouse macrophages was computed; in most instances it was 5–15 min. and substantially the same for virulent and avirulent strains. The mean times of intracellular survival obtained by these workers agreed well with the direct microscopical observations of Wilson, Wiley & Bruno (1957) on the intracellular survival of Gram-positive organisms.

The rapidity with which bacterial multiplication in the extracellular phase can overshadow and obscure results in systems of cells and bacteria maintained *in vitro* is convincingly revealed in several studies (Rowley & Whitby, 1959; Jenkin & Benacerraf, 1960; Mackaness, 1960). Undoubtedly this is the main reason for the contradictory results obtained by workers who used experimental procedures that did not dissociate the ingestion and post-ingestion phases of the inactivation of organisms by phagocytic cells (Rogers & Tompsett, 1952; Baker, 1954; Kapral & Shayegani, 1959; Furness, 1958; Furness & Ferreira, 1959).

Some workers have used antibiotics to circumvent the problems created by the presence of unknown numbers of extracellular bacteria growing at an unknown rate. Streptomycin is commonly employed in the belief that it does not readily penetrate cells. It should be used at a concentration which inhibits extracellular bacteria without interfering with the metabolic activities of bacteria within cells. Even if an appropriate concentration of a drug is selected, a precaution seldom taken, there are still serious objections to its use (Rowley, 1962). Drugs cannot be applied until after phagocytosis has taken place, and by this time bacterial inactivation is often well advanced. Cells may have been damaged by antigens of ingested organisms, and the penetration of such cells by drugs could produce a spurious effect on bacterial survival. The same would be true if any egestion and re-ingestion of organisms took place in the experimental system.

Other workers have sought to control extracellular multiplication by repeated or copious washing of cell monolayers in order to remove redundant organisms (Kapral & Shayegani, 1959; Mackaness, 1962). It is virtually impossible to eliminate all uningested organisms by this means, especially if the contours of the culture vessel are at all complex and difficult to irrigate.

The measurement of phagocytosis in vivo

Despite the difficulties, a number of observations made *in vitro* seem to be consistent with what is known of the fate of certain organisms *in vivo*. Before any comparison of behaviour in the two environments can be made, however, we require quantitative methods for measuring phagocytosis *in vivo*. Only two procedures are available.

Blood clearance as a measure of phagocytosis

An adaptation of the method of measuring the rate of clearance of indian ink from the circulation (Halpern, Benacerraf & Biozzi, 1953; Benacerraf, Biozzi, Halpern & Stiffel, 1957) has been used for estimating the rate of phagocytosis of organisms *in vivo*. The simplest procedure employs bacteria labelled with ^{32}P. Clearance rates are expressed as an exponential function (K), known as the phagocytic index. Although the method has limitations, it has provided a clearer picture of the early phases of the interaction between micro-organisms and the phagocytic cells of the reticulo-endothelial system (Rowley, 1962). Unlike the behaviour of inert particulate material, the removal of organisms such as *Escherichia coli* or *Staphylococcus aureus* is exponential for only a short period following injection (Benacerraf, Sebestyan & Schlossman, 1959) because other factors, such as bacterial multiplication, rapidly come into play when the method is applied to viable organisms. None the less, a good correlation has been found between the rate of clearance and the level of circulating antibody: pre-opsonization of organisms increases the clearance rate in a manner which is consistent with its effect in increasing the rate of phagocytosis *in vitro*. It is not, however, possible to make any closer comparison than this, for phagocytosis depends among other things upon the frequency of the random contacts that occur between cells and bacteria; we have no control over this parameter *in vivo*. Moreover, an increased rate of clearance is not necessarily an expression of increased resistance. This is borne out by the observations of Howard (1959), who showed that although Thorotrast and bacterial endotoxin both caused an increase in the rate of clearance of bacteria, against certain organisms the former increased the level of antibacterial resistance whereas the latter depressed it.

Clearance rates provide information concerning only the *initial* phase of interaction between host cells and organisms. They tell us nothing about subsequent events. To examine the events that follow phagocytosis, more complicated procedures are required.

Quantitative studies on peritoneal infections

The peritoneal cavity has long been used for the study of infectious processes. This anatomical compartment has been successfully exploited by Cohn (1962 *a*, *b*, *c*) for the quantitative analysis of the interaction between cells and organisms *in vivo*. The methods were similar to those used by Cohn & Morse (1959, 1960) to examine the influence of various factors on the interaction between bacteria and suspensions of rabbit polymorphonuclear leucocytes *in vitro*.

The peritoneal cavity of the mouse has the special merit that it contains a large and relatively constant population of resident macrophages free in the peritoneal fluid. These appear sufficient to cope with an approximately equivalent number of staphylococci. Even virulent, coagulase-positive strains under these conditions are efficiently ingested and destroyed. In this respect the peritoneal cavity appears to provide more favourable conditions for phagocytosis than are commonly encountered *in vitro*. In shaken suspensions at a density of 10^7 per ml., a cell density which was not substantially different from that in the peritoneal cavity, rabbit macrophages were relatively inefficient in the phagocytosis of unopsonized, coagulase-positive staphylococci (Mackaness, 1960). Specific opsonization of organisms abolished this impediment to phagocytosis showing that cells *in vitro* were physiologically intact, but unable for physico-chemical reasons to make effective contact with the organism.

In Cohn's studies the injection of larger doses of staphylococci into the peritoneal cavity revealed the necessity for adequate opsonization of the organism and the importance of phagocyte numbers. At a bacterium–phagocyte ratio of 10:1, effective control over the bacterial population had to await the diapedesis of many immigrant granulocytes. However, pre-opsonization of the organism appeared to abolish this requirement for additional cells. It is possible, therefore, that the inflammatory response not only provides cells, but augments the supply of phagocytosis-promoting factors as well. It is known, for instance, that normal complement levels in the peritoneal cavity of mice are low (Amos, 1962).

The importance of the available supply of phagocytic cells was emphasized by observations made in animals treated with agents that interfere with the host's cellular response. Cortisone produces a gross impairment of antibacterial resistance which is not caused by any effect on the phagocytic or bactericidal activity of leucocytes, or on the level of serum opsonins (Hirsch & Church, 1961). During staphylococcal infection in the peritoneal cavity of mice treated with cortisone, Cohn (1962b) observed a gross depression in the number of granulocytes entering the peritoneal fluid despite their presence in undiminished numbers in the peripheral blood. X-irradiation caused both a fall in the level of circulating leucocytes and a deficiency in the cellular response during peritoneal infection. In each case the effectiveness of peritoneal clearance was correspondingly reduced.

The studies of Whitby & Rowley (1959) on the inactivation of Gram-negative organisms in the peritoneal cavity of mice show the dependence of cellular defences on serum opsonins. Moreover, the non-specific resistance produced by substances such as endotoxin was due in part to an augmented supply of opsonic factors (Rowley, 1962). The source and nature of these naturally occurring opsonins is the subject of current inquiry.

It is an eloquent justification for the use of models *in vitro* that the dynamics of a peritoneal infection in mice could have been predicted with considerable accuracy on the basis of evidence previously obtained from studies performed outside the body.

THE REACTIONS OF PHAGOCYTES WITH FACULTATIVE INTRACELLULAR PARASITES

The relationship between host phagocytes and organisms that are able to survive and multiply after ingestion has proved more difficult to analyse. Some of the problems are technical in nature; but others stem from the peculiarities of the mechanism of acquired resistance to intracellular parasites. There is, moreover, the further complication that certain microbial parasites not only survive in cells, but also resist ingestion. The special problems created by organisms of this sort will be discussed later.

The behaviour of organisms that survive in cells can only be studied in an experimental system that remains stable over relatively long periods. So far as we know, the cells that such organisms commonly inhabit *in vivo* are the mononuclear phagocytes of the reticulo-endothelial system. The only representatives of these phagocytic cells that can be readily obtained in sufficient numbers for experimental purposes are the free phagocytes obtained from the peritoneal cavity or lung. It has been asked whether these are truly representative of reticulo-endothelial phagocytes at large (Pavillard, 1963). The mononuclear cells which populate a tubercle or which invade the connective tissues during an inflammatory response are mainly of haematogenous origin; hence the cells obtained from a post-inflammatory exudate in the peritoneal cavity have at least as much relevance to the problem of host–parasite relations, as do the fixed phagocytic cells of the reticulo-endothelial system.

Intracellular growth rates

Mycobacterium tuberculosis is an extreme example of a facultative intracellular parasite. Under artificial cultural conditions *in vitro*, strains of varying virulence grow at similar rates (Fenner & Leach, 1953). In the tissues, however, the initial growth rate and the ultimate size reached by the bacterial population vary according to the virulence of the strain (Pierce, Dubos & Schaefer, 1953). Studies on infected organs such as the spleen and liver may not provide a true picture of the growth characteristics of all bacterial species *in vivo*. They may however do so in the case of the tubercle bacillus, which is extraordinarily resistant, even in the tissues of immune animals. For instance, Rees & Hart (1961) showed in chronically infected mice that the total stainable bacterial population in the lungs was equal to the total viable population, and remained almost constant for many weeks. Furthermore, the number of stainable organisms was not reduced during therapy with drugs. Tubercle bacilli neither died nor left the lungs in significant numbers at any stage of infection. Hence the growth characteristics of tubercle bacilli *in vivo* can be determined with considerable accuracy on the basis of viable counts. Moreover, as the tubercle bacillus offers little or no resistance to phagocytosis, and occasions no necrosis in the tissues of the mouse, it is probable that the growth rates measured *in vivo* are those of intracellular bacilli.

The growth characteristics of tubercle bacilli have been observed in cultures of mononuclear phagocytes *in vitro*. The intracellular growth rate is higher for virulent than for attenuated strains (Mackaness, 1954; Berthrong & Hamilton, 1958). The methods used were relatively crude. The long generation time of the tubercle bacillus (*c*. 24 hr.) necessitates that cells be maintained in undiminishing numbers for protracted periods in order to obtain significant data. At present we are unable to meet this exacting requirement but gross differences between growth rates of different strains are detectable *in vitro*, and they reflect the behaviour of the same strains *in vivo*.

The rate at which organisms multiply *in vivo* has a profound influence on the course of an infectious process. This is illustrated by comparing the behaviour of three facultative intracellular parasites—*Listeria monocytogenes*, *Brucella abortus* and *Mycobacterium tuberculosis*.

Listeria monocytogenes increases in the spleens of infected mice with a mean generation time of *c*. 4–8 hr. (Mackaness, 1962). Within mouse macrophages *in vitro*, it appears to grow at the same rate, and to kill the host cell when the intracellular population has increased to about 16. Hence cells which are initially infected with a single organism will die and begin to release their content of *L. monocytogenes* after *c*. 20 hr. What happens during the second intracellular growth cycle would depend upon the number of organisms ingested by phagocytes in the vicinity. Microscopically, organisms appear to be well distributed among cells of the lesion, so that the cycling time between cells possibly continues at only a slightly faster rate.

Brucella abortus behaves differently. Its growth rate *in vivo* cannot be accurately determined but the rate of increase of the vaccine strain (strain 19) in the spleen of mice corresponded to a generation time of approximately 24 hr. (Mackaness, 1963). However, infected macrophages maintained *in vitro* remained viable until they were virtually choked with organisms (Holland & Pickett, 1958). They did not die until the 4th or 5th day of incubation (Mackaness, unpublished). Hence the time required for *B. abortus* to cycle from one host cell to the next is longer than for a virulent strain of *Listeria monocytogenes*. Although accurate data are not available it appears from the figures of Pierce *et al.* (1953), that the growth characteristics of *Mycobacterium tuberculosis* in tissues are similar to those of *B. abortus*. The significance of these different growth characteristics of pathogenic organisms will be considered later in relation to the evolution of the host's immune response.

THE ROLE OF CELLS IN ACQUIRED RESISTANCE

Resistance against extracellular parasites

We have already discussed the importance to the host of rapidly transferring parasites from an extracellular to an intracellular location. The efficiency of anti-bacterial mechanisms of phagocytes is usually so high, and the opsonic activity of serum so good, that the normal host can deal with most microbial intrusions. However, serum does not usually have opsonic activity against all potential pathogens. In the absence of appropriate opsonic factors, phagocytosis may proceed too slowly for effective control over parasite numbers. It is commonly supposed that in such cases the tide is turned after the host has been immunologically activated to produce specific antibodies capable of neutralizing anti-phagocytic or cytotoxic properties of the bacterial cell. Although we have undoubted evidence of the enhancing effect of specific antiserum on phagocytosis rates *in vitro*, clearance rates *in vivo*, and on resistance to a challenge infection, we have not yet satisfactorily assessed the importance of normal and induced opsonins in contributing to the resolution of an infectious process. This should not surprise us, for opsonization is only one factor that influences the rate of phagocytosis. An increase in density of the phagocyte population may compensate for any deficiency in the supply of opsonic factors (Cohn, 1962 *a*, *b*). Indeed for any given rate of phagocytosis there may be an inverse relationship between phagocyte density and opsonic requirements.

A proportion of the bacteria taken up by phagocytic cells is not immediately killed. As many as 10 % of ingested staphylococci were not killed within rabbit macrophages; when the survivors were extracted and exposed to fresh cells they were as susceptible to inactivation as the original population (Mackaness, 1960). Apparently all phagocytic cells are not equally equipped, and some are deficient in bacteriolytic activity. This defect in cell function is unimportant if the percentage of organisms that survive the first cycle of ingestion is relatively small, and the survivors kill the cell before much multiplication occurs. Under these conditions the number of bacteria surviving, by virtue of the incompetent cells in the phagocyte population, would decline progressively during successive cycles of ingestion. If, however, the proliferation within cells was extensive, bacterial numbers would increase and the organisms would be more appropriately classified as facultative intracellular parasites.

Our concepts of the mechanism of acquired resistance to extracellular parasites has been influenced by observations on the interaction between cells and organisms *in vitro*. Benefits that would attend the elaboration of antibodies which neutralize antiphagocytic components present at the bacterial surface, or which neutralize bacterial products with cytotoxic properties that interfere with phagocyte function, has beguiled us into thinking that this is often the essential process whereby the host achieves ascendancy over invading microbial parasites. The promotion of phagocytosis by neutralizing microbial products is only one way in which the host's immune response prevents an increase in bacterial numbers; there are several others, of which two are conveniently discussed at this stage. The nature of acquired resistance to intracellular parasites will be discussed in subsequent sections.

The chemotactic effect of antigen–antibody complexes

Bacterial products can influence the migration of phagocytic cells (Harris, 1954). So far as we know this phenomenon is not immunologically determined. However, when antigen and antibody react in the presence of fresh serum, a chemotactic agent is released (Boyden, 1962). This was demonstrated by a novel and quantitative technique; the chemotactic agent was placed on one side of a porous membrane and leucocytes on the other; net migration of cells through the membrane gave a measure of the chemotactic response. This model system bears the closest possible analogy to the situation *in vivo*, where cells must pass an endothelial membrane to contact a provoking stimulus in the tissues. There are many examples of the accumulation of leucocytes near antigen–antibody reactions *in vivo*. Germuth, Maumenee, Senterfit & Pollack (1962) showed the marked tendency for granulocytes to migrate towards the line of precipitate formed during counter-diffusion of antigen and antibody in the rabbit cornea. Phagocyte density influences bacterial survival, and once antibody has been produced in infection, the complexes that it forms with bacterial products would influence, by chemotaxis, the cell content of infected tissues. The immunological reaction influences the behaviour of host cells rather than the susceptibility of the parasite.

Phagocyte stimulation by antigen–antibody complexes

The ingestion of particles by phagocytic cells alters the metabolic rate of phagocytes (Stähelin, Suter & Karnovsky, 1956; Sbarra & Karnovsky, 1959), the rate of phagocytosis of particles (Cohn & Morse, 1959; Rowley, 1960) and possibly the rate of killing of ingested bacteria. Antigen–antibody complexes are phagocytozed by cells of the reticulo-endothelial system (Benacerraf, Sebestyan & Cooper, 1959), and this ingestion results in a similar increase in metabolic activity (Strauss & Stetson, 1960). Hence antigen–antibody complexes arising in infection could lead to an advantageous increase in the functional capacity of the reticulo-endothelial system.

Resistance against intracellular parasites

Although antibodies are present in the serum of animals and man infected with tubercle bacilli or a variety of other facultative intracellular parasites, we have failed to demonstrate a role for them in specific acquired resistance. Indeed, an absence of correlation between the level of circulating antibody and of acquired resistance is one of the notable features of infections caused by these parasites (Dubos, 1954). They produce chronic infection which militates against the demonstration of passive protection by serum. This accounts for the paucity of efforts to confer resistance passively with the serum of tuberculous animals; even in acute infections, such as that caused by *Listeria monocytogenes* the passive transfer of serum of highly resistant convalescent animals may be quite ineffective (Osebold & Sawyer, 1957; Mackaness, 1962).

The inability to demonstrate humoral immunity, and the fact that tubercle bacilli were commonly located within cells in lesions that showed evidence of regression, suggested that immunity against tuberculosis was cellular (Rich, 1951). The first direct evidence supporting the speculative view that phagocytic cells of animals immunized against certain micro-organisms possessed properties not found in cells of normal animals came from the memorable experiments of Lurie (1942). Monocytes from normal or immunized rabbits were parasitized with tubercle bacilli and implanted in the anterior chamber of the eye. From histological appearances and viable counts, it seemed that growth had been restricted in the presence of cells from immunized donors. The experiment, however, did not answer the important question of whether the transplanted phagocytes had themselves exerted this restraining influence over the organisms contained within them (Mackaness, 1954). The cells obtained from peritoneal exudates were similar to those used by Chase (1945) to transfer tuberculin sensitivity to a normal recipient. They were also a mixed cell population and probably contained immunologically competent cells which, from immunized donors, would be capable of responding to antigens present in the infecting tubercle bacilli. These qualifications do not gainsay the significance of Lurie's observations; they merely draw attention to the fact that conditions *in vivo* can be more complex than they seem.

Numerous attempts have been made to reproduce Lurie's findings in a simpler system *in vitro*, reviewed by Jenkin & Rowley (1963). Many observers found that the mononuclear phagocytes obtained from animals immunized with such organisms as *Mycobacterium tuberculosis*, *Brucella abortus*, *Salmonella enteritidis*, *Listeria monocytogenes* and *Histoplasma capsulatum*, when washed and isolated *in vitro*, exerted some restraining influence over the intracellular growth of the corresponding organism. Sometimes the antibacterial activity of immune cells was so great as to exclude artifact (Mackaness, 1962). The demonstration that cells from immune animals possessed properties not found in the cells of normal animals was, however, only one aspect of the problem. Perhaps the more important question concerns the mechanism whereby this change in cells was brought about. Jenkin & Rowley (1963) pointed out that antibody has not yet been excluded as the essential mediator of this form of acquired resistance. A satisfactory explanation of cellular immunity must encompass specificity, and there seems to be a conspicuous lack of specificity in the antibacterial activity of 'immune' phagocytes (Auzins & Rowley, 1962; Mackaness, 1963).

The immunological basis of acquired cellular resistance

The mechanism of acquired resistance to intracellular parasites cannot be rationally explained until we learn more about the immunological response that such organisms provoke. Practically all organisms which survive and multiply within mononuclear phagocytes of infected animals produce delayed hypersensitivity. A relationship between the hypersensitive state and acquired antibacterial immunity has frequently been observed. However, immunity may exist in the apparent absence of hypersensitivity (Rich, 1951). An understanding of the true relationship between them must await more knowledge of delayed hypersensitivity itself. In the meantime, a related problem can be considered, that of specificity in acquired resistance to intracellular parasites.

The specificity of acquired resistance to
facultative intracellular parasites

Considerable cross-resistance develops between animals that have been infected with organisms producing an immunity that cannot be transferred with serum. Thus, tuberculous animals are partially resistant to *Brucella abortus* (Pullinger, 1936), to *Pasteurella pestis* (Girard & Grumbach, 1958), to *Salmonella enteritidis* (Howard *et al.* 1959), and to *Staphylococcus aureus* (Dubos & Schaedler, 1957), while brucella-infected mice are highly resistant to *Listeria monocytogenes*

(Mackaness, 1963). This awkward phenomenon of cross-resistance between antigenically unrelated organisms is regarded usually as 'non-specific immunity'; but there has been a notable disregard for the immunological aspects of the phenomenon. An examination of this problem has been made in mice immunized with *L. monocytogenes, B. abortus* (strain 19) and *M. tuberculosis* (B.C.G.) (Mackaness, 1963).

During infection with *Listeria monocytogenes*, peritoneal macrophages of mice developed marked anti-listeria activity when tested *in vitro* (Mackaness, 1962). This property appeared in host cells on the 4th day of infection. It coincided with the development of delayed sensitivity to listeria antigens, and also with the development of an active mechanism for the destruction of the infecting organisms in the spleen and liver. For a period of 2–3 weeks after infection, mice were completely resistant to re-infection. During this period the peritoneal cavity contained a gradually diminishing population of macrophages which could actively destroy *L. monocytogenes in vitro*, but the serum contained no humoral factor which would passively protect normal mice. While mice were in this phase of resistance to *L. monocytogenes*, *Brucella abortus* was also incapable of multiplying in the tissues. Later in convalescence, when resistant macrophages disappeared from the tissues, a challenge inoculum of *L. monocytogenes* grew briefly in spleen and liver, then ceased to multiply in the face of a rapid anamnestic response. *B. abortus*, however, was found to be incapable of provoking an accelerated immune response in convalescent listeria-immune mice; it grew in them exactly as it did in the tissues of normal animals.

In a reciprocal study (Fig. 1), *Brucella abortus* grew at a constant rate in the spleens of normal animals for 10–12 days. During this time *Listeria monocytogenes* multiplied at a normal rate, in the organs of the brucella-infected host and survived in the peritoneal macrophages of these animals. During the next 12–14 days, however, the brucella population in the spleen ceased to multiply and underwent a partial inactivation. When *L. monocytogenes* was injected during this phase, day 18, it was rapidly destroyed in the spleen. Macrophages recovered at this stage and tested *in vitro* showed a corresponding increase in anti-listeria activity. More than 99·9 % of them were able to kill ingested listeria. The brucella infection then entered a third, or latent, phase in which the numbers of organisms in the spleen remained almost constant. *L. monocytogenes*, at this time, day 35, multiplied in the spleen, but at a slower rate than normal. The peritoneal macrophages also changed, for the proportion showing anti-listeria activity had fallen to a relatively low level. Furthermore, *L. monocytogenes* grew in the spleens during the latent phase of brucella infection without eliciting an anamnestic response.

These facultative intracellular parasites induce an immunological response associated with delayed type hypersensitivity and a change in the functional capacity of the host's macrophages. The modified cells which arise during the immunological response persist for only a limited

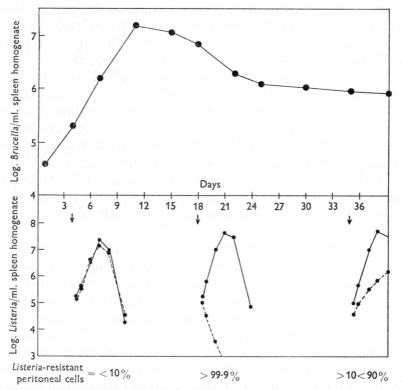

Fig. 1. Upper: growth curve of *Brucella abortus* (strain 19) in spleens of normal mice. Lower: growth curves of *Listeria monocytogenes* in the spleens of normal (●———●), and brucella-infected mice (●----●). Percentages of peritoneal macrophages with anti-listeria activity on day of challenge (arrowed) are recorded below.

period. After they have disappeared they can be rapidly recalled, but only by injecting the homologous organism. Since the recall mechanism is immunologically specific, it appears that the functional change in host cells is the result of a specific immunological reaction. This conclusion is warranted even though the antibacterial properties of the altered cells are manifestly non-specific in their effect on ingested organisms. A similar lack of specificity in the antibacterial activity of macrophages from immunized animals has been observed by Elberg, Schneider & Fong (1957). The susceptibility of *Listeria monocytogenes* to inactivation by cells of immunized animals provided a sensitive indicator of bactericidal capacity of phagocytic cells of the reticulo-endothelial system. The organism's behaviour in spleens and particularly in isolated macrophages of brucella-infected mice reflected in an exaggerated way the influences to which the brucella population was itself subject at successive phases of the infection. The conspicuous variation that occurred in host resistance in this experiment is an aspect of host–parasite relations that has not been adequately explored. Why had the antibacterial efficiency, so high on the 18th day of infection, dwindled to such a low ebb on the 35th day when the spleens still contained relatively large numbers of viable brucellae? The only obvious factor which could account for this difference was the difference in numbers of bacteria present in the tissues at these times. The prevailing antigenic mass may control the level of host resistance by exerting its influence in at least two ways: by stimulating antibody production, or by reacting with tissues that have become allergically sensitized to it. An examination of the resistance of B.C.G.-infected mice to subsequent infection with *L. monocytogenes* was undertaken in the hope of distinguishing between these two possibilities (Mackaness, 1963).

A group of mice was injected intravenously with a suspension of a vaccine strain of B.C.G., 0·25 mg. wet weight. Fourteen weeks later half of them, and an equal number of normal mice, received more living B.C.G., 0·65 mg. wet weight. Three days later all mice and an additional group of untreated animals were challenged intravenously with a virulent strain of *Listeria monocytogenes*, one LD 50. The fate of this inoculum was followed by viable spleen counts performed at intervals on groups of 5 mice from each of the 4 groups. The second injection of B.C.G. produced a marked effect on the behaviour of listeria only in the spleens of animals which were already sensitized to it (Fig. 2). This effect of a second injection of B.C.G. cannot be satisfactorily explained in terms of a secondary antibody response unless we ignore the fact that here again the antibacterial mechanism was nonspecific. We are left with the alternative explanation that increased antibacterial activity in the spleen was due to the action of B.C.G. antigens on allergically sensitive tissues. It is suggested that host macrophages become sensitized during infection, perhaps by the antibodies responsible for delayed hypersensitivity. When sensitized in this way, the cells respond to further contact with specific antigen by changing their metabolic habit and this change accounts for the nonspecific antibacterial activity of the host's macrophage population.

There is a certain amount of evidence to support this hypothesis. First the phagocytes of tuberculous animals are more active than cells of normal animals (Lurie, 1939); they show an increased metabolic activity when isolated *in vitro* (Allison, Zappasodi & Lurie, 1962), and are

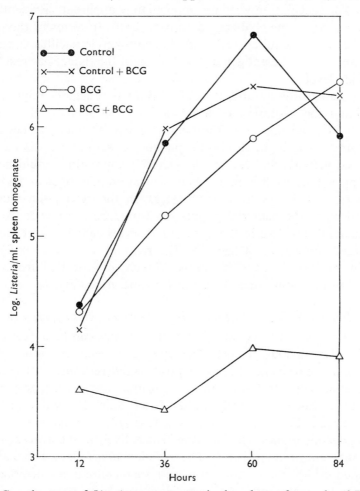

Fig. 2. Growth curves of *Listeria monocytogenes* in the spleens of normal and B.C.G.-infected mice, some of which had received additional antigen (B.C.G.) 3 days before challenge.

richer in enzymes (Grogg & Pearse, 1952). In addition, other features of acquired resistance to the tubercle bacillus are consistent with the foregoing hypothesis. The protective effects of vaccination with B.C.G. can be revealed in a number of ways: by a difference in survival time of the host and a reduction in the number of virulent organisms in the tissues at an interval following challenge, or by a difference in severity of the lesions. However, when quantitative studies were made immediately after challenge, the number of organisms recovered from organs was the same in normal as in vaccinated animals and there were equivalent numbers of microscopic lesions (Lévy *et al.* 1961). The benefits of vaccination were not immediately apparent. Evidently they are activated by the challenge organism itself. In terms of the present hypothesis, the acquired resistance to the tubercle bacillus is not fully expressed unless the antigen concentration in the tissues has reached an adequate level.

This view of the nature of acquired cellular resistance might help in explaining the evolution of a latent infection such as that seen in brucella-infected mice. At the height of the infection, when antigen concentration is high, the host's phagocytes would be stimulated to increased activity, and the viable bacterial population would begin to fall. This would result in reduced concentration of antigen and a falling-off in the activity of host phagocytes. In theory a point would be reached when the bacterial population would be too low to excite an antibacterial response in the host's phagocytes sufficient to ensure the continued elimination of bacteria. The reduced number of phagocytes showing antibacterial activity on the 35th day of brucella infection is an indication that some such mechanism is operating (Fig. 1).

Immunological induction of reticulo-endothelial hyperplasia

We have discussed three ways in which a change can be induced in the functional behaviour of host phagocytes as a result of immunological processes. The fourth concerns the proliferative response of the reticulo-endothelial system following antigenic stimulation of sensitized animals.

Antigens can destroy cells of sensitized animals (Waksman, 1959; Amos, 1962). For example, the beta toxin of *Clostridium welchii* will lyse guinea pig monocytes that have been actively or passively sensitized with antiserum (Allan, 1963). The reaction takes place in the absence of added complement, and with sera that contain no demonstrable precipitating antibody. Antigens, however, can have effects upon sensitized cells that are in no way destructive. These are less familiar, but of greater significance in the present context.

Our interest in the proliferation of host macrophages was aroused during an investigation of the lesions caused by *Listeria monocytogenes* in mice (Mackaness, 1962). On the 4th day of a primary infection, when delayed sensitivity was first in evidence, the focal lesions in the spleen and liver showed evidence of mitosis involving mononuclear phagocytes. During reinfection of hypersensitive animals the mitoses occurred within 24–48 hr. The intravenous injection of listeria culture filtrate into convalescent mice caused proliferation among the free cells of the peritoneal cavity. The effect was not seen in normal animals. The immunological significance of this cellular proliferation has been examined in more detail (Forbes & Mackaness, 1963). It is a general phenomenon equally evident in animals that have been sensitized to non-bacterial antigens such as bovine serum albumin or rabbit serum.

The subcutaneous injection of a small eliciting dose of antigen into previously sensitized animals causes marked changes in the free cells of the peritoneal cavity. After an interval of 12–16 hr. the pre-existing macrophages begin to synthesize DNA. The number of synthesizing cells rapidly increases until virtually all of them are involved. The cell cytoplasm remains morphologically unchanged, but by 24 hr. the nucleus has enlarged, becomes ovoid and finely reticulated, and contains several prominent nucleoli. From this time on the cells begin to divide. The maximum mitotic rate occurs 30–36 hr. after injection when as many as 25 % of cells may be in division at one time (Pl. 1, figs. 1, 2). In some mice, presumed to be more sensitive, the pre-existing macrophages show toxic changes and may disappear almost completely from free fluid in the peritoneal cavity. In such cases the lymphocyte population begins to synthesize DNA and to undergo morphological changes (Pl. 2, figs. 3, 4). The cytoplasm enlarges and becomes basophilic, the nucleus loses its pachychromatic structure, acquires nucleoli and increases in size. These changes take longer to develop so that mitosis does not usually occur until later (60 hr. or more) when the transformed cells have assumed the cytoplasmic characteristics of macrophages. It may, however, occur at an earlier stage while the cell cytoplasm is still undifferentiated (Pl. 2, fig. 4). In all cases, however, the ultimate cellular products of this proliferative response are new macrophages. Waksman & Matoltsy (1958) described the induction *in vitro* of a similar series of cell changes when tuberculin was introduced into cultures of peritoneal cells from tuberculin-sensitive guinea pigs. The derivation of macrophages from lymphocytes has been argued for many years (Rebuck, 1960; Yoffey, 1962), but still opinions differ on this point (Rabinowitz & Schrek, 1962; Goldman & Walker, 1962). Mitosis in mature macrophages, on the other hand, has seldom been encountered. Recently Kelly, Brown & Dobson (1962) showed conclusively that endotoxin causes mitosis in the Kupffer cells of the liver. It had a similar effect on the free cells of the peritoneal cavity of normal mice (Forbes & Mackaness, 1963).

From the viewpoint of host–parasite relations there can be little doubt that the intense mitotic stimulation caused by a bacterial antigen to which the host has become sensitized in the course of an infection is an important mechanism of defence. This would be true even if an increased number of phagocytes were the only benefit that accrues. It is possible, though unproved, that the stimulus which provokes cells to divide also induces the increase in antibacterial activity that was discussed in the previous section. These aspects of the host's cellular response have been discussed at some length in order to emphasize the dynamic nature of the infectious process.

PHAGOCYTOSIS-RESISTANT PARASITES THAT SURVIVE INGESTION

The problem of analysing the host–parasite relationship is formidable in infections caused by organisms with attributes that enable them not only to resist phagocytosis but to survive ingestion: for example, certain enteric pathogenic organisms particularly *Salmonella typhimurium* and *S. enteritidis*.

Salmonella typhimurium is especially interesting because of its natural pathogenicity for mice and the fact that strains of varying virulence are available for comparative studies. Virulent strains of *S. typhimurium* are relatively resistant to phagocytosis *in vitro* (Whitby & Rowley, 1959; Jenkin & Benacerraf, 1960) and are correspondingly slow to disappear from the circulation after intravenous injection (Jenkin & Rowley, 1961). Prior opsonization of virulent organisms increases their phago-cytosis rate *in vitro* and their clearance rate *in vivo* (Jenkin & Rowley, 1959), rendering them indistinguishable from avirulent strains under the same conditions of test. Hence, *S. typhimurium* does not appear to differ from other organisms that resist ingestion. However, there is one significant anomaly: the rapid clearance that results from opsoniz-ing a virulent inoculum of *S. typhimurium* is not attended by any signi-ficant increase in the survival rate of challenged animals (Jenkin & Rowley, 1963) except at minimal challenge doses (Jenkin & Rowley, 1959). Since opsonization with specific antiserum renders virulent strains of *S. typhimurium* as readily phagocytozed as avirulent strains and equally susceptible to intracellular inactivation, why is there no significant influence on lethality? Evidently *S. typhimurium* behaves differently from virulent strains of *Staphylococcus aureus* (Cohn, 1962*a, b*) which is perhaps even more resistant to ingestion (Whitby & Rowley, 1959). The chief characteristic of *S. typhimurium* which accounts for this absence of an effect on lethality, and explains the organism's extreme pathogenicity for mice is an ability to survive ingestion. It is difficult, however, to assess the exact proportion of ingested organisms that survive within cells. When ingested in the presence of limiting amounts of opsonin it is probably not less than 40–50 % (Jenkin & Benacerraf, 1960); when fully opsonized the survival rate is much lower. However, in dealing with a parasite for which the LD 50 is only 10 organisms, even a single intracellular survivor is a potential threat. Its intracellular growth could result not only in a bacterial population of lethal proportions, but also in a phenotypic change associated with an increased ability to survive intracellularly. Passage of other organisms, such as *Listeria monocytogenes*, through mononuclear phagocytes significantly increases their intracellular survival rate (Mackaness, 1962). Similar instances of phenotypic change are well known (Meynell, 1961).

Having mentioned the salient characteristics of a typical enteric pathogenic organism it would perhaps be worth while to make a comparison between experimental infections in mice caused by *Salmonella typhimurium* on the one hand, and *Staphylococcus aureus* (a typical extracellular parasite) and *Listeria monocytogenes* (a typical intracellular parasite) on the other. Table 1 contains a qualitative evaluation of several factors which bear upon host cell–parasite relations in mice experimentally infected with these three organisms.

The few bacterial cells that suffice to kill mice infected with *Salmonella typhimurium* portray an aggressiveness that can be attributed to three characteristics of the parasite: rapid growth rate, resistance to phagocytosis, and ability to survive ingestion. On the other hand, the large LD 50 of *Listeria monocytogenes* can be explained by its relatively slow growth rate in relation to its susceptibility to the host's mechanisms of acquired resistance. Following an inoculum of 10^5, the bacterial population in the tissues does not exceed 5×10^8 after 3 days of a primary infection. Hence acquired resistance can develop before the parasite population has reached unmanageable proportions. A further fact of equal importance, however, is the efficiency and bactericidal nature of acquired resistance to this organism (Mackaness, 1962). It will be recalled that quite the opposite obtains in the case of the tubercle bacillus (Rees & Hart, 1961; Lévy *et al.* 1961). Although *Staphyloccus aureus* grows rapidly, and can to some extent resist ingestion, it is relatively avirulent for mice (Table 1). There is little likelihood of a staphylococcus establishing itself in the tissues of this host because opsonic factors are normally present in adequate concentration, and the

234

Table 1. Significant characteristics of representatives of three main classes of microbial parasites for mice

Organism	LD 50 for mice	Normal growth rate in vivo	Resistance to phagocytosis	Intracellular survival	Effect of opsonins on Phagocytosis	Effect of opsonins on Intracellular survival	Development of acquired cellular resistance
Salmonella typhimurium	10	Fast	++	++	Marked increase	Marked decrease	No
Listeria monocytogenes	10⁵	Relatively slow	Very low	+++	Little change	Nil	Yes
Staphylococcus aureus	10⁷	Fast	++	±	Marked increase	Nil	No

organism survives poorly within cells (Cohn, 1962 a). Moreover, organisms which do survive probably cannot proliferate (Mackaness, 1960). It is the latter property which distinguishes *S. aureus* from *S. typhimurium*. Even when fully opsonized, *S. typhimurium* is still highly lethal for the mouse. We shall consider briefly the special features of acquired resistance to this organism.

Recent studies have revealed that specific antibodies to certain Gram-negative bacteria have, in addition to opsonic activity, the property of increasing intracellular killing of the bacteria (Rowley, 1958; Jenkin & Benacerraf, 1960; Rowley & Jenkin, 1962). The most persuasive evidence for this was provided by Jenkin (1963 a); *Salmonella typhimurium* was opsonized by an antibody directed against phage particles adsorbed to the bacterial surface. Organisms which had been ingested under these conditions were found to survive much better than organisms which had been ingested after treatment with an antibody directed against the bacterial cell wall. This introduced the concept that humoral factors may be important in the case of *S. typhimurium* as much for their effect on intracellular survival as for their effect on the rate of phagocytosis. Jenkin (1963 b) pointed out that the ultimate behaviour of *S. typhi-murium* in the tissues of infected mice depends to a large extent on the availability of opsonins that ensure not only the rapid ingestion but also the subsequent death of ingested bacteria. Antibodies with these properties have recently been found in the serum and on the cells of mice suffering from latent *S. typhimurium* infection. Such animals are highly resistant to challenge with *S. typhimurium* (Jenkin & Rowley, 1963) and repeated doses of their serum will passively protect normal mice.

It is clear that antibody is the essential mediator of cellular immunity to enteric organisms. This does not seem to be so for the class of parasites to which *Listeria monocytogenes* belongs. Here the serum of convalescent mice (Mackaness, 1962), even in repeated doses (Miki & Mackaness, unpublished results), was without effect on the growth of *L. monocytogenes* in normal recipients. Without pre-judging the fundamental mechanism responsible for acquired resistance to this and other facultative intracellular parasites, the fact remained that the level of acquired resistance varied according to the number of antibacterial macrophages induced in the tissues of immune animals. Claims have been made that similar, but immunologically specific changes occurred in the cells of animals immunized against *Salmonella enteritidis* (Ushiba, Saito, Akiyama & Sugiyama, 1959). While this may be true, certain doubts arise concerning the interpretation of these findings. Cells from

mice immunized against *S. enteritidis* and *S. typhimurium* had adsorbed
to them a trypsin-labile antibody which could be extracted with 2M
urea and promoted the phagocytosis and killing of *S. enteritidis* in the
peritoneal cavity of mice (Jenkin & Rowley, 1963). It appears that the
cellular changes observed by the Japanese workers were due to the
presence of antibody, particularly in view of their specific nature.
Hence it is difficult to argue that acquired resistance to enteric pathogenic organisms depends upon a change in cell function rather than the
presence of a cell-bound and specific opsonin. For the same reason it
would perhaps be wise to reserve judgement on the mechanism of
resistance in other reputed instances of cell-mediated immunity.

This review of work on the behaviour of microbial parasites in relation
to phagocytic cells fits well with the theme of this Symposium: it shows
how our present knowledge has come from a balanced exploitation of
experimental situation *in vivo* and those with living cells *in vitro*.

REFERENCES

ALLAN, D. (1963). Cytotoxic effects produced by beta toxin of *Clostridium welchii*
on guinea-pig monocytes. *Immunology*, 6, 3.
ALLISON, M. J., ZAPPASODI, P. & LURIE, M. B. (1962). Metabolic studies on mono-
nuclear cells from rabbits of varying genetic resistance to tuberculosis: II.
Studies on cells from B.C.G.-vaccinated animals. *Amer. Rev. resp. Dis.* 85,
364.
AMOS, D. B. (1962). The use of simplified systems as an aid to the interpretation of
mechanisms of graft rejection. *Progr. Allergy*, 6, 468.
AUZINS, I. & ROWLEY, D. (1962). On the question of specificity of cellular immunity.
Aust. J. exp. Biol. med. Sci. 40, 283.
BAKER, H. J. (1954). Effects of penicillin and streptomycin on staphylococci in
cultures of mononuclear phagocytes. *Ann. N.Y. Acad. Sci.* 58, 1232.
BENACERRAF, B., BIOZZI, G., HALPERN, B. N. & STIFFEL, C. (1957). Physio-
pathology of the reticulo-endothelial system: a Symposium. Ed. by B. Bena-
cerraf & J. F. Delafresnaye. Springfield: C. C. Thomas.
BENACERRAF, B., SEBESTYAN, M. M. & COOPER, N. S. (1959). The clearance of
antigen-antibody complexes from the blood by the reticulo-endothelial system.
J. Immunol. 82, 131.
BENACERRAF, B., SEBESTYAN, M. M. & SCHLOSSMAN, S. (1959). A quantitative
study of the kinetics of blood clearance of ^{32}P-labelled *Escherichia coli* and
Staphylococcus by the reticulo-endothelial system. *J. exp. Med.* 110, 27.
BERTHRONG, M. & HAMILTON, M. A. (1958). Tissue culture studies on resistance
in tuberculosis. I. Normal guinea-pig monocytes with tubercle bacilli of
different virulence. *Amer. Rev. Tuberc.* 77, 436.
BOYDEN, S. V. (1962). The chemotactic effect of mixtures of antibody and antigen
on polymorphonuclear leucocytes. *J. exp. Med.* 115, 453.
CHASE, M. W. (1945). The cellular transfer of cutaneous hyper-sensitivity to
tuberculin. *Proc. Soc. exp. Biol., N.Y.* 59, 134.

COHN, Z. A. (1962a). Determinants of infection in the peritoneal cavity. I. Response to and fate of *Staphylococcus aureus* and *Staphylococcus albus* in the mouse. *Yale J. Biol. Med.* **35**, 12.

COHN, Z. A. (1962b). Determinants of infection in the peritoneal cavity. II. Factors influencing the fate of *Staphylococcus aureus* in the mouse. *Yale J. Biol. Med.* **35**, 29.

COHN, Z. A. (1962c). Determinants of infection in the peritoneal cavity. III. The action of selected inhibitors on the fate of *Staphylococcus aureus* in the mouse. *Yale J. Biol. Med.* **35**, 48.

COHN, Z. A. & MORSE, S. I. (1959). Interaction between rabbit polymorpho-nuclear leucocytes and staphylococci. *J. exp. Med.* **110**, 419.

COHN, Z. A. & MORSE, S. I. (1960). Functional and metabolic properties of poly-morphonuclear leucocytes. II. The influence of a lipopolysaccharide endotoxin. *J. exp. Med.* **111**, 689.

DUBOS, R. J. (1954). *Biochemical Determinants of Microbial Diseases*. Cambridge: Harvard University Press.

DUBOS, R. J. & SCHAEDLER, R. W. (1957). Effects of cellular constituents of myco-bacteria on the resistance of mice to heterologous infections. *J. exp. Med.* **106**, 703.

ELBERG, S. A., SCHNEIDER, P. & FONG, J. (1957). Cross-immunity between *Brucella melitensis* and *Mycobacterium tuberculosis*. Intracellular behaviour of *Brucella melitensis* in monocytes from vaccinated animals. *J. exp. Med.* **106**, 545.

FENNER, F. & LEACH, R. H. (1953). The growth of mammalian tubercle bacilli in Tween-albumin medium. *Amer. Rev. Tuberc.* **68**, 321, 342.

FISHMAN, M. (1961). Antibody formation *in vitro*. *J. exp. Med.* **114**, 837.

FISHMAN, M. & ADLER, F. L. (1963). Antibody formation *in vitro*. II. Antibody synthesis in X-irradiated recipients of diffusion chambers containing nucleic acid derived from macrophages incubated with antigen. *J. exp. Med.* **117**, 595.

FORBES, I. J. & MACKANESS, G. B. (1963). Immunological induction of reticulo-endothelial hyperplasia. Unpublished.

FURNESS, G. (1958). Interaction between *Salmonella typhimurium* and phagocytic cells in tissue culture. *J. infect. Dis.* **103**, 272.

FURNESS, G. & FERREIRA, I. (1959). The role of macrophages in natural immunity to *Salmonellae*. *J. infect. Dis.* **104**, 203.

GERMUTH, F. G., MAUMENEE, A. E., SENTERFIT, L. B. & POLLACK, A. D. (1962). Immunohistologic studies on antigen-antibody reactions in the avascular cornea. I. Reactions in rabbits actively sensitised to foreign protein. *J. exp. Med.* **115**, 919.

GIRARD, G. & GRUMBACH, F. (1958). L'infection tuberculeuse de la souris entraîne sa résistance à l'infection pesteuse expérimentale. *C.R. Soc. Biol. Paris*, **152**, 280.

GOLDMAN, A. S. & WALKER, F. (1962). The origin of cells in the infiltrates found at the sites of foreign protein injections. *Lab. Invest.* **11**, 808.

GROGG, E. & PEARSE, A. G. E. (1952). The enzyme and lipid histo-chemistry of experimental tuberculosis. *Brit. J. exp. Path.* **33**, 567.

HALPERN, B. N., BENACERRAF, B. & BIOZZI, G. (1953). Quantitative study of the granulopectic activity of the reticulo-endothelial system. I. The effect of the ingredients present in India ink and of substances affecting blood clotting on the fate of carbon particles administered intravenously in rats, mice and rabbits. *Brit. J. exp. Path.* **34**, 426.

HARRIS, H. (1954). Role of chemotaxis in inflammation. *Physiol. Rev.* **34**, 529.

HIRSCH, J. G. & CHURCH, A. B. (1961). Adrenal steroids and infection: the effect of cortisone administration on poly-morphonuclear leukocytic functions and on serum opsonins and bactericidins. *J. clin. Invest.* **40**, 794.

HOLLAND, J. J. & PICKETT, M. J. (1958). A cellular basis of immunity in experimental *Brucella* infection. *J. exp. Med.* **108**, 343.

HOWARD, J. G. (1959). M.D. thesis quoted by Rowley (1962).

HOWARD, J. G., BIOZZI, G., HALPERN, B. N., STIFFEL, C. & MOUTON, D. (1959). The effect of *Mycobacterium tuberculosis* (B.C.G.) infection on the resistance of mice to bacterial endotoxin and *Salmonella enteritidis* infection. *Brit. J. exp. Path.* **40**, 281.

JENKIN, C. R. (1963a). The effect of opsonins on the intra-cellular survival of bacteria. *Brit. J. exp. Path.* **44**, 47.

JENKIN, C. R. (1963b). The immunological basis for the carrier state in mouse typhoid. *Proceedings Symposium on Bacterial Endotoxins, Institute of Microbiology.* Rutgers University: Rutgers University Press.

JENKIN, C. R. & BENACERRAF, B. (1960). *In vitro* studies on the interaction between mouse peritoneal macrophages and strains of *Salmonella* and *Escherichia coli. J. exp. Med.* **112**, 403.

JENKIN, C. R. & ROWLEY, D. (1959). Opsonins as determinants of survival in intraperitoneal infections in mice. *Nature, Lond.* **184**, 474.

JENKIN, C. R. & ROWLEY, D. (1961). The role of opsonins in the clearance of living and inert particles by cells of the reticulo-endothelial system. *J. exp. Med.* **114**, 363.

JENKIN, C. R. & ROWLEY, D. (1963). The basis for immunity to typhoid in mice and the question of 'cellular immunity'. *Bact. Rev.* (in the Press).

KANTOCH, M. (1961). The role of phagocytes in virus infections. *Arch. Immunol. Ter. dośw.* **9**, 261.

KAPRAL, F. A. & SHAYEGANI, M. G. (1959). Intracellular survival of staphylococci. *J. exp. Med.* **110**, 123.

KELLY, L. S., BROWN, B. A. & DOBSON, E. L. (1962). Cell division and phagocytic activity in liver reticulo-endothelial cells. *Proc. Soc. exp. Biol., N.Y.* **110**, 555.

LÉVY, F. M., CONGE, G. A., PASQUIER, J. F., MAUSS, H., DUBOS, R. J. & SCHAEDLER, R. W. (1961). The effect of B.C.G. vaccination on the fate of virulent tubercle bacilli in mice. *Amer. Rev. resp. Dis.* **84**, 28.

LURIE, M. B. (1939). The mobilisation of mononuclear phagocytes in normal and immunised animals and their relative capacities for division and phagocytosis. *J. exp. Med.* **69**, 579.

LURIE, M. B. (1942). Studies on the mechanism of immunity in tuberculosis: the fate of tubercle bacilli ingested by mono-nuclear phagocytes from normal and immunised animals. *J. exp. Med.* **75**, 247.

MAALØE, O. (1946). *On the Relation between Alexin and Opsonin.* Copenhagen: Einar Munksgaard.

MACKANESS, G. B. (1954). The growth of tubercle bacilli in monocytes from normal and vaccinated rabbits. *Amer. Rev. Tuberc.* **69**, 495.

MACKANESS, G. B. (1960). The phagocytosis and inactivation of staphylococci by macrophages of normal rabbits. *J. exp. Med.* **112**, 35.

MACKANESS, G. B. (1962). Cellular resistance to infection. *J. exp. Med.* **116**, 383.

MACKANESS, G. B. (1963). The specificity of cellular resistance to infection. Unpublished.

MEYNELL, G. G. (1961). Phenotypic variation and bacterial infection. In *Microbial Reaction to Environment. Symp. Soc. gen. Microbiol.* **11**, 174.

MUDD, S., MCCUTCHEON, M. & LUCKÉ, B. (1934). Phagocytosis. *Physiol. Rev.* **14**, 210.

OSEBOLD, J. N. & SAWYER, M. T. (1957). Immunisation studies on listeriosis in mice. *J. Immunol.* **78** 262.

PAVILLARD, E. R. J. (1963). *In vitro* phagocytic and bactericidal ability of alveolar and peritoneal macrophages of normal rats. *Aust. J. exp. Biol. med. Sci.* **41**, 265.

PIERCE, C. H., DUBOS, R. J. & SCHAEFER, W. B. (1953). Multiplication and survival of tubercle bacilli in the organs of mice. *J. exp. Med.* **97**, 189.

PULLINGER, E. J. (1936). The influence of tuberculosis upon the development of *Brucella abortus* infection. *J. Hyg., Camb.* **36**, 456.

RABINOWITZ, Y. & SCHREK, R. (1962). Studies on cell source of macrophages from human blood in slide chambers. *Proc. Soc. exp. Biol., N.Y.* **110**, 429.

REBUCK, J. W. (1960). *The Lymphocyte and Lymphocytic Tissue.* New York: Hoeber.

REES, R. J. W. & HART, P. D. (1961). Analysis of the host–parasite equilibrium in chronic murine tuberculosis by total and viable bacillary counts. *Brit. J. exp. Path.* **42**, 83.

RICH, A. R. (1951). *The Pathogenesis of Tuberculosis.* Springfield: C. C. Thomas.

ROGERS, D. E. & TOMPSETT, R. (1952). The survival of staphylococci within human leucocytes. *J. exp. Med.* **95**, 209.

ROWLEY, D. (1958). Bactericidal activity of macrophages *in vitro* against *Escherichia coli*. *Nature, Lond.* **181**, 363.

ROWLEY, D. (1960). The role of opsonins in non-specific immunity. *J. exp. Med.* **111**, 137.

ROWLEY, D. (1962). Phagocytosis. *Advanc. Immunol.* **2**, 241.

ROWLEY, D. (1963). Endotoxin-induced changes in susceptibility to infections. *Proceedings Symposium on Bacterial Endotoxins, Institute of Microbiology.* Rutgers University: Rutgers University Press.

ROWLEY, D. & JENKIN, C. R. (1962). Antigenic cross-reaction between host and parasite as a possible cause of pathogenicity. *Nature, Lond.* **193**, 151.

ROWLEY, D. & WHITBY, J. L. (1959). The bactericidal activity of mouse macrophages *in vitro*. *Brit. J. exp. Path.* **40**, 507.

SBARRA, A. J. & KARNOVSKY, M. L. (1959). The biochemical basis of phagocytosis. I. Metabolic changes during the ingestion of particles by polymorphonuclear leucocytes. *J. biol. Chem.* **234**, 1355.

STÄHELIN, H., SUTER, E. & KARNOVSKY, M. L. (1956). Studies on the interaction between phagocytes and tubercle bacilli. I. Observations on the metabolism of guinea-pig leucocytes and the influence of phagocytosis. *J. exp. Med.* **104**, 121.

STEVENS, K. M. & McKENNA, J. M. (1958). Studies on antibody synthesis initiated *in vitro*. *J. exp. Med.* **107**, 537.

STRAUSS, B. S. & STETSON, C. A. (1960). Studies on the effect of certain macromolecular substances on the respiratory activity of the leucocytes of peripheral blood. *J. exp. Med.* **112**, 653.

USHIBA, D., SAITO, K., AKIYAMA, T. & SUGIYAMA, T. (1959). *Mechanisms of Infection and Immunity of Cytopathogenic Bacteria. Symposium, 11th Kanto Branch Meeting of the Japan Bacteriological Association.* Tokyo: Yamamoto Shoten.

WAKSMAN, B. H. (1959). The toxic effects of the antigen-antibody reaction on the cells of hypersensitive reactors. In *Cellular and Humoral Aspects of the Hypersensitive States*, p. 123. Ed. by H. S. Lawrence. New York: Hoeber-Harper.

WAKSMAN, B. H. & MATOLTSY, M. (1958). The effect of tuberculin on peritoneal exudate cells of sensitized guinea-pigs in surviving cell culture. *J. Immunol.* **81**, 220.

WHITBY, J. L. & ROWLEY, D. (1959). The role of macrophages in the elimination of bacteria from the mouse peritoneum. *Brit. J. exp. Path.* **40**, 358.

WILSON, A. T., WILEY, G. G. & BRUNO, P. (1957). The fate of non-virulent group A streptococci phagocytised by human and mouse neutrophils. *J. exp. Med.* **106**, 777.

WOOD, W. B., SMITH, M. R. & WATSON, B. (1946). Studies on the mechanism of recovery in pneumococcal pneumonia. IV. The mechanism of phagocytosis in the absence of antibody. *J. exp. Med.* **84**, 387.

YOFFEY, J. M. (1962). The present status of the lymphocyte problem. *Lancet*, i, 206.

EXPLANATION OF PLATES

PLATE 1

Fig. 1. Smear taken directly from the peritoneal fluid of a mouse sensitized by intramuscular injection of bovine serum albumin in Freund's complete adjuvant and injected 25 hr. before with bovine serum albumin (1·0 mg.) given subcutaneously. The field shows an array of macrophages at the onset of the mitotic response. Leishman, × 250.

Fig. 2. A field from fig. 1 showing two prophase nuclei and the primitive nuclear structure of every macrophage in the field. Leishman, × 1250.

PLATE 2

Fig. 3. Smear of peritoneal fluid from a sensitized mouse given rabbit serum (0·1 ml.) 45 hr. before. Pre-existing macrophages, normally accounting for about 75 % of the free cells, have largely disappeared. In their place there is an unphased differentiation of lymphoid cells to a cell of larger and more primitive type. These cells subsequently mitose and differentiate to form macrophages. Leishman, × 1250.

Fig. 4. A later stage of a reaction similar to that shown in fig. 3. The field was chosen to show that differentiation of lymphoid cells is still in progress. The presence of 3 pairs of daughter cells indicates that mitosis is occurring, but at an earlier stage of differentiation. Leishman, × 1250.

PLATE 1

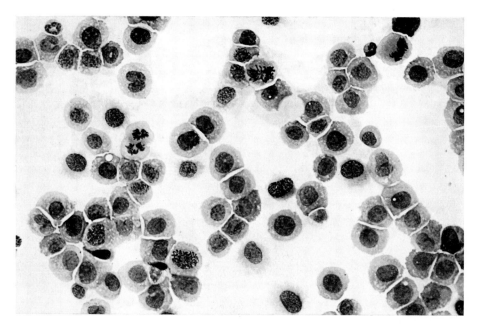

Fig. 1

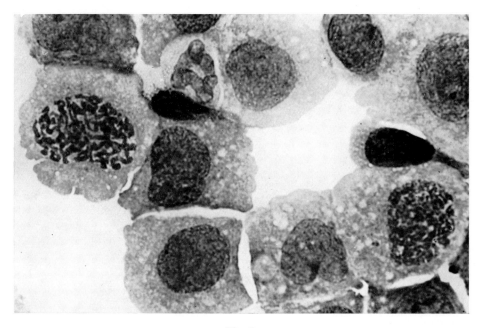

Fig. 2

(*Facing page* 240)

PLATE 2

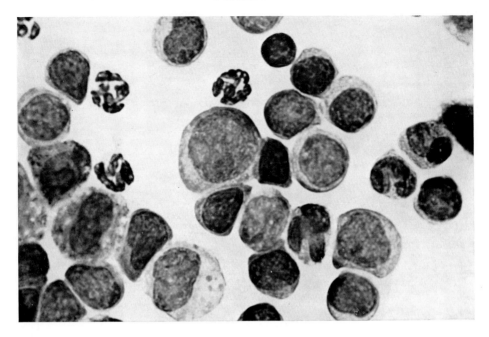

Fig. 3

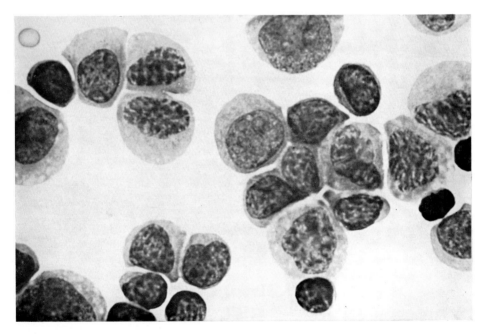

Fig. 4

MIXED POPULATIONS *IN VIVO* AND *IN VITRO*

D. W. HENDERSON

Microbiological Research Establishment, Porton, Salisbury, Wiltshire

In this paper attention is confined to the reaction of animal hosts to invasion by mixed populations of various pathogens and to related experiments *in vitro*.

EXPERIMENTS *IN VIVO*

Induced non-specific resistance

There is a vast literature on the subject of mixed microbial populations *in vivo*, most of which is concerned with protection induced by one organism against infection with another which is later superimposed on the host. As early as 1893 Klein, challenging the specificity of Haffkine's anti-cholera vaccine, inoculated mice and guinea pigs intraperitoneally or subcutaneously with various unrelated organisms such as *Escherichia coli*, *Serratia marcescens* and a *Proteus* species. These animals proved as resistant to a lethal dose of the vibrio as those inoculated with the cholera vaccine. He therefore denied the specificity of Haffkine's preventive method. Similar types of result have been obtained in the virus field. For example, Schlesinger (1959) lists about thirty references to interference between heterologous viruses in animal species ranging from man to mouse.

In more recent times, work on induced non-specific resistance has been concerned with the 'systems' or chemical substances produced by the host in response to primary challenge, which may be responsible for the reaction. Thus the properdin system of Pillemer *et al.* (1954) has been extensively studied in relation to bacterial interference, and interferon, first described by Isaacs & Lindenmann (1957), has evoked great interest in studies of viruses. These are academic studies of much importance, but it is too early to pronounce on the part such substances may play in the control of naturally occurring diseases.

There are other examples of non-specific resistance to infection which are not due to the reaction of the host to primary infection, but to a direct anti-bacterial action of one organism on another, for example, the action of the 'normal' flora of the intestine on a potential secondary invader. Bohnhoff & Miller (1962) showed that after mice had been treated orally with streptomycin to eliminate or reduce the bacterial population of the gastro-intestinal tract, they succumbed to certain salmonella infections given orally to which normally they were resistant.

There is evidence from clinical observations in naturally occurring disease in man that suppression of one infection by another can occur. For example, inhibition of *Treponema pallidum* in late neuro-syphilis may be brought about by an acute attack of malaria (Meagher, 1929). That this results from the severe pyrexia induced in the host by malaria is corroborated by a similar effect being produced by some pyrogens.

Our own work on this subject (Henderson, 1960) has been concerned with organisms of high virulence. The hosts were the guinea pig and the mouse. The respiratory route of infection was used and the time between infection by primary and secondary invader was varied. Attention was first directed to the pathogenesis of infection by each invader acting separately.

Infections in the guinea pig

An estimated lung retention of about 50 single organisms of *Brucella suis* leads to infection of approximately 90 % of animals. The disease produces an extensive lymphoid hyperplasia with no evidence of necrosis at the height of infection and no invasion of lung parenchyma. The peak of infection is reached about the 28th day and the disease progresses steadily until the 5th week with great increase in lymphoid tissue (Pl. 1, fig. 1). In addition, in bronchial lymph nodes there is abscess formation with peripheral necrosis. By about the 16th week the whole lymphatic system is rapidly regenerating and few organisms can be recovered from the spleen or other parts of the reticulo-endothelial system.

Guinea pigs so infected can be shown to possess a high degree of non-specific resistance to a secondary infection or none, according to the secondary invader or the route of its entry into the host. If one chooses as secondary invader an organism such as *Bacillus anthracis*, given by the respiratory route so that it invades the lung by a similar pathway to *Brucella suis*, then for a period of some weeks a high degree of non-specific resistance to anthrax is observed (Table 1). This first appears about the 3rd week after the primary infection and lasts until about the 8th week. By the time there is evidence of lymphatic repair, all non-specific resistance has gone. This type of result occurs no matter by what route the preliminary infection with *B. suis* is given because in the guinea pig this disease, however incurred, is characterized by a general-

Table 1. *Inhibitory action of brucellosis on respiratory infection with* Bacillus anthracis

	No. of guinea pigs		Autopsy of survivors
Exposed to	Tested	Died	
B	30	0	27 (+B)
A	60	54 (+A)	Not examined
B and A	60	10 (+B, +A)	42 (+B)
.		5 (+A, −B)	2 (−B)
.		1 (+B, −A)	.

Animals exposed to anthrax spores 21 days after respiratory infection with *Brucella suis*. B = *B. suis*; A = *B. anthracis*; + = positive culture; − = negative culture.

ized infection of the lymphatic system with hyperplasia and marked monocyte and lymphocyte infiltration. This does not apply to the secondary invader, *B. anthracis*. If the portal of entry is other than by the respiratory route, for example intracutaneously, from which a nidus of continued multiplication develops, the non-specific resistance is overcome and the animals are no more resistant to *B. anthracis* than control animals (Table 2).

Table 2. *Specific immunity to anthrax in guinea pigs with brucellosis after surviving exposure to anthrax spores*

	Secondary challenge with A* No. of animals		Re-challenge of survivors with A† No. of animals	
	Tested	Survived	Tested	Survived
B+Resp. A	20	19	19	11
B+intracut. A	20	1		
Resp. A control	20	3	.	
Intracut. A control	20	2	.	.

B = *Brucella suis*; A = *Bacillus anthracis*.
* Twenty-one days after exposure to *B. suis*.
† Intracutaneously 35 days later; of 20 control animals injected at the same time 3 survived.

It was interesting to find that when anthrax spores were inhaled by animals infected with brucellosis, they made a sufficiently serious attempt to invade the lymphatic system as to induce a specific immunity to anthrax. Thus, some of these animals could later be shown to be specifically immune to an intracutaneous challenge with anthrax spores (Table 2). No such immunity develops in animals without brucellosis but which survive exposure to a cloud of anthrax spores.

If *Pasteurella pestis* is the secondary invader, non-specific resistance induced by brucellosis can be demonstrated only if the organisms are

presented to the animal in a cloud of large, even-sized particles (say
12μ diameter). These are known to be retained in the upper respiratory
tract and to invade through the upper lymphatics. This leads in normal
animals to septicaemia and death with no trace of pneumonia. On the
other hand, if *P. pestis* is presented to the host as a cloud of single
particles which penetrate to the deeper parts of the lung, a bronchopneumonia is induced which rapidly leads to septicaemia and death in
normal animals, and primary infection with *Brucella suis* does not alter
the result (Table 3).

Table 3. *Effect of brucellosis by respiratory route on subsequent
infection with* Pasteurella pestis

	No. of guinea pigs		Autopsy of survivors
Exposed to	Tested	Died	
B	17	0	15 (+B)
B and P (single cell particles)	30	26 (+P, +B)	1 (+B, −P)
		3 (+P, −B)	.
B and P (12μ diameter particles)	30	5 (+P, +B)	21 (+B, −P)
		2 (+P, −B)	2 (−B, −P)
P (single cell particles)	25	25 (+P)	.
P (12μ diameter particles)	25	23 (+P)	2 (−P)

Animals exposed to *P. pestis* 21 days after infection with *Brucella suis*.
B = *B. suis*; P = *P. pestis*; + = positive culture; − = negative culture.

Similar experiments were made using *Mycobacterium tuberculosis*
as the secondary invader. A bovine strain (Vallée) was chosen because
it could be grown in such a way that suspensions of single organisms
and units of them about $2–3\mu$ in diameter could be obtained. One to
ten of these particles retained in the lung infected all mice. Macroscopic
tubercles were apparent in the lungs 10–14 days after infection, at the
sites of primary infection, in the alveolar wall and associated lymphoid
tissue. Secondary spread occurred via the peribronchial lymphatics to
the bronchial lymph nodes. When guinea pigs with active brucellosis
were exposed to a cloud of *M. tuberculosis* a marked delay was observed
in the development of the secondary infection but there was no evidence of total suppression of tuberculosis. However, when *M. tuberculosis* was the primary invader and *Brucella suis* the secondary one,
given 21 days after primary infection, a pronounced non-specific resistance
to infection with *B. suis* resulted (Table 4). Similarly, when anthrax
was the secondary invader after *M. tuberculosis* a high degree of resistance was observed.

Similar findings to those we have recorded have already been
reported by Pullinger (1938) on the influence of infection with *Myco-*

bacterium tuberculosis in reducing the severity of secondary infection with *Brucella abortus* in guinea pigs, and Mica, Goodlow, Victor & Braun (1954) also showed that guinea pigs infected with brucellosis were insusceptible to secondary infection with *Coxiella burnetii*.

Table 4. *Effect of pulmonary tuberculosis on development of brucellosis*

Guinea pigs examined; weeks after exposure to *Brucella suis*

Exposed to	2 weeks	3 weeks	4 weeks	5 weeks
TB	10/10 (+TB)	10/10 (+TB)	10/10 (+TB)	10/10 (+TB)
TB and B	10/10 (+TB)	10/10 (+TB)	10/10 (+TB)	10/10 (+TB)
	1/10 (+B)	1/10 (+B)	1/10 (+B)	2/10 (+B)
B	.	.	.	17/20 (+B)

Animals exposed to *B. suis* 21 days after infection with *Mycobacterium tuberculosis*. TB = *M. tuberculosis*; B = *B. suis*; + = positive culture.

A few experiments were made with guinea pigs using the same groups of organisms given simultaneously by the respiratory route. There was no evidence of suppression of one infection by another nor any evidence of synergic action.

These results suggested that it is solely the response to primary infection of the lymphatic tissue, which the secondary invader attacks, that determines the fate of the secondary organisms and therefore that of the host. If so, the outcome ought to be different in a host with a different reaction to the primary invader or in which the secondary infection takes another course. This speculation led us to similar experiments with mice.

Infections in the mouse

The pathogenesis of primary respiratory infection in the mouse with *Mycobacterium tuberculosis* and *Pasteurella pestis* follows closely that occurring in the guinea pig but the reaction of the mouse to infection with *Brucella suis* or *Bacillus anthracis* is quite different.

With *Brucella suis* the respiratory infective dose is similar in both species but the course of the infection is different. Peribronchial or perivascular lymphoid hyperplasia is minimal or absent in the mouse. Instead there is a rapid infection of lung parenchyma commencing in the alveolar spaces and by about the 4th day of infection small abscesses are well developed (Pl. 1, fig. 2). Cellular change reaches a peak about the 14th day and by 6 weeks all that remain are small fibrotic foci. The respiratory infective dose (LD90) with *Bacillus anthracis* in the mouse is high (about 10^5 spores). The first evidence of infection is found about 24 hr. after exposure, when vegetative organisms are seen in the alveolar spaces. There is no evidence of infection of lymphoid or lymphatic tissue. At death, there are no obvious changes in lung tissue but capillaries are packed with bacilli (Pl. 2, fig. 3). The essential difference, therefore, in infection with these organisms in the guinea pig and the mouse, is that in the former lymphoid and lymphatic invasion is dominant; in the latter it is minimal or absent.

When *Brucella suis* was the primary invader in the mouse there was no evidence of development of non-specific resistance to secondary exposure to *Bacillus anthracis*, *Pasteurella pestis* or *Mycobacterium tuberculosis* at any stage after the primary infection. Similar results were obtained when *M. tuberculosis* was the first invader. Further, there was no evidence of synergic action between any two invaders or of increased susceptibility to the secondary one. In fact, each disease ran a normal course or, if say, *B. anthracis* and *P. pestis* acted together, all that resulted was an additive effect.

We have been unable to demonstrate non-specific resistance in the mouse. This is in contrast to previous work with this species as, for example, the studies of the 'properdin system' (Pillemer *et al.* 1954) or the earlier work of Klein (1893). What may account for our failure to demonstrate this type of reaction? The choice of organisms of high virulence in our tests and their portal of entry into the mouse could be responsible. The organisms used by other workers have been of low virulence. Ledingham (1931), reviewing an already large literature on this subject, stated that if 'an animal has the power to deal successfully with small numbers of an organism which only in large doses and at a particular site is capable of producing a lethal effect there is no doubt that the sum total of the local and general defences can definitely be enhanced by artificial stimuli'. In respiratory infections such as we have used, the portal of entry in the mouse is through the alveolar spaces and not, as in the guinea pig, directly through lymphatic tissue. This could be considered a situation similar to that obtaining in the guinea pig with brucellosis when it is challenged intracutaneously with anthrax spores and no evidence of non-specific immunity is found.

Synergic action between invaders

A classical example of this type of action was first described by Shope (1931). He studied the aetiology of swine influenza, which is characterized by the onset of high fever followed by lobar pneumonia and frequently death. It was already known that an extract of macerated lung from affected animals could produce the typical disease in normal swine when instilled intranasally. Shope isolated *Haemophilus influenzae* from diseased lung tissue but pure cultures of this organism failed to produce the disease. Later he isolated from similar tissue a virus which, by itself, could not produce the typical disease although it induced a mild pyrexia and subclinical lung lesions. However, when swine were infected first with the virus and then with *H. influenzae*, characteristic symptoms of swine influenza followed. The combination led to lobar pneumonia, septicaemia and death. The development of lobar pneumonia suggests that the site of bacterial invasion is in the alveolar region as in the experiments described above in the mouse. Had it been otherwise, an associated bronchopneumonia would have been expected.

Another interesting example of this type of combined infection is mouse hepatitis, first described by Gledhill & Andrewes (1951) and further explored by Gledhill, Dick & Niven (1955). It transpired that certain strains of mouse carried a latent virus which could be activated by a parasite present in some other strains of mice. The latter is *Eperythrozoon coccoides*, 'related to *Haemobartonella* and variously recorded as a bacterium or a protozoon'. The virus given alone produced only small areas of hepatic necrosis and the disease was self-limiting with the virus remaining in a latent form. The protozoon produced no macroscopic pathological changes. When the two agents were given to mice normally free of them, a rapid and highly fatal disease resulted. Massive necrotic hepatic lesions appeared and there was a great increase in both virus and protozoon in practically all tissues until death supervened.

There is also circumstantial evidence from clinical observations in naturally occurring disease in man that excitation of one infection by another can occur. To take one example, patients carrying a latent virus of *Herpes simplex* not infrequently suffer from its eruption during the latter stages of some upper respiratory infections. Whether this is a specific reaction is doubtful, but it is of sufficiently frequent occurrence to be suggestive, although the mechanism of release of the *Herpes* virus remains unknown.

We have observed similar types of reaction in experiments in the mouse using influenza virus as a primary or secondary invader. For most of the experiments, a mouse-adapted strain of the Asiatic type was used. One or two egg-infecting units of the virus retained in the lung resulted in approximately 100 % infection. In the present experiments, however, 12–20 egg-infecting units were mostly used. The disease ran a relatively mild course although about 10 % of the animals died. The first pathological changes were seen about the second day; there was polymorph infiltration of the main bronchi and some degeneration of the epithelium in the region. Viral antigen was detected in these areas by fluorescent labelled antiserum. By the 3rd day infection had reached the alveoli. Some pneumonic lesions had appeared and in these areas alveoli were filled with oedema fluid, macrophages, a few poly-morphs and cell debris. By this time the epithelium of the main bronchi was regenerating and nearly back to normal. Around the 8th day the areas of lung consolidation seemed maximal and there was prominent peribronchial and perivascular lymphatic infiltration. Thereafter the disease most usually regressed and around the 14th day the epithelium of the alveolar walls had regenerated.

Table 5. *Effect on mice of anthrax spores given at intervals after exposure to influenza virus*

No. of animals dying and their median death time (M.D.T.)* when 40 of them were challenged with anthrax at the stated time after exposure to influenza virus

Animals exposed to	0 hr.		4 days		2 weeks		3 weeks		5 weeks	
	Died	M.D.T.	Died	M.D.T.	Died	M.D.T.	Died	M.D.T.	Died	M.D.T.
Virus + *Bacillus anthracis*	35	7	24	3	13	4	16	3	19	4
Anthrax (controls)	8	3	7	3	7	5	3	4	7	8
Virus (controls)	2	9	0	.	0	.	0	.	0	.

Both infections given by the respiratory route.
* In days.

Experiments were made in which anthrax spores were given by the respiratory route either simultaneously with influenza virus or at varying intervals after it (Table 5). It is clear that the lung damage produced by the virus caused a great increase in the death rate from anthrax. The effect remained, although in decreasing degree, until at least the 35th day after exposure to influenza. Histologically the picture in combined infection followed closely that seen when the organisms were given separately; that is to say there was no evidence of increased severity of the influenza lesions or increase in virus titre, and the anthrax followed the normal course in mice, with germination and early multiplication confined to the alveoli. It is surprising, therefore, that the enhancing effect should have been evident until at least the 35th day, for histologically it was shown that by about the 14th day alveolar repair was practically complete, with regenerated epithelium.

It was known that in mice which survived exposure to anthrax spores by the respiratory route, spores could be recovered from the lung for some weeks after deaths from anthrax had ceased. However, superinfection of such animals with influenza virus re-awakened the latent spores and death frequently followed (Table 6).

Table 6. *Effect on mice of influenza virus given at intervals after
exposure to anthrax spores*

No. of animals dying and their median death time (M.D.T.)*
when challenged with influenza virus at the stated time after
exposure to anthrax

Animals exposed to	2 weeks			3 weeks			5 weeks		
	Tested	Died	M.D.T.	Tested	Died	M.D.T.	Tested	Died	M.D.T.
Anthrax + virus	100	43	8–9	50	22	7	50	6	10
Anthrax (controls)	30	0	.	30	0	.	30	0	.
Virus (controls)	50	2	8–9	50	5	10–11	50	4	11

Both infections given by the respiratory route.
* In days.

The latency of anthrax spores in the lung had led us, in earlier
experiments, to test the effect of penicillin in preventing or delaying
death. If penicillin was given daily for 5 days commencing the day after
exposure to anthrax, there was a marked reduction in the death rate and
a marked delay in death time. We tested the effect of superimposing
anthrax spores on an established influenza infection and then treating
mice with penicillin as described above (Table 7). The treatment was
still effective, but less so than in control animals given only anthrax
spores and penicillin.

Table 7. *Effect of penicillin on mice with 4-day influenza
exposed to respiratory anthrax*

Animals exposed to	No. of mice		M.D.T. (days)	Autopsy of survivors 36 days after A
	Tested	Died		
V+A+P	40	10 (+A)	11	24 (−A)
V+A	40	30 (+A)	3	10 (−A)
A	40	7 (+A)	2	33 (−A)
A+P	40	4 (+A)	9	36 (−A)
V	40	0	.	.

V = Influenza virus; A = *Bacillus anthracis*; P = penicillin; + = positive culture;
− = negative culture.
M.D.T. = Median death time.
Penicillin given daily for 5 days commencing 1 hr. after respiratory exposure to *B. anthracis*.

Experiments in which *Brucella suis* was the primary invader provided evidence that the infection could be aggravated when influenza
virus was superimposed at various times after infection. This was shown
by about a 100-fold increase in the number of organisms recoverable
from the lungs. Interestingly, there was no detectable increase in the
severity of lesions produced by the *B. suis* over that occurring when the
organism was given alone.

EXPERIMENTS *IN VITRO*

Many attempts have been made *in vitro* to analyse the part that cellular and humoral immunity may play in the specific control of various diseases, as well as in the type of non-specific resistance I have described. This has been done by introducing infective organisms into various combinations of normal and immune serum and phagocytes, and observing the rate of growth of the phagocyte-ingested organisms and the resulting damage to the phagocytes. The results have been discordant. Much of the earlier work was abortive, showing no difference in growth of *Mycobacterium tuberculosis*, for example, in an immune serum-cell mixture or a similar normal serum-cell complex (Rich & McCordock, 1929). Some of the more recent data follow the same pattern. For example, Mackaness (1954) not only found no difference in the growth rate of *M. tuberculosis* in immune and normal serum-cell mixtures, but noted that 'at all levels of infection immune monocyte cultures always died sooner than those of normal cells if the two had been infected with equivalent numbers of bacilli'. He thought that this might be a result of cellular hypersensitivity. Clearly these failures to demonstrate a difference in growth rates of bacteria within the immune and normal systems might be attributed to preparative and cultural procedures *in vitro*, for experiments *in vivo* had proved otherwise. Lurie (1942), for example, summarizing his earlier work on immunization against tuberculosis, states 'that bacilli of re-infection are either destroyed at once, if small numbers penetrate the tissues, or fail to multiply in the immune animal if large numbers invade the body, whereas in the normal individual the bacilli, whether many or few, at first grow unhindered'. Recently a large number of papers have appeared demonstrating inhibition of growth of pathogenic organisms by immune systems *in vitro*. For example, Suter (1953) found that monocytes from animals vaccinated against the tubercle bacillus markedly inhibited the growth of *M. tuberculosis in vitro*, and Fong, Schneider & Elberg (1956) made a similar observation. Later Elberg (1960) described the same phenomenon in work with *Brucella melitensis*. These workers also found that non-specific resistance between brucellosis and tuberculosis can be shown *in vitro*, as we have shown it *in vivo*. Serum and cells from animals immune to brucellosis inhibited the growth of *M. tuberculosis*, and a similar system taken from animals immunized against tuberculosis suppressed the growth of *B. melitensis*.

We became interested in these experiments in the hope of explaining non-specific resistance observed in mixed infections *in vivo*. The general details of the techniques we employed were as described by Suter (1953) and Elberg (1960). However, we decided not only to examine serum-cell systems *in vitro* by microscopic techniques but to make accurate assessment of bacterial numbers by cultural methods. The latter necessitated serum-cell mixtures in quantities of about 2 ml. and Carrel flasks were used instead of the chambers described by Suter and Elberg. For microscopic examination of cells and bacteria similar flasks were used, each containing several cover slips on to which the cells became evenly deposited. The coverslips were removed at intervals for microscopic examination of the deposited cells.

In the first instance, we examined *in vitro* the behaviour of *Brucella suis* ingested by monocytes from normal guinea pigs and rabbits and those from animals immunized against further attack by *B. suis*. The infected monocytes were re-suspended and incubated in 20–40 % (v/v) homologous serum or plasma in a balanced salt mixture. The immune monocyte reacted differently from normal cells even at relatively low levels of ingestion (e.g. 1 *B. suis* per monocyte). The intracellular organisms multiplied at a greater rate in immune than in normal phagocytes and the immune cells also disintegrated more rapidly (Table 8). In many respects the results were similar to those described by Rich & McCordock (1929) and Mackaness (1954).

Table 8. *Growth of* Brucella suis *in immune and normal* monocytes in vitro

Time (hr.) after ingestion	Type of monocyte	Average no. B. suis per monocyte
0	Immune	0·95
	Normal	0·50
24	Immune	5·66
	Normal	0·79
48	Immune	276
	Normal	39·4

Monocytes suspended in mixtures of homologous serum and balanced salt solution.

This failure to demonstrate *in vitro* the immunity clearly shown *in vivo* prompted us to combine techniques *in vitro* with those *in vivo*. Lurie (1942) showed that if monocytes from animals vaccinated with *Mycobacterium tuberculosis*, and those from normal animals were parasitized *in vitro* with virulent tubercle bacilli, and injected into the anterior chamber of rabbits' eyes, immune monocytes in one eye and normal monocytes in the other, the immune monocytes suppressed the multiplication of their ingested organisms; the normal monocytes did not. We used a modification of Lurie's technique with encouraging results.

First, guinea pigs were infected with *Brucella suis* by the respiratory route. After the infection had subsided and when the organisms could no longer be recovered from any tissues or fluids other than the lung, monocytes were collected from the peritoneum by the techniques we had used in the earlier experiments. They were parasitized as in our tests *in vitro*, and then injected intraperitoneally into normal guinea pigs. The dose of intracellular organisms was approximately that needed to produce 90–100 % infection with unphagocytozed organisms (*c.* 200 organisms). To achieve this, various adjustments of bacterial population for ingestion had been found necessary in earlier experiments. Immune monocytes in normal serum ingest 80–90 % of the organisms presented to them, while normal cells in normal serum ingest only 10–20 % of the bacilli. This necessitated an increase in the bacterial population presented to normal monocytes so that a like dose of ingested organisms was injected into the test animals. On the other hand, no change in bacterial concentration for ingestion was required when normal monocytes were suspended in immune serum, for again about 90 % ingestion occurred. After ingestion, organisms remaining extracellularly were eliminated by incubation of the system in a mixture of serum and salt solution containing streptomycin. The monocytes were then well washed with the same serum mixture without streptomycin and finally resuspended for injection in such a mixture. Typical findings including those from various combinations of serum and monocytes are given in Tables 9–11.

Whatever relation these findings may have to the disease process as a whole, the immediate results are statistically unequivocal. Table 9 shows the dramatic difference between the immune and normal system. Table 10 gives the combined results of the three experiments. It is seen first that immune monocytes do not appear to require immune serum either for ingestion or for effecting destruction of bacilli; secondly, that the immune serum not only influences the uptake of bacilli by normal monocytes but plays some part in intracellular destruction. Table 11 shows that immune serum alone can probably play a small part in protection. This may be due to the influence it has on the normal phagocytic system of the animal injected, *vide infra*, rather than to direct destruction of organisms. The growth *in vitro* of *Brucella suis* in normal and immune serum was examined. A mixture (5 ml.) of 20 % serum in the balanced salt solution was inoculated with about 200 organisms. This inoculum was used because it had provided the ID 90 dose for the guinea pig by the intraperitoneal route and was similar to that used in the monocyte experiments. Quite a marked difference between normal and immune serum cultures was observed. The growth in immune serum proceeded more slowly than in normal serum so that after 6 days' incubation the immune serum culture reached the peak of 10^6 organisms per ml., whereas the normal serum cultures rose to about 10^9 organisms per ml.

Table 9. *Infectivity of* Brucella suis *ingested by monocytes* in vitro *then injected into guinea pigs*

System		% B. suis ingested *in vitro*	No. of intracellular organisms injected	No. of guinea pigs	
Monocyte	Serum			Tested	Infected
Immune	Immune	85	329	20	0
Normal	Normal	16	380	20	18
B. suis infection (control)		.	125*	20	17

* Not intracellular.

Table 10. *Effect of serum on ingestion of* Brucella suis *in immune and normal monocytes and on subsequent infection in the guinea pig*

System		% B. suis ingested *in vitro*	No. of intracellular organisms injected	No. of guinea pigs	
Monocyte	Serum			Tested	Infected
Immune	Immune	91	216	60	9
Immune	Normal	90	183	60	9
Normal	Normal	15	259	60	54
Normal	Immune	97	276	60	36
B. suis infection (control)		.	126*	60	54

* Not intracellular.

Table 11. *Infectivity of* Brucella suis *in immune and normal monocyte-serum systems when injected into normal guinea pigs*

System		% B. suis ingested *in vitro*	No. of intracellular organisms injected	No. of guinea pigs	
Monocyte	Serum			Tested	Infected
Immune	Immune	100	294	20	3
.	Immune	.	190*	20	13
Normal	Normal	17	265	20	17
.	Normal	.	225*	20	19
B. suis infection (control)		.	135*	20	19

* Not intracellular.

It was interesting to find in all these experiments that, when infection followed the administration of the pathogenic organism in an immune system, the disease was much less acute than in control animals or in those receiving the organisms in a normal system. This was observed by gross and microscopic examination of the lesions and by the number of organisms recovered from them. Animals were normally examined 28 days after injection of the test material. It is possible, therefore, that the disease in the instances noted was slower in developing. However, no significant difference in the picture was seen when such animals were examined 56 days after injection.

The immune monocyte appears to play a dominant role in protection. Immune serum seems of only secondary importance. However, one point remains unresolved. It is known that even after 'thorough washing' of monocytes it is practically impossible to remove all traces of normally associated serum. If this does remain adherent it is possible that its effect is greater than the experiments indicate. Further, the fact that immune serum has a retarding effect on the growth of *Brucella suis in vitro* may indicate that when given intraperitoneally (Table 11) its effect may be similar and it might also enable the normal monocytes to ingest developing organisms and destroy them.

The main object of this type of experiment was to examine the role that humoral and cellular factors played in preventing secondary infection with a pathogenic organism unrelated to the primary invader. So far we can report only negative results. When *Bacillus anthracis* was phagocytozed by monocytes from guinea pigs with active brucellosis and injected intraperitoneally in an immune serum–salt solution into normal guinea pigs no evidence of protection against anthrax was observed. This failure may be due to the ease with which *B. anthracis* can multiply extracellularly. Similar experiments are now planned with organisms which are more strictly intracellular in their growth.

I have chosen to discuss in some detail only one aspect of mixed populations *in vivo* and *in vitro* for the reason that it is the only one with which I have been actively engaged. Clearly the fields for study are legion and may vary from academic exercises to attempts to understand the factors involved in suppression or exacerbation of the activities of one organism by another. In any circumstance a first essential is the isolation of the organisms in pure culture. Thereafter the choice of the menstruum in which the mixed population might best be studied clearly depends on the object of the exercise. If it is to understand the factors involved in naturally occurring processes the object should be to use a menstruum as near the natural state as can be devised. If it is likely that direct action of one organism on another is responsible for the effect as observed in nature, direct observations *in vitro* might best be the first choice and here the classical example is in antibiotic reactions. However, if there is reason to suspect that it is the reaction of the 'host' to 'invasion' by one organism that determines the fate of the second it might be best to carry out part of the reaction *in vitro* and part *in vivo* as in the example given in this paper.

The experimental work described was done in close collaboration with my colleagues Mr S. Peacock, Mr W. J. Randles and Mr L. P. Packman.

REFERENCES

BOHNHOFF, M. & MILLER, C. P. (1962). Enhanced susceptibility to *Salmonella* infection in streptomycin-treated mice. *J. infect. Dis.* **111**, 117.

ELBERG, S. S. (1960). Cellular immunity. *Bact. Rev.* **24**, 67.

FONG, J., SCHNEIDER, P. & ELBERG, S. S. (1956). Studies on tubercle bacillus-monocyte relationship. 1. Quantitative analysis of effect of serum of animals vaccinated with B.C.G. upon bacterium-monocyte system. *J. exp. Med.* **105**, 455.

GLEDHILL, A. W. & ANDREWES, C. H. (1951). A hepatitis virus of mice. *Brit. J. exp. Path.* **32**, 559.

GLEDHILL, A. W., DICK, G. W. A. & NIVEN, J. S. F. (1955). Mouse hepatitis virus and its pathogenic action. *J. Path. Bact.* **69**, 299.

HENDERSON, D. W. (1960). Bacterial interference. *Bact. Rev.* **24**, 167.

ISAACS, A. & LINDENMANN, J. (1957). Virus interference. 1. The interferon. *Proc. roy. Soc.* B, **147**, 258.

KLEIN, E. (1893). The anticholera vaccination: an experimental critique. *Brit. med. J.* i, 632.

LEDINGHAM, J. C. G. (1931). *A System of Bacteriology in Relation to Medicine*, 6, 45. London: H.M. Stationery Office.

LURIE, M. B. (1942). Studies on the mechanism of immunity in tuberculosis. The fate of tubercle bacilli ingested by mononuclear phagocytes derived from normal and immunized animals. *J. exp. Med.* **75**, 247.

MACKANESS, G. B. (1954). Growth of tubercle bacilli in monocytes from normal and vaccinated rabbits. *Amer. Rev. Tuberc.* **69**, 495.

MEAGHER, E. T. (1929). *Board of Control on General Paralysis and its Treatment by Induced Malaria*. London: H.M. Stationery Office.

MICA, L. A., GOODLOW, R. J., VICTOR, J. & BRAUN, W. (1954). Studies on mixed infections. 1. Brucellosis and Q fever. *Proc. Soc. exp. Biol., N.Y.* **87**, 500.

PILLEMER, L., BLUM, L., LEPOW, I. H., ROSS, D. A., TODD, E. W. & WARDLAW, A. C. (1954). The properdin system and immunity. *Science*, **120**, 279.

PULLINGER, E. J. (1938). Induced tissue resistance to *Brucella abortus* infection. *J. Path. Bact.* **47**, 413.

RICH, A. R. & MCCORDOCK, H. A. (1929). An enquiry concerning the role of allergy, immunity and other factors of importance in the pathogenesis of human tuberculosis. *Johns Hopk. Hosp. Bull.* **44**, 273.

SCHLESINGER, R. W. (1959). Interference between animal viruses. In *The Viruses*, 3, 157. Ed. by F. M. Burnet & W. M. Stanley. New York: Academic Press.

SHOPE, R. E. (1931). Swine influenza. III. Filtration experiments and aetiology. *J. exp. Med.* **54**, 373.

SUTER, E. (1953). Multiplication of tubercle bacilli within mononuclear phagocytes in tissue cultures derived from normal animals and animals vaccinated with B.C.G. *J. exp. Med.* **97**, 235.

256   D. W. HENDERSON

EXPLANATION OF PLATES

PLATE 1

Fig. 1. Lung of guinea pig 5 weeks after respiratory infection with *Brucella suis*, showing massive peribronchial lymphoid hyperplasia. There is no evidence of alveolar abscesses or invasion of lung parenchyma. Haematoxylin-eosin stain, × 80.

Fig. 2. Lung of mouse 4 days after respiratory infection with *Brucella suis*. Abscess formation in lung parenchyma is commencing in alveolar spaces. Haematoxylin-eosin stain, × 80.

PLATE 2

Fig. 3. Lung of mouse dead of anthrax, showing a bronchial lymph node with no evidence of bacterial invasion. Vegetative organisms indicated by arrow are confined to blood capillaries. Haematoxylin-eosin stain, × 660.

PLATE 1

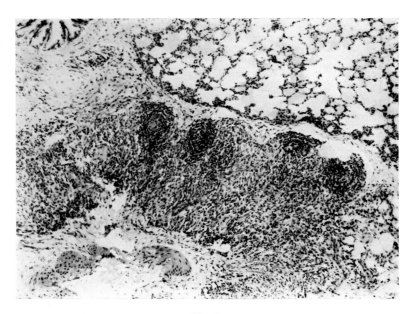

Fig. 1

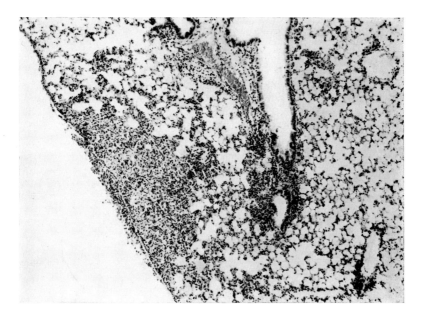

Fig. 2

PLATE 2

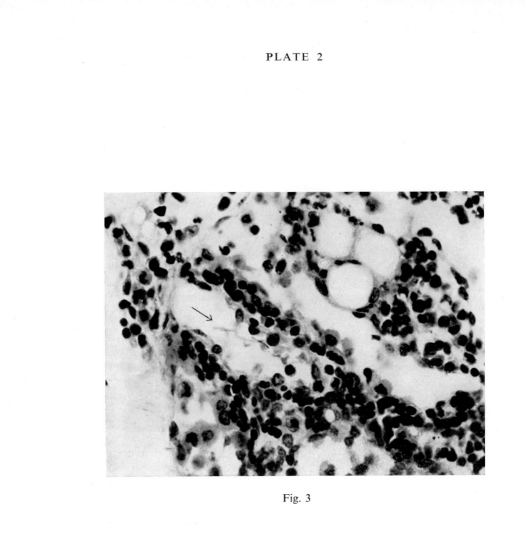

Fig. 3

VIRUSES IN ANIMALS AND IN CELL CULTURE

J. J. HOLLAND

Department of Microbiology, University of Washington, Seattle, Washington

It would seem that discussion of true viruses is to a large extent outside the content of this Symposium since viral activities are confined, with but occasional exceptions, to genetic redirection of host–cell metabolism. Studies of the mechanisms of virus infection of animal cells in cell-free systems have only recently begun to gain momentum. Therefore, this review will be restricted mainly to consideration of virus disease as elucidated by recent studies using animal cells cultured and infected *in vitro*, and to comparison with findings *in vivo*. Great breadth of coverage is not attempted in this review due to the proliferation of literature in animal virology, to the diversity of host–virus systems and more specifically to the narrow range of competence of the reviewer. Rather, a number of studies are cited which appear to clarify certain aspects of host–virus interaction. Heavy emphasis is given to poliovirus studies because these viruses have been so thoroughly investigated and because this is an excellent model system.

POSSIBLE MECHANISMS OF VIRUS-INDUCED CELL PATHOLOGY IN CELLS GROWN *IN VITRO*

Before attempting to discuss virus-mediated pathology at the level of tissues, organs, and intact animals, it would be well to consider the mechanisms by which viruses alter and destroy animal cells grown *in vitro*.

A molecular basis for cytopathogenic effects has not been clearly established for any animal virus infection of cultured cells. Boyer, Leuchtenberger & Ginsberg (1957) showed that heat-inactivated adenovirus preparations contained a toxic factor which caused early, reversible clumping of HeLa cells. Pereira (1958) separated this 'toxin' from the virus particles by centrifugation, and later it was purified and characterized chromatographically and immunologically (Wilcox & Ginsberg, 1961; Klemperer & Pereira, 1959). Ackermann, Payne & Kurtz (1958) reported preliminary evidence for a weakly toxic material elaborated by poliovirus-infected HeLa cells, but no thorough characterization of a poliovirus 'toxin' has been reported subsequently. There are a number of reports of 'virus toxicity' in cells exposed to a variety of viruses which are incapable of forming complete infectious virus within these cells, but these are cases of abortive cycles of infection which fail to culminate in production of mature infectious virus.

There is little evidence that animal viruses cause cytopathogenic effects mainly as a result of toxin production in infected cells. By analogy with bacterial virus systems it is more likely that most virulent animal viruses injure and destroy their host cells by interfering with essential metabolic functions. Lysis of infected bacteria appears to be mediated by endolysins produced, or activated, by virus infection. But the relatively slow release of most animal viruses from infected cells suggests that these viruses do not promote the production of enzymes causing dramatic lysis (Dulbecco & Vogt, 1954; Howes & Melnick, 1957; Sheek & Magee, 1961). Recent biochemical studies indicate that virulent animal viruses, like bacterial viruses, substitute virus genetic control for host cell gene activity, thereby profoundly altering macromolecular synthesis in infected mammalian cells.

Salzman, Lockhart & Sebring (1959) reported that poliovirus infection of HeLa cells caused inhibition of net synthesis of DNA, RNA, and protein within 6 hr. Holland (1961 a, 1962 a) found that RNA synthesis continued at near normal rates for about 5 hr. following poliovirus infection, but that the RNA produced was not normal cell RNA and resembled purified poliovirus RNA in base composition. This RNA was produced at near normal rates in infected cells treated with actinomycin to prevent host RNA synthesis. Reich, Franklin, Shatkin & Tatum (1961) had shown that RNA viruses multiply normally in actinomycin-treated cells despite inability of these cells to produce normal cellular RNA. Thus, it appeared that RNA animal viruses substitute virus-directed RNA synthesis for host cell directed RNA synthesis soon after infection. Baltimore & Franklin (1962 a) and Holland (1962 b, 1963) demonstrated that host DNA-directed RNA synthesis is in fact suppressed by animal virus infections both in the living cells and in cell-free DNA-primed RNA synthesizing systems. The mechanism of this suppression of HeLa cell genetic activity during poliovirus infection was not determined, but DNA from infected cells was not inhibited in its ability to prime RNA synthesis in a reaction mixture with RNA polymerase of *Escherichia coli* (Holland, 1962 b). Therefore it appears that poliovirus infection renders HeLa cell genes non-functional by some mechanism which does not incapacitate the genetic potential of cell DNA. Franklin & Baltimore (1962) showed that puromycin inhibits this depression of host DNA-primed RNA synthesis, thereby implicating protein synthesis in the process. However, puromycin will prevent virus and virus RNA synthesis at very early stages of infection so its effect may be due to a general inhibition of virus activity. Nomura, Okamoto & Asano (1962) have reported that T4 bacteriophage infection of *E. coli* leads to rapid inhibition of ribosomal and soluble RNA synthesis which can be prevented by stopping protein synthesis with chloramphenicol.

Zimmerman, Heeter & Darnell (1963) and Fenwick (1963) showed that nearly all of the RNA synthesized during poliovirus infection is about the size (37 S) of purified poliovirus RNA. Warner, Madden & Darnell (1963) established that purified poliovirus RNA can act as messenger in stimulating ribosomes to produce protein, at least part of which is antigenically related to virus capsid protein. Finally, Baltimore & Franklin (1962*b*) found a new RNA polymerase in cells infected by RNA animal viruses. This polymerase appears to be associated with its template RNA in a particulate aggregate of some kind.

Thus, it appears that the small virulent RNA animal viruses depress host RNA synthesis and redirect RNA synthesis to production of RNA of viral size and composition through the action of a new polymerase. This virus RNA acts as a template for producing virus capsid protein and possibly other proteins. With such redirection of host RNA synthesis, and with strong inhibition of DNA and protein synthesis in cells infected with RNA viruses (Franklin & Baltimore, 1962), it is obvious that cell death might not require specific 'toxin' production.

Cellular macromolecular synthesis may not be as rapidly altered in mammalian cells infected with many DNA viruses, but there is increasing evidence that these viruses also cause considerable reorientation of metabolic activity analogous to that caused by DNA phages in bacteria. Ginsberg & Dixon (1959), and Green (1962) showed accumulation of abnormal DNA, probably virus DNA, and increase in protein in adenovirus-infected HeLa cells. Sheek & Magee (1961) reported unusual cytoplasmic DNA synthesis or accumulation in vaccinia-infected HeLa cells. Ben-Porat & Kaplan (1963) showed gross synthesis of virus-like DNA in pseudorabies virus-infected cells in stationary phase cultures. Furthermore, a number of recent studies indicate that DNA animal viruses induce new DNA polymerases and nucleoside and nucleotide kinases (Magee, 1962; Hanafusa, 1962; McAuslan & Joklik, 1962; Green, 1962; Kit & Dubbs, 1963).

In short, animal virus nucleic acid like that of phages may cause extensive metabolic deviation in host cells so that death or pathologic behaviour of cells is to be expected. Death of cells may not always ensue, as in the case of cells altered by tumour viruses.

Turning to the effect of viruses *in vivo*, it is a reasonable assumption that virus-induced destruction of infected cells involves essentially the same mechanisms regardless of whether the cells are in the body or cultured *in vitro*. That virus infection of cells *in vivo* causes pathology in these cells similar to that occurring *in vitro* is attested by the similarity of inclusion bodies, giant cells and other morphological changes characteristic of certain viruses infecting cells both inside and outside the body (Enders, 1954). Cheatham (1959) has discussed the similarity between varicella inclusions in human cells *in vivo* and *in vitro*, and the tendency for measles virus to form giant cells in lung, thymus and trachea, analogous to those seen in cell culture. However, it is recognized that not all cellular pathology in virus diseases can be attributed to virus infection of affected cells. Cellular pathology might occur indirectly in uninfected cells after virus infection of other cells has caused disturbance of blood circulation, of innervation, of endocrine secretions, or other changes. Cellular damage may also result from inflammatory or immunological responses affecting virus-infected cells and adjacent uninfected cells. An observation *in vivo* which as yet is not supported by findings *in vitro* is that motor neurons infected by poliovirus *in vivo* may recover from early damage (Bodian, 1949). This observation is hard to reconcile with the metabolic consequences of poliovirus infection of cells *in vitro* which is referred to above.

IMMUNOLOGICAL MECHANISMS OPERATIVE AGAINST VIRUS INFECTION IN ANIMALS

In contrast to phage infection of bacteria and to virus infection of cells cultured *in vitro*, the outcome of virus infection of animals is determined to a great extent by host immune reactions. Generalizations about immunity to virus infections are on uncertain ground since there are variations in virus architecture, tropism and multiplication rate; also differences exist in onset and duration of effective immunity. Duration of immunity ranges from lifelong against measles, mumps and others, to weak, transient immunity against viruses such as the common cold viruses. Also, effective immunity may develop reasonably quickly, or as for rabies virus it may occur so late during a long incubation period as to be uniformly unsuccessful in preventing death unless developed artificially before, or early in the course of infection.

Mode of antibody neutralization

There is little doubt that antiserum will, to some extent, passively protect animals against infection with some viruses; the mode of action of antibody in this respect has been indicated by studies *in vitro*. As with phage systems, antibody in great excess prevents neutralized virus from adsorbing to cells or from entering cells if added soon after adsorption. However, virus neutralized by low concentrations of antiserum is capable of adsorbing but not of 'eclipsing' or 'penetrating' (Rubin, 1957; Mandel, 1958). The electron micrographs of Hummeler, Anderson & Brown (1962) demonstrate that the surface of poliovirus particles may be so extensively coated by specific antibody that adsorption to host cells would be highly improbable. Antibody acts only on virus capsid protein and has no effect on extracted infectious nucleic acid (Alexander, Koch, Morgan-Mountain & Van Damme, 1958). Furthermore, in some cases virus inactivated by antibody can be reactivated by dissociation of antibody (Hummeler & Kettler, 1958). Reactive sites on neutralizing antibody are directed against an antigenic configuration found only when virus capsid subunits are arranged in the form of the intact virus or at least as an incomplete virus capsid. Dispersed capsomeres of tobacco mosaic virus are antigenically distinct from intact virus rods, but reaggregation of the subunits into rods results in the acquisition of original virus antigenicity (Aach, 1959). A similar situation apparently prevails with foot and mouth disease virus which changes antigenicity upon disruption of the capsid (Brown, Cartwright & Stewart, 1961). Poliovirus changes its antigenicity even without disruption of the capsid. Mild heating, drying, ultraviolet light irradiation and a variety of chemical treatments can change poliovirus capsid to a different antigenic specificity (Roizman, Mayer & Roane, 1959; LeBouvier, 1959; Hummeler *et al.* 1962). Although an ideal inactivated virus vaccine would be a concentrated purified capsid protein, preparation of such a vaccine would be greatly complicated by the necessity for maintaining capsid subunits in such juxtaposition as to maintain virus antigenicity.

Antibody neutralization of some animal viruses requires a heat-labile serum accessory factor for optimal virus inactivation. These requirements for a labile accessory factor have been reported mainly for lipid-containing viruses such as arbor viruses (Morgan, 1945; Sabin, 1950), mumps (Leymaster & Ward, 1949) and measles (Adams & Imagawa, 1957). This accessory substance may not be complement but it shows similar characteristics (Leymaster & Ward, 1949; Dozois,

Wagner, Chemerda & Andrew, 1949) and probably acts in concert with antibody to damage virus lipid membrane much as antibody and complement damage cellular membranes.

Antibody neutralization of viruses appears to be exclusively extracellular, presumably because immune globulins do not enter cells except possibly in pinocytosis vacuoles, and because intracellular stages of virus replication would not be susceptible to anti-capsid globulin action. The work of innumerable investigators has established the resistance of intracellular virus to antibody, whereas reports of significant intracellular antiviral action of antibody are not definitive (Furusawa, Hagiwara & Kamahora, 1956).

An observation on infected cells *in vitro* which might have important implications *in vivo* is as follows. Despite the ability of neutralizing antibody to prevent attachment and penetration by animal viruses, some viruses can spread from cell to cell and form plaques in cell monolayers under medium containing neutralizing antibody. If antibody is present when these viruses are added to the cell cultures, plaque formation is prevented, but if antibody is added after virus has adsorbed and penetrated, plaque formation will proceed. Herpes virus spreads from cell to cell in the presence of antibody, and poliovirus is a virus which does not (Black & Melnick, 1955). The ability of viruses to pass from cell to cell in the presence of immune serum may play a large part in the pathogenesis of such diseases as recurrent herpetic lesions which develop despite high levels of humoral antibody. The fact that such lesions do not progress indefinitely but usually regress raises important questions concerning non-antibody mechanisms of virus resistance. The means by which these viruses accomplish reinfection in the presence of antibody remains to be determined, but it is interesting that most viruses exhibiting this behaviour are viruses with a lipid envelope. These lipid viruses often cause giant cell formations, syncitia and other evidence for disturbance of cell membrane structure and function (Roizman, 1962). Thus, cell-to-cell transmission by general membrane fusion or by fusion of small cytoplasmic extrusions might occur without necessity for virus release to the outside.

Mechanisms of virus immunity other than by antibody

The studies by Good (1957) and others of hypogammaglobulinaemic patients demonstrated that they recover from and remain immune to many virus infections about as well as do normal persons, despite their inability to form detectable amounts of antiviral antibody. This response to virus infection is in contrast to the greatly enhanced susceptibility of hypogammaglobulinaemic patients to infections with Gram-positive bacteria. Furthermore, repeated immunization of these patients with viral antigens nearly always failed to evoke detectable levels of neutralizing antibody (Good, 1957). Along similar lines, Friedman & Baron (1961) showed that rabbits irradiated with X-rays and unable to make detectable antibody recovered normally from vaccinia virus infection. On the other hand, measles symptoms may progress despite high levels of neutralizing antibody in the patient's serum (Enders, 1960). Such findings emphasize the importance of little-understood factors other than humoral antibody which play a role in controlling virus infections. Some studies have been made on the nature of these factors using techniques *in vivo* and *in vitro*.

Nearly all cell culture virologists eventually encounter inhibitory effects of 'normal' animal sera on virus growth. There are numerous reports of virus inhibitory substances in normal animal and human sera (Allen, Finkelstein & Sulkin, 1958). While some of these are globulin and may be antibody, many others are not. Various protein and lipid serum components exhibit antiviral activity. The best studied is the 'Francis inhibitor,' a mucoprotein found in serum and other tissues (Francis, 1947). This material contains groupings identical with, or closely related to, the receptor substance of cells able to adsorb influenza virus (Gottschalk, 1957). This inhibitor has proved invaluable for delineation of the structure of host cell receptors for influenza virus, but little is known of the role that this and other virus-inhibitory serum factors may play in host resistance.

The possible participation of neutrophiles, monocytes and other leucocytes in resistance to virus infection has received much attention since Todd (1928) showed association of fowl plague virus with chicken leucocytes *in vivo*. Sabin (1935) demonstrated association of vaccinia virus with phagocytes, and Nungester, Gordon & Collins (1950) reported attachment of influenza virus to guinea-pig leucocytes. Ginsberg & Blackmon (1956) found that influenza virus adsorbed by polymorphonuclear leucocytes quickly lost infectivity but no new virus was produced. No material capable of inactivating virus infectivity was recovered from cell homogenates. Leucocytes which adsorbed virus were metabolically damaged and this was attributed to virus-induced derangement of energy-yielding mechanisms (Fisher & Ginsberg, 1956). Boand, Kempf & Hanson (1957) showed that human and rabbit leucocytes ingest and inactivate influenza virus more efficiently in the presence of immune serum than normal serum. Also peritoneal phagocytes of immunized mice ingested influenza virus more rapidly and cleared the peritoneum of infectious virus more quickly than did those of normal control mice. From these and other studies it appears that leucocytes may take up and inactivate viruses with or without the aid of antibody.

On the other hand, leucocytes might tend to disseminate virus by elution of adsorbed virus after migration across architectural impediments such as the blood–brain barrier. Inglot & Davenport (1962) suggested that leucocytes transport influenza virus to extrapulmonary sites in the body. Some viruses might even multiply within leucocytes. Florman & Enders (1942) reported that vaccinia virus multiplied to low titres in monocytes in culture, although others have not observed this. Cultured monocytes have been shown to support growth of a number of viruses such as poliovirus (Barski & Robbe-Fossat, 1957), mouse hepatitis (Bang & Warwick, 1960), and arbor viruses (Goodman & Koprowski, 1962). Gresser & Chany (1963) reported isolation of measles virus from washed leucocytes taken from the blood of 10 of 11 measles patients in the first day of examination. Berg & Rosenthal (1961) had previously shown that human leucocytes support growth of measles virus *in vitro*.

Production of interferon by virus-infected leucocytes has recently been reported and may be a major component of virus resistance (Gresser, 1961; Glasgow & Habel, 1962). Infected leucocytes were shown to produce significant titres of interferon with no multiplication of virus.

Duration of immunity to virus infection

Until more information is available, the factors determining duration of immunity can only be speculated upon, using as a basis the general behaviour of diseased animals and some pertinent indications from experiments with infected cell cultures. It appears that virus infections of superficial epithelium (e.g. upper respiratory tract infections) do not usually confer solid, long-lasting immunity. This may be because humoral immune mechanisms are less important here, with immunity depending on the local levels of antibody and numbers of leucocytes in the fluids bathing the affected epithelial surfaces. It is logical that in such superficial infections the probability of large amounts of viral antigen reaching immunologically competent cells, and of large amounts of antibody reaching the site of infection, is much lower than in a deep-seated infection.

Conversely, viruses such as measles, mumps and German measles, which may confer lifelong immunity, are often characterized by having tropisms for internal organs and tissues, and by having long incubation periods during which the virus reaches these tissues or organs by way of viraemia. Obviously, this type of pathogenesis offers maximal opportunity for strong generalized antigenic stimulation and, after immunity to the first infection has developed, for subsequent neutralization of reinfecting virus either by preformed circulating antibody or by antibody quickly produced in an anamnestic response which could be effective before the end of a long incubation period.

However, such an explanation for wide variations in duration of immunity cannot account for the patterns of immunity seen in many virus diseases. It cannot explain the poor immunity against serum hepatitis for example. Also, there are many good alternatives to account for differences in immunity. The continued production or persistence of virus or virus antigens throughout life could be the basis for lifelong immunity. In persons who experience recurrent herpetic 'fever blisters' the virus appears to persist in some form for decades, although in this case immunity is not strong enough to prevent recurrences.

Cell culture studies have shown a number of mechanisms by which virus and/or virus antigens might continue to be produced at low levels for indefinite periods. Virus carrier cultures are cell cultures which are chronically infected with a virus in which there is an equilibrium between cell multiplication and virus multiplication (Ginsberg, 1958). There are at least three mechanisms by which the carrier state can be achieved after infection of cell cultures with virulent viruses. Low levels of antibody in the cell-culture medium can promote the carrier state (Ackermann & Kurtz, 1955); production of interferon can also allow a cell-virus equilibrium to be attained (Henle, Henle, Deinhardt & Bergs, 1959; Ho & Enders, 1959) or virus can saturate or alter cell receptors (Crowell & Syverton, 1961). Any one or all of these mechanisms might operate in the body to establish chronic virus infection analogous to cell culture carrier states.

A more intriguing possibility to explain persistence of virus and immunity would be some form of integration of animal virus genes into host cells in a relationship resembling lysogeny in phage-bacterial systems. No conclusive demonstration of animal virus lysogeny has yet taken place. However, there is increasing evidence, among the tumour viruses particularly, that some temperate relationships between animal cell and animal virus genes may exist. Tumour virus-cell relationships will be discussed below, but it is pertinent here to mention that cells may be transformed to malignancy by several DNA-containing tumour viruses and may continuously produce virus-specific antigens without producing mature virus or damaging the cells (Sjögren, Hellström & Klein, 1961; Habel, 1962; Sachs, 1961). There is recent evidence that Rous sarcoma virus, an RNA tumour agent, is able to integrate at least part of its genome into the host cell that it transforms to malignancy (Hanafusa, Hanafusa & Rubin, 1963; Temin, 1963). An interesting phenomenon in human cell cultures inoculated with measles virus has been described by Rustigian (1962). He isolated surviving, multiplying cells from the infected cultures which were completely free of detectable infectious measles virus but which produced distinctive measles virus inclusions and readily detectable quantities of measles virus antigen continuously. If such a situation can occur *in vivo*, its contribution to lifelong anti-measles immunity would be obvious. The entire field of latent infection with animal viruses requires more extensive investigation before its significance *in vivo* can be evaluated.

The methods required for the first isolations of adenoviruses from human adenoids (Rowe *et al.* 1953) suggested that infectious mature adenovirus was not present, but appeared only after cultivation *in vitro* of adenoid cells had triggered its maturation from a latent state. Indeed, it is not impossible that a 'purpose' (i.e. a reason for evolutionary survival) of tonsils and adenoids lies in their capacity to become easily or chronically infected by many viruses and bacteria and thereby to provide rapid or prolonged antigenic stimulus for immune responses. Further work may show that latency is as widespread and as important to the biology of animal viruses as it is to bacteriophages and to plant viruses. Hildebrand (1958) has reviewed knowledge of the important role of latency in plant virus diseases.

Deleterious effects in vivo *of immune response to viruses*

As with other antigens, virus antigens might be expected to elicit tissue-damaging inflammatory infiltrates and hypersensitivity, and numerous pathologists and clinicians have suggested a role for these effects in human virus lesions and symptoms. Studies by Rowe (1956) and Hotchin (1962) indicate that lymphocytic choriomeningitis virus infection of mice may be extensive, chronic and symptomless for years when the virus is given to immunologically incompetent animals such as neonatal or irradiated mice. Yet death occurred in about 1 week after infection of immunologically mature mice. Despite the fact that virus multiplied to the same titre in the two groups, extensive lymphocytic infiltration of viscera and central nervous systems was seen in immunologically mature mice at death, but not in symptomless X-rayed mice. It appeared that disease and death were due to immunologic reactions associated with the lymphocytic invasion, rather than to virus destruction of cells.

Specific immune serum has been shown to be capable of accelerating arbor virus pathology of the mouse central nervous system (Berg *et al.* 1961). Apparently antibody–antigen interaction stimulated perivascular cuffing, glial proliferation and lymphocytic infiltration similar to that which would have occurred much later in the normal virus infection. It is interesting that antihistaminics can protect mice against lethal effects of mouse hepatitis virus, and in organ culture can protect mouse liver parenchymal cells from virus-induced necrosis (Judah, Bjotvedt & Vainio, 1960).

INTERFERENCE BETWEEN ANIMAL VIRUSES

It has long been known that concurrent or sequential infection of the same cells by two related or unrelated animal viruses may lead to depressed multiplication of one or both of the viruses. This interference has been extensively studied (Henle, 1950; Schlesinger, 1959). Even virus inactivated with ultraviolet light and by other means is capable of interfering with infectious virus. Interference is of importance not only because of its usefulness in studying phases of virus replication, but also because it may be a prime factor determining pathogenesis of a virus disease. Also, interference can be an impediment to mass immunization with live vaccines.

Interferon

Production and action of interferon in vitro

A significant advance in the understanding of interference phenomena occurred with the discovery by Isaacs & Lindemann (1957) of an interfering protein produced by virus-infected cells. This substance, interferon, was capable of inhibiting virus growth and cytopathic effects in cells which were exposed to it but had no effect on free virus. It is not certain nor even likely that all viral interference is mediated by interferon, but it has been found in many diverse virus-cell systems. There have been recent reviews (Isaacs, 1962; Ho, 1962; Wagner, 1963), so extensive coverage will not be attempted here.

The original studies of Isaacs and Lindemann provided clear evidence that interferon was protein in nature, but recent attempts by other workers at purification and characterization of interferon have not been in agreement as to molecular weight, isoelectric point and other characters, so further work is necessary before the nature of interferon proteins is established.

The mode of action of interferon is likewise not clearly delineated to date. Isaacs, Klemperer & Hitchcock (1961) suggested that interferon acts by uncoupling oxidative phosphorylation, but Zemla & Schramek (1962) showed it to be active in an anaerobic environment. It is of theoretical and practical importance to learn whether interferon acts in some general way to depress cellular metabolism and therefore virus metabolism, or whether it is a specific repressor of virus function. If interferon inhibits virus replication by interfering with oxidative phosphorylation or some other host cell function, then these effects should be detectable. In fact, concentrated interferon has been reported (Paucker, Cantell & Henle, 1962) to inhibit cell multiplication. The reviewer (Holland, unpublished data) observed slight inhibition of cellular RNA synthesis in normal chick cells treated with concentrated interferon, but there were no such effects at lower concentrations still able to inhibit virus replication. It is clear that interferon acts intracellularly to inhibit replication of virus nucleic acid rather than by preventing some early stage of infection such as adsorption or penetration (DeSomer, Prinzie, Denys & Schonne, 1962; Ho, 1961; Mayer, Sokol & Vilcek, 1961; Grossberg & Holland, 1962). Cocito, DeMaeyer & DeSomer (1962) found that interferon completely prevented virus-induced RNA synthesis, but had little effect on normal cell RNA synthesis. A slight depression of nucleic acid synthesis in normal cells was attributed to impurities in the interferon preparation. It would be most attractive to find that interferons are molecules capable of specifically recognizing and repressing virus genomes, but the lack of virus specificity of interferons, coupled with their host cell specificity, indicate that they are molecules produced under the control of host genes activated by virus infection. It seems, therefore, more likely that they act on cells in some manner that affects virus synthesis more than cell-directed synthesis.

Recent studies (DeMaeyer & Enders, 1961; Isaacs, 1962; Wagner, 1963) of the conditions favouring virus and interferon synthesis indicate that they are independent events dissociated in time, in cells which produce both as a result of virus infection. By controlling temperature it was possible to allow selective synthesis of either interferon or virus, and it appeared that at any given time synthesis of significant quantities of one mitigates against synthesis of the other.

Production and action of interferon in vivo

There has been considerable interest in interferon as a significant factor in recovery from, and resistance to, virus diseases, and in possible therapeutic implications. There have been many reports of protection of animals from lethal effects of one virus by interference due to another virus inoculated into the same area. Recently, attempts have been made to demonstrate interference 'at distant sites' *in vivo* (Grossberg, Hook & Wagner, 1962; Wagner, 1960; Hitchcock & Isaacs, 1960) both to determine possible protection by a diffusible substance, and to see whether local virus multiplication is capable of providing generalized resistance by interference. These studies indicated slight interference in areas of the body other than those in which the interfering virus multiplied, suggesting that interferon was carried throughout the body.

A number of investigators have shown that interferon is produced *in vivo* using several different viruses and animal hosts. Interferon-like material was extracted from tissue or organ homogenates and from allantoic fluid and blood (Gledhill, 1959; Isaacs & Hitchcock, 1960; Hitchcock & Isaacs, 1960; Hitchcock & Porterfield, 1961; Isaacs & Barron, 1960; Grossberg, Hook & Wagner, 1962; Vainio, Gwatkin & Koprowski, 1961). There is no reason to expect that these interferons formed *in vivo* would not protect cells *in vivo* as they do in cell culture. Interferons produced *in vitro* are capable of conferring slight protection on animals and chicken embryos (Isaacs & Westwood, 1959; Hitchcock & Isaacs, 1960; Wagner, 1960). Interferons were most effective when given locally but some systemic protection was observed. More work is required using purified, highly concentrated interferons in large quantities to determine its potency, toxicity, allergenicity and stability *in vivo* before its therapeutic potential can be adequately assessed. Possibly, active stimulation of interferon by inoculation of attentuated or inactivated viruses offers greater promise than passive administration. Schulman & Kilbourne (1963) reported definite interference protection of mice against influenza pneumonitis by intranasal aerosols of influenza virus inactivated with ultraviolet light. Interferon was not detected in the lungs of treated mice so it was not certain whether interference was due to interferon or to some other mechanism.

FACTORS ASSOCIATED WITH CELL SUSCEPTIBILITY OR RESISTANCE TO VIRUSES

Animal viruses vary widely in host specificity from poliovirus which normally infects only certain cells or tissues of man and other higher primates, to arbor viruses, some of which can infect mammals, birds, reptiles and insects. Several plant viruses multiply both in plants and insects. Some viruses, such as rabies virus, are able to infect almost any mammal, but are restricted *in vivo* to multiplication only in relatively few well-defined tissues or organs. It is obvious that an understanding of viral tropisms would help elucidate the basis for disease patterns seen with different viruses both at the level of tissue and organ pathology and at the level of host species or strain susceptibility.

Viral tropisms in vitro

Relatively few attempts have been made to elucidate the factors controlling animal virus host range and tissue tropisms. It has been known for decades that virus mutants with altered tissue affinities or host range can be selected by empirical means, but little has been known of the basis for this alteration. It has also been known (Stone, 1948) that cells susceptible to influenza virus can be rendered relatively resistant by treatment with receptor-destroying enzyme from *Vibrio cholerae*. This enzyme cleaves N-acetyl neuraminic acid from cell surface receptor sites, thereby rendering the cells unable to adsorb virus (Gottschalk, 1957).

Using poliovirus as a model system, the reviewer and colleagues have examined the differences between susceptible primate cells and insusceptible non-primate cells in culture. Since it would be a more complex task to look for significant metabolic differences between susceptible and insusceptible cells, the early stages of virus cell interaction were studied. The development of animal virus plaque techniques and of methods for measuring animal virus adsorption to cultured cell (Dulbecco, 1952) made such a study feasible. Cultured human and monkey cells, which are susceptible to poliovirus, adsorbed virus reproducibly with first-order kinetics; cultured non-primate cells, which fail to respond to poliovirus infection, also failed to adsorb detectable quantities of virus. Cells from a wide variety of non-primates, dogs, cats, pigs, calves, chicks, rabbits, guinea pigs and mice, all failed to produce detectable amounts of virus progeny after inoculation, to show cytopathic changes or to adsorb poliovirus (McLaren, Holland & Syverton, 1959). When susceptible cells were disrupted, material adsorbing virus was demonstrated on insoluble membranes of the cells (Holland & McLaren, 1959). No such virus-adsorbing material could be detected in homogenates of insusceptible non-primate cells. Attempts to render these receptor materials soluble were unsuccessful; they appeared to be proteins associated with insoluble lipoproteins of cell membrane (Holland & McLaren, 1961). This membrane-bound receptor showed the same virus specificities and cofactor requirements for adsorption as whole cells (McLaren, Holland & Syverton, 1960).

If the major determinant of poliovirus species specificity was simply the presence or absence of receptor sites on the cell membrane, non-primate cells, insusceptible to intact virus, should be infected with infectious viral nucleic acid since the protein coat would determine attachment specificity of the intact virus. It proved possible to infect cell cultures, organ cultures, and tissues of living animals of every warm-blooded species inoculated with poliovirus RNA (Holland, McLaren & Syverton, 1959). Also Mountain & Alexander (1959) and DeSomer, Prinzie & Schonne (1959) found that poliovirus RNA replicated in avian and rodent cells. Thus many insusceptible cells have the metabolic capacity to replicate poliovirus but lack those cell membrane components necessary for adsorbing virus and initiating infection. However, virus nucleic acid would not infect fish cells, frog cells, plant cells, bacterial protoplasts or cells of other cold-blooded species tested, and it is clear the host range of viral nucleic acid is limited. Understanding the basis for this limitation may be of great aid in studying virus re-direction of cell synthesis.

Viral tropisms in vivo

Studies similar to those described above for cells cultivated *in vitro* were performed using homogenates of tissues and organs from humans and rhesus monkeys to investigate tissue tropisms in susceptible species (Holland, 1961 b). Again a correlation was found between susceptibility and presence or absence of virus-binding material resembling the receptors found in cultivated cells. Homogenates of human and monkey brain, spinal cord and intestine showed considerable virus-binding activity, but tissues and organs in which virus does not replicate extensively *in vivo* failed to exhibit significant receptor activity, with the unexplained exception of certain liver homogenates. Kunin & Jordan (1961) reported essentially similar findings but found virus-binding material in insusceptible tissues and organs which we have not encountered. For some reason their virus infectivity decayed even in a salt solution containing no tissues. Possibly their virus was unstable or was adsorbing non-specifically to glassware. Quersin-Thiry (1961) found no virus-binding activity in non-cultivated monkey tissue. Kunin & Jordan (1961) observed that certain human breast, stomach and rectal carcinomata adsorbed poliovirus, indicating that malignant change had altered the virus adsorptive capacity of the cell membranes. McLaren (personal communication) found an excellent correlation between poliovirus tissue tropisms and presence or absence of receptor activity in human and rhesus tissues and organs. Therefore, it seems that in poliovirus and probably other virus infections the nature of the disease process depends to a large extent on cell ability to adsorb virus. Obviously, many other factors such as age, trauma, immune response, hormones and body temperature, may play a role in determining the outcome of any particular infection.

If cell resistance to virus is defined simply as inability to support virus multiplication without regard to the fate of the cell, then there are some interesting cases of cell resistance at an intracellular level. Several myxovirus-cell systems have been reported in which virus adsorbs to and enters cells and causes their ultimate death with production of virus subunits, but in which mature infectious virus is not formed. Influenza virus in HeLa cells forms viral haemagglutinins without mature virus formation (Henle, Girardi & Henle, 1955). Newcastle disease virus kills Ehrlich ascites tumour cells but does not multiply in them (Prince & Ginsberg, 1957) and fowl plague virus also undergoes an abortive cycle of infection in L strain mouse cells (Franklin & Breitenfeld, 1959). Franklin & Breitenfeld (1959) clearly defined the basis for the abortive infection of L cells by demonstrating that these cells supported normal production of viral nucleoprotein in the nucleus, and usual amounts of haemagglutinins accumulated in the cytoplasm, but viral nucleoprotein was not released from the nucleus into the cytoplasm so that normal maturation could not take place in these 'insusceptible' but moribund cells.

Cell 'resistance' of this kind could be of great importance in preventing virus multiplication and spread *in vivo* despite the fact that the cells are destroyed by these abortive infections. Such abortive infections take place *in vivo*, and with sufficient viral inocula they can cause gross pathological changes in the host, as shown with influenza virus in mouse brain (Schlesinger, 1950), in mouse lung (Ginsberg, 1954) and in rabbit corneal endothelium (Wilcox *et al.* 1958).

Susceptibility of cells in vivo *as compared with the same cells cultivated* in vitro

Since the signal discovery of Enders, Weller & Robbins (1949) that non-neural human tissues cultivated *in vitro* supported multiplication of poliovirus, it seemed anomalous that the same cells in the body are insusceptible. Kaplan (1955) demonstrated that injection of poliovirus directly into the kidney of living monkeys led neither to virus multiplication nor to detectable organ damage. However, cells from the same kidneys when cultivated as monolayers on glass became completely susceptible to poliovirus multiplication and cytopathic effects. When this phenomenon was subsequently examined from the standpoint of cellular receptor activity, susceptibility was correlated with presence or absence of virus-binding material in homogenates (Holland, 1961 b). Mono-layer cultures of human and monkey kidney cells or human amnion cells acquired receptor activity within several days of cultivation *in vitro*, whereas no receptor could be detected in homogenates of these cells before cultivation. This 'dedifferentiation' of cells to receptor produc-tion and virus susceptibility was not triggered *in vitro* by the culture medium, but rather by loss of normal contact relationships between cells and their substrate membrane, because organ culture of the intact human amnion membrane in the same medium failed to lead to acquisition of virus susceptibility or receptor activity (Holland, 1961 b).

There are other indications that interruption of normal tissue associa-tions leads to acquisition of virus susceptibility. Hermodssen & Wesslin (1958) described extensive multiplication of poliovirus in human embryonic tissue grafted into rats. Chaproniere & Andrewes (1958) showed that myxoma virus and adenovirus multiplied in normally insusceptible tissues after grafting. Quersin-Thiry (1961) reported poliovirus-binding receptor activity in monkey kidney tissues grafted subcutaneously into other monkeys. In agreement with our findings, she could find no virus receptor in these tissues before either grafting or culture *in vitro*. As mentioned previously, Kunin & Jordan (1961) found that human carcinoma cells acquired poliovirus receptor activity. This might be a necessary component of malignant change, but may simply represent concomitant dedifferentiation as a result of the cells being freed from normal tissue relationships in tumours. Koprowska & Koprowski (1957) demonstrated that benzpyrene-induction of cervical carcinomas in mice led to ability of the vaginal tract to support multi-plication of mengo and vesicular stomatitis viruses, whereas no virus was produced in the normal vaginal tract not even after irritation induced by benzpyrene. Considering the cell membrane alterations which may accompany malignant change, it would not be surprising if membrane receptors for viruses were gained or lost during carcinogenesis.

Culture techniques *in vitro* which preserve normal tissue relationships also allow retention of differentiated cell characteristics and virus resistance. Barski, Kourilsky & Cornefert (1957) reported that inocula of poliovirus or adenovirus which quickly destroyed dedifferentiated human or monkey cells *in vitro* had little or no cytopathic effect on differentiated ciliated respiratory epithelium maintained as explant cultures. Herpes virus and vaccinia virus showed greater ability to attack differentiated epithelium explants (Barski, Cornefert & Wallace, 1959. These viruses in general show less host range specificity and less specific tropisms than poliovirus. It may be that these and other lipid-containing viruses, such as arbor viruses, can adsorb to a wide variety of cell membranes without requirement for highly specific receptors. The requirement by enteroviruses for specific cell receptors rests not only on necessity for efficient adsorption, but also on necessity for some alteration of the virus capsid by plasma membrane receptors (Holland, 1962c; Philipson & Bengtsson, 1962; Holland & Hoyer, 1962).

Leucocyte susceptibility as a determinant of virus resistance

Bang & Warwick (1960) reported a selective destructive effect of mouse hepatitis virus for mouse macrophages cultivated *in vitro* as compared with fibroblasts and epithelial cells. Susceptible strains of mice yielded susceptible reticulo-endothelial cells, whereas insusceptible strains yielded macrophages which were resistant in cell culture. Interstrain mating and back-crosses between mice showed genetic segregation of animal (and macrophage) susceptibility which suggested that susceptibility was dominant over resistance and that a single gene might be involved. Goodman & Koprowski (1962) found similar agreement between arbor B virus susceptibility in mouse strains and ability of cultured peritoneal macrophages to support virus replication. Again, segregation of genes for susceptibility or resistance was reflected at the level of the whole animal and at the cellular level in cultivated macrophages. It would be of real value to study the molecular basis for the effects of these genes. It is interesting that a continuous cell line from the liver of a mouse strain resistant to mouse hepatitis virus is susceptible to this virus (Hartley & Rowe, 1963). This may be due to cellular changes accompanying dedifferentiation in culture, similar to those described above.

FACTORS INFLUENCING VIRUS VIRULENCE

Virus virulence *in vivo* is complex and involved with interaction of multiple host and virus functions. With no animal virus–host system is it possible to define all the parameters controlling virulence, although with the extensively studied polioviruses it is possible to speculate about the influence of some well-established properties. Empirical methods have been employed since the time of Jenner for selecting and testing avirulent variants of virulent viruses. Little is known of the basic biological or molecular differences between virulent and avirulent strains, but some differences observed *in vitro* will be listed and discussed in relation to findings *in vivo*. It is assumed that the common methods for obtaining avirulent variants by passage in foreign host cells (Li, Schaeffer & Nelson, 1955; Wenner, 1955) merely select mutant

viruses of reduced virulence, but the rapidity with which virus populations may shift on passage (Lehmann-Grube & Syverton, 1961) often
leads to use of the words 'variant' and 'adaptation' in discussing these
shifts.

Plaque size

Vogt, Dulbecco & Wenner (1957) observed that strains of poliovirus
avirulent for monkeys often produced smaller plaques in cell culture
than did virulent strains. Hsiung (1960) found similar reduction of
mouse virulence in small plaque mutants of Coxsackie A9 virus, as did
McClain & Hackett (1959) with vesicular exanthema virus of swine and
Mussgay & Suarez (1962) with Venezuelan equine encephalitis virus in
mice. This correlation between virulence and rapidity of lesion spread
in vitro is easily understandable if the assumption is made that the rate of
cell-to-cell advance of lesions *in vivo* generally reflects the rate observed
in vitro. A slower spread of lesions would obviously allow greater
mobilization of host defence mechanisms before tissue destruction
became critical. However, smaller plaque size is probably itself a manifestation of some basic alteration in the virus, such as increased capacity to elicit interferon production, or reduced capacity to cause early
cell lysis and virus release, or reduced growth rate.

Temperature and pH sensitivity and virulence

Dubes & Wenner (1957) found that poliovirus strains selected for
ability to multiply at 23° showed a reduced neurovirulence and suggested
that they were avirulent because they were unable to multiply at the
temperature of the primate body. Lwoff (1959, 1962) has shown an
excellent correlation between ability of poliovirus to develop at high
temperature and neurovirulence, and that the temperature-sensitive
step occurs during a short interval of the virus cycle. Other viruses also
show a correlation between growth at high temperature and virulence
(Isaacs, 1962). Hoggan, Roizman & Turner (1960) demonstrated that
optimal spread of Herpes simplex plaques by giant cell formation in cell
monolayers under antiserum occurred at 37°. They suggested that in
man febrile episodes, which elevate the skin temperature to about 37°,
may promote 'fever blisters' of the recurrent herpetic type by activation
in vivo of giant cell formation which is not affected by host antibody.
It seems likely from results *in vitro* that fever might also depress
virus replication *in vivo*. Walker & Boring (1958) protected mice from
Coxsackie B1 multiplication and pathology by keeping the mice at 37°.
When they were kept at 4° virus multiplication and death were rapid.
Lwoff (1959) obtained similar effects with mouse-adapted poliovirus,

and pointed out earlier experiments by Thompson (1938) which had shown thermal protection of rabbits from myxoma and fibroma viruses.

Lwoff (1959) also reviewed the effects which pH exerts on poliovirus replication. Lowering the pH of the cell culture medium to about 6·6 quickly depressed virus synthesis. Vogt *et al.* (1957) and Sabin (1957) showed that most avirulent strains of poliovirus were more sensitive to a low pH than were virulent strains. However, not all avirulent strains were sensitive to low pH, so that pH insensitivity was not restricted to virulent strains. As Lwoff (1959) has pointed out, it would be of interest to know the pH within a mammalian neuron, since in molluscs the internal pH of neurons is sufficiently low (pH 6·6 and below) as to be capable of inhibiting acid-sensitive avirulent strains of poliovirus.

Receptor affinities and virulence

As described above, receptor distribution in tissues and organs seems to determine poliovirus tropisms in the body, and it seems reasonable to investigate what role, if any, receptor specificities might play in virulence. Virulent strains of poliovirus are those which multiply in and destroy cells of the central nervous system *in vivo*, whereas avirulent strains grow in pharynx and gut, but grow poorly or not at all in central nervous system cells and cause little or no damage to these cells (Sabin, 1957). Schwerdt & Pardee (1952) and Francis & Chu (1953) showed that brain homogenates contained an inhibitor which bound virulent poliovirus. Sabin (1957) found that an avirulent strain of type 1 poliovirus was not bound by central nervous system tissue homogenates under conditions in which virulent type 1 poliovirus was bound. It seems likely that the 'inhibitors' found by the above investigators were receptors. Holland (1961 b) repeated Sabin's work and found little or no receptor for avirulent poliovirus in brain or spinal cord homogenates despite considerable binding of virulent strains by the same homogenates. However, avirulent strains were adsorbed by intestinal homogenates, indicating that attenuated strains may have lost their affinity for central nervous system receptors without losing affinity for and ability to multiply in intestinal cells. Kunin (1962), however, reported that an avirulent strain of poliovirus was bound by central nervous system tissue homogenates and the reason for this difference is not clear. Sabin (1957) found that under certain conditions attachment of attenuated virus to some spinal cord and basal ganglion homogenates could occur. He also reported limited multiplication of attenuated virus at the local site of intraspinal injection. If polioviruses differ in their affinity for tissue receptors, then there must be different species of

receptors for polioviruses. Sabin (1957) described differential susceptibility of brain stem and lumbar cord neurons to different strains of
poliovirus. This may suggest that receptors on various cells differ in
their affinity for each virus strain. Virulent and avirulent strains differ
chromatographically (Hodes, Zepp & Ainbender, 1960) and can be
separated by countercurrent distribution (Bengtsson & Philipson,
1963), so the capsid protein surface must be considerably changed
during mutation to avirulence, and altered receptor affinities might
result from such change.

Interferon and virulence

De Maeyer & Enders (1961) reported that avirulent strains of poliovirus
induced significant amounts of interferon whereas virulent strains
stimulated no detectable quantities. Isaacs (1962) encountered a
similar correlation between virulence and inability to stimulate interferon among a number of other viruses, although there were several
exceptions. Isaacs (1962) also found a correlation between virulence,
temperature of optimal growth and insensitivity to the action of exogenous interferon. Not only were virulent viruses in general poor
producers of interferon, and relatively insensitive to its action, but they
also were capable of interfering with its production in cells simultaneously infected with another, avirulent virus, depending on temperature of incubation and the temperature optima of each virus. The
relationship between temperature and interferon production may be
more than fortuitous, since it appears (Isaacs, 1962) that higher
temperatures favour interferon production at the expense of virus
production.

In addition to factors discussed above, many other variables may
influence virus virulence, such as age of the host or hormonal balance.
Many viruses are highly virulent for young mice but relatively avirulent
in older mice (Sigel, 1952; McLaren & Sanders, 1959), and this is not
always due to differences in immunological competence. Likewise,
cortisone treatment can transform a mild herpetic keratitis into a
fulminating lesion (Jawetz, Okumoto & Sonne, 1959), and adrenalin
can activate silent herpes infection of rabbits causing a fatal herpetic
encephalitis (Schmidt & Rasmussen, 1960).

To sum up these factors a virulent virus must be capable of rapid
spread from cell to cell which precedes the onset of adequate immune
responses. It must be capable of adsorbing to cells whose destruction
would cause disease. It must be capable of multiplying at the pH and
temperature usually found in the host and possibly even at the pH and

temperatures resulting from local inflammatory and febrile responses to infection. Finally, it must not stimulate production of large amounts of interferon; if it does, it must be relatively resistant to interferon action or have a predilection for cells which are refractory to interferon.

TUMOUR VIRUSES

Although it is not possible to discuss them at length, no review of virus effects *in vivo* and *in vitro* would be complete without mentioning the tumour viruses. Since Stewart *et al.* (1957) cultivated polyoma virus in monkey cells *in vitro*, it has become increasingly clear that transformations of animal cells *in vitro* by tumour viruses may serve as excellent models for malignant transformation *in vivo*. Vogt & Dulbecco (1960) demonstrated that polyoma virus transformed normal hamster cells to malignancy *in vitro*, and this has been confirmed and widely studied by others (Stoker & Abel, 1962) in mouse and hamster cells. It appears that cells transformed by polyoma virus and grown either in cell culture or as tumour transplants *in vivo* may lose the ability to produce virus, but retain malignant properties and continue to produce a specific antigen peculiar to polyoma-transformed cells (Sjögren *et al.* 1961; Habel, 1962; Sachs, 1961). Much of the pathogenesis and epidemiology of polyoma virus infection has been elucidated by studies *in vitro* (Habel, 1962). Shein & Enders (1962) have observed a similar malignant transformation of hamster and human cells *in vitro* with SV 40 virus from monkeys. It is still not clear how polyoma virus so quickly and regularly causes malignant transformation *in vitro* or *in vivo*, but the persistence of a specific antigen suggests that all or part of the virus genome may be integrated into the transformed cells.

With avian tumour viruses also, transformation *in vitro* has been observed (Temin & Rubin, 1958; Baluda, 1962). Temin (1962, 1963) showed that chicken cells transformed by Rous sarcoma may fail to produce virus, and Hanafusa *et al.* (1963) made the remarkable discovery that cells transformed *in vitro* by Rous sarcoma virus carry the virus genome in a masked form for many generations with no production of infectious virus. The tumour cells seem to carry the virus in a defective form, so that the presence of the virus genome can only be detected by superinfecting the cells with the closely related, virulent Rous-associated virus to cause the production of both types of virus. Study of the integrated state of Rous sarcoma virus in tumour cells should yield valuable information about silent infection of animal cells.

SUMMARY

Recent knowledge of virus-cell and virus-animal interactions are discussed. It is implicit throughout that most animal virus-cell interactions observed *in vitro* reflect virus-cell relations *in vivo*, although the latter may be considerably influenced by the differentiated state of the animals' cells, and by the multitude of physiologic and immunologic responses which can take place. It is suggested that with both virulent animal viruses and tumour viruses, cellular alterations observed in the whole animal can usually be successfully reproduced in cultured cells.

REFERENCES

AACH, H. G. (1959). Serologische untersuchungen zur struktur des tabakmosaikvirus. *Biochim. biophys. Acta*, **32**, 140.

ACKERMANN, W. W. & KURTZ, H. (1955). Observations concerning a persisting infection of HeLa cells with poliomyelitis. *J. exp. Med.* **102**, 555.

ACKERMANN, W. W., PAYNE, F. E. & KURTZ, H. (1958). Concerning the cytopathogenic effect of poliovirus; evidence for an extraviral toxin. *J. Immunol.* **81**, 1.

ADAMS, J. M. & IMAGAWA, D. T. (1957). Immunological relationship between measles and distemper virus. *Proc. Soc. exp. Biol., N.Y.* **96**, 240.

ALEXANDER, H. E., KOCH, G., MORGAN-MOUNTAIN, I. & VAN DAMME, O. (1958). Infectivity of ribonucleic acid from poliovirus in human cell monolayers. *J. exp. Med.* **108**, 493.

ALLEN, R., FINKELSTEIN, R. A. & SULKIN, E. S. (1958). Viral inhibitors in normal animal sera. *Tex. Rep. Biol. Med.* **16**, 472.

BALTIMORE, D. & FRANKLIN, R. M. (1962*a*). The effect of mengovirus infection on the activity of the DNA-dependent RNA polymerase of L-cells. *Proc. nat. Acad. Sci., Wash.* **48**, 1383.

BALTIMORE, D. & FRANKLIN, R. M. (1962*b*). Preliminary data on a virus-specific enzyme system responsible for the synthesis of viral RNA. *Biochem. biophys. Res. Comm.* **9**, 388.

BALUDA, M. A. (1962). Properties of cells infected with avian myeloblastosis virus. *Cold Spr. Harb. Sym. quant. Biol.* **27**, 415.

BANG, F. B. & WARWICK, A. (1960). Mouse macrophages as host cells for the mouse hepatitis virus and genetic basis of their susceptibility. *Proc. nat. Acad. Sci., Wash.* **46**, 1065.

BARSKI, G., CORNEFERT, F. & WALLACE, R. E. (1959). Response of ciliated epithelia of different histological origin to virus infections *in vitro*. *Proc. Soc. exp. Biol., N.Y.* **100**, 407.

BARSKI, G., KOURILSKY, R. & CORNEFERT, F. (1957). Resistance of respiratory ciliated epithelium to action of polio- and adeno-viruses *in vitro*. *Proc. Soc. exp. Biol., N.Y.* **96**, 386.

BARSKI, G. & ROBBE-FOSSAT, F. (1957). Multiplication du virus poliomyétique dans les cultures *in vitro* de cellules exsudatives de singe sans effet cytopathogène généralisé. *Ann. Inst. Pasteur*, **92**, 301.

BENGTSSON, S. & PHILIPSON, L. (1963). Countercurrent distribution of poliovirus type 1. *Virology*, **20**, 176.

BEN-PORAT, T. & KAPLAN, A. S. (1963). The synthesis and fate of pseudorabies virus DNA in infected mammalian cells in the stationary phase of growth. *Virology*, **20**, 310.

BERG, T. O., GLEISER, C. A., GOCHENOUR, W. S., JR., MIESSE, M. L. & TIGERTT, W. D. (1961). Studies on the virus of Venezuelan equine encephalomyelitis. II. Modification by specific immune serum of response of central nervous system of mice. *J. Immunol.* **87**, 509.

BERG, R. B. & ROSENTHAL, M. S. (1961). Propagation of measles virus in suspensions of human and monkey leucocytes. *Proc. Soc. exp. Biol., N.Y.* **106**, 581.

BLACK, F. L. & MELNICK, J. L. (1955). Micro-epidemiology of poliomyelitis and Herpes-B infections. Spread of the viruses within tissue cultures. *J. Immunol.* **74**, 236.

BOAND, A. V., JR., KEMPF, J. E. & HANSON, R. J. (1957). Phagocytosis of influenza virus. I. *In vitro* observations. *J. Immunol.* **79**, 416.

BODIAN, D. (1949). Histopathologic basis of clinical findings in poliomyelitis. *Amer. J. Med.* **6**, 563.

BOYER, G. S., LEUCHTENBERGER, C. & GINSBERG, H. S. (1957). Cytological and cytochemical studies of HeLa cells infected with adenoviruses. *J. exp. Med.* **105**, 192.

BROWN, F., CARTWRIGHT, B. & STEWART, D. L. (1961). Mechanism of infection of pig kidney cells by foot and mouth disease virus. *Biochim. biophys. Acta*, **47**, 172.

CHAPRONIERE, D. M. & ANDREWES, C. H. (1958). Factors involved in the susceptibility of tissues of various species to myxoma virus. *Virology*, **5**, 120.

CHEATHAM, W. J. (1959). A comparison of *in vitro* and *in vivo* characteristics as related to the pathogenesis of measles, varicella, and Herpes zoster. *Ann. N.Y. Acad. Sci.* **81**, 6.

COCITO, C., DEMAEYER, E. & DESOMER, P. (1962). Synthesis of messenger RNA in neoplastic cells treated *in vitro* with interferon. *Life Sci.* **12**, 759.

CROWELL, R. L. & SYVERTON, J. T. (1961). The mammalian cell-virus relationship. VI. Sustained infection of HeLa cells by Coxsackie B3 virus and effect on super-infection. *J. exp. Med.* **113**, 419.

DE MAEYER, E. & ENDERS, J. F. (1961). An interferon appearing in cell cultures infected with measles virus. *Proc. Soc. exp. Biol., N.Y.* **107**, 573.

DE SOMER, P., PRINZIE, A., DENYS, P., JR. & SCHONNE, E. (1962). Mechanism of action of interferon. I. Relationship with viral ribonucleic acid. *Virology*, **16**, 63.

DE SOMER, P., PRINZIE, A. & SCHONNE, E. (1959). Infectivity of poliovirus ribonucleic acid for embryonated eggs and unsusceptible cell lines. *Nature, Lond.* **184**, 652.

DOZOIS, T. F., WAGNER, J. C., CHEMERDA, C. M. & ANDREW, V. M. (1949). The influence of certain serum factors on the neutralization of Western equine encephalitis virus. *J. Immunol.* **62**, 319.

DUBES, G. R. & WENNER, H. A. (1957). Virulence of polioviruses in relation to variant characteristics distinguishable on cells *in vitro*. *Virology*, **4**, 275.

DULBECCO, R. (1952). Production of plaques in monolayer tissue cultures by single particles of an animal virus. *Proc. nat. Acad. Sci., Wash.* **38**, 747.

DULBECCO, R. & VOGT, M. (1954). One step growth curve of Western equine encephalomyelitis virus on chicken embryo cells grown *in vitro* and analysis of virus yields from single cells. *J. exp. Med.* **99**, 183.

ENDERS, J. F. (1954). Cytopathology of virus infections. (Particular reference to tissue culture studies.) *Annu. Rev. Microbiol.* **8**, 473.

ENDERS, J. F. (1960). A consideration of the mechanisms of resistance to viral infection based on recent studies of the agents of measles and poliomyelitis. *Trans. Coll. Phys. Philad.* **28**, 68.

ENDERS, J. F., WELLER, T. H. & ROBBINS, F. C. (1949). Cultivation of the Lansing strain of poliomyelitis virus in cultures of various human embryonic tissues. *Science*, **109**, 85.

FENWICK, M. L. (1963). The influence of poliovirus infection on RNA synthesis in mammalian cells. *Virology*, **19**, 241.

FISHER, T. N. & GINSBERG, H. S. (1956). The reaction of influenza viruses with guinea pig polymorphonuclear leucocytes. III. Studies on the mechanism by which influenza viruses inhibit phagocytosis. *Virology*, **2**, 656.

FLORMAN, A. L. & ENDERS, J. F. (1942). The effect of homologous antiserum and complement on multiplication of vaccinia virus in roller-tube cultures of blood mononuclear cells. *J. Immunol.* **43**, 159.

FRANCIS, T., JR. (1947). Dissociation of hemagglutinating and antibody-measuring capacities of influenza virus. *J. exp. Med.* **85**, 1.

FRANCIS, T., JR. & CHU, L. W. (1953). The interaction *in vitro* between poliomyelitis virus and nervous tissue. *Abstr. VIth Int. Congr. Microbiol.* **3**, 162.

FRANKLIN, R. M. & BALTIMORE, D. (1962). Patterns of macromolecular synthesis in normal and virus-infected mammalian cells. *Cold Spr. Harb. Sym. quant. Biol.* **27**, 175.

FRANKLIN, R. M. & BREITENFELD, P. M. (1959). The abortive infection of Earle's L-cells by fowl plague virus. *Virology*, **8**, 293.

FRIEDMAN, R. M. & BARON, S. (1961). The role of antibody in recovery from infection with vaccinia virus. *J. Immunol.* **87**, 379.

FURUSAWA, E., HAGIWARA, K. & KAMAHORA, J. (1956). The effect of antibody on intracellular virus multiplication. *Med. J. Osaka Univ.* **7**, 551.

GINSBERG, H. S. (1954). Formation of noninfectious influenza virus in mouse lungs. Its dependence upon extensive pulmonary consolidation initiated by the virus inoculum. *J. exp. Med.* **100**, 581.

GINSBERG, H. S. (1958). The significance of the viral carrier state in tissue culture systems. *Prog. med. Virol.* **1**, 36.

GINSBERG, H. S. & BLACKMON, J. R. (1956). Reactions of influenza viruses with guinea pig polymorphonuclear leucocytes. I. Virus-cell interactions. *Virology*, **2**, 618.

GINSBERG, H. S. & DIXON, M. K. (1959). Deoxyribonucleic acid and protein alterations in HeLa cells infected with type 4 adenovirus. *J. exp. Med.* **113**, 283.

GLASGOW, L. A. & HABEL, K. (1962). Interferon production by mouse leukocytes *in vitro* and *in vivo*. *J. exp. Med.* **117**, 149.

GLEDHILL, A. W. (1959). The interference of mouse hepatitis virus with Ectromelia in mice and a possible explanation of its mechanism. *Brit. J. exp. Path.* **40**, 291.

GOOD, R. A. (1957). Morphologic basis of the immune response and hypersensitivity. In *Host–Parasite Relationships in Living Cells*, p. 79. Springfield: C. C. Thomas.

GOODMAN, G. T. & KOPROWSKI, H. (1962). Macrophages as a cellular expression of inherited natural resistance. *Proc. nat. Acad. Sci., Wash.* **48**, 160.

GOTTSCHALK, A. (1957). Virus enzymes and virus templates. *Physiol. Rev.* **37**, 66.

GREEN, M. (1962). Studies on the biosynthesis of viral DNA. *Cold Spr. Harb. Sym. quant. Biol.* **27**, 219.

GRESSER, I. (1961). Production of interferon by suspensions of human leukocytes. *Proc. Soc. exp. Biol., N.Y.* **108**, 799.

GRESSER, I. & CHANY, C. (1963). Isolation of measles virus from the washed leucocyte fraction of blood. *Proc. Soc. exp. Biol., N.Y.* **113**, 695.

GROSSBERG, S. E. & HOLLAND, J. J. (1962). Interferon and viral ribonucleic acid. Effect on virus-susceptible and insusceptible cells. *J. Immunol.* **88**, 708.

GROSSBERG, S. E., HOOK, E. W. & WAGNER, R. R. (1962). Hemorrhagic encephalopathy in chicken embryos infected with influenza virus. III. Viral interference at a distant site induced by prior allantoic infection. *J. Immunol.* **88**, 1.

HABEL, K. (1962). Antigenic properties of cells transformed by polyoma virus. *Cold Spr. Harb. Sym. quant. Biol.* **27**, 433.

HANAFUSA, H. (1962). Factors involved in the initiation of multiplication of vaccinia virus. *Cold Spr. Harb. Sym. quant. Biol.* **27**, 209.

HANAFUSA, H., HANAFUSA, T. & RUBIN, H. (1963). The defectiveness of Rous sarcoma virus. *Proc. nat. Acad. Sci., Wash.* **49**, 572.

HARTLEY, J. W. & ROWE, W. P. (1963). Tissue culture cytopathic and plaque assays for mouse hepatitis viruses. *Proc. Soc. exp. Biol., N.Y.* **113**, 403.

HENLE, W. (1950). Interference phenomena between animal viruses. *J. Immunol.* **64**, 203.

HENLE, G., GIRARDI, A. & HENLE, W. (1955). A non-transmissible cytopathogenic effect of influenza virus in tissue culture accompanied by formation of non-infectious hemagglutinins. *J. exp. Med.* **101**, 25.

HENLE, W., HENLE, G., DEINHARDT, F. & BERGS, V. V. (1959). Studies on persistent infections of tissue cultures. IV. Evidence for the production of an interferon in MCN cells by myxovirus. *J. exp. Med.* **110**, 525.

HERMODSSEN, S. & WESSLIN, T. (1958). Multiplication of poliomyelitis virus in heterotransplanted human embryonic tissue. *Acta path. microbiol. scand.* **42**, 371.

HILDEBRAND, E. M. (1958). Masked virus infection in plants. *Annu. Rev. Microbiol.* **12**, 441.

HITCHCOCK, G. & ISAACS, A. (1960). Protection of mice against the lethal action of an encephalitis virus. *Brit. med. J.* **2**, 1268.

HITCHCOCK, G. & PORTERFIELD, J. S. (1961). Production of interferon in brains of mice infected with an athropod-borne virus. *Virology*, **13**, 363.

HO, M. (1961). Inhibition of the infectivity of poliovirus ribonucleic acid by an interferon. *Proc. Soc. exp. Biol., N.Y.* **107**, 639.

HO, M. (1962). Interferons. *New Engl. J. Med.* **266**, 1367.

HO, M. & ENDERS, J. F. (1959). Further studies on an inhibitor of viral activity appearing in infected cell cultures and its role in chronic viral infections. *Virology*, **9**, 446.

HODES, H. L., ZEPP, H. D. & AINBENDER, E. (1960). A physical property as a virus marker. Difference in avidity of cellulose resin for virulent (Mahoney) and attenuated (L Sc 2ab) strain of type 1 poliovirus. *Virology*, **11**, 306.

HOGGAN, M. D., ROIZMAN, B. & TURNER, T. B. (1960). The effects of the temperature of incubation on the spread of Herpes simplex virus in an immune environment in cell culture. *J. Immunol.* **84**, 152.

HOLLAND, J. J. (1961a). Altered base ratios in HeLa cell RNA during poliovirus infection. *Biochem. biophys. Res. Comm.* **6**, 196.

HOLLAND, J. J. (1961b). Receptor affinities as major determinants of enterovirus tissue tropism in humans. *Virology*, **15**, 312.

HOLLAND, J. J. (1962a). Altered base ratios in RNA synthesized during enterovirus infection of human cells. *Proc. nat. Acad. Sci., Wash.* **48**, 2044.

HOLLAND, J. J. (1962b). Inhibition of DNA-primed RNA synthesis during poliovirus infection of human cells. *Biochem. biophys. Res. Comm.* **9**, 556.

HOLLAND, J. J. (1962c). Irreversible eclipse of poliovirus by HeLa cells. *Virology*, **16**, 163.

HOLLAND, J. J. (1963). Depression of host-controlled RNA synthesis in human cells during poliovirus infection. *Proc. nat. Acad. Sci., Wash.* **49**, 23.

HOLLAND, J. J. & HOYER, B. H. (1962). Early stages of entero-virus infection. *Cold Spr. Harb. Sym. quant. Biol.* **27**, 101.

HOLLAND, J. J. & McLAREN, L. C. (1959). The mammalian cell-virus relationship. II. Adsorption, reception, and eclipse of polio-virus by HeLa cells. *J. exp. Med.* **109**, 487.

HOLLAND, J. J. & McLAREN, L. C. (1961). The location and nature of enterovirus receptors in susceptible cells. *J. exp. Med.* **114**, 161.

HOLLAND, J. J., McLAREN, L. C. & SYVERTON, J. T. (1959). The mammalian cell-virus relationship. IV. Infection of naturally insusceptible cells with enterovirus nucleic acid. *J. exp. Med.* **110**, 65.

HOTCHIN, J. (1962). The biology of lymphocytic choriomeningitis infection: Virus-induced immune disease. *Cold Spr. Harb. Sym. quant. Biol.* **27**, 479.

HOWES, D. W. & MELNICK, J. L. (1957). The growth cycle of polio-virus in monkey kidney cell. I. Maturation and release of virus in monolayer culture. *Virology*, **4**, 97.

HSIUNG, G. D. (1960). Studies on variation in Coxsackie A-9 virus. *J. Immunol.* **84**, 285.

HUMMELER, K., ANDERSON, T. F. & BROWN, R. A. (1962). Identification of poliovirus particles of different antigenicity by specific agglutination as seen in the electron microscope. *Virology*, **16**, 84.

HUMMELER, K. & KETTLER, A. (1958). Dissociation of poliomyelitis virus from neutralizing antibody. *Virology*, **6**, 297.

INGLOT, A. & DAVENPORT, F. M. (1962). Studies on the role of leukocytes in infection with influenza virus. *J. Immunol.* **88**, 55.

ISAACS, A. (1962). Production and action of interferon. *Cold Spr. Harb. Sym. quant. Biol.* **27**, 343.

ISAACS, A. & BARRON, S. (1960). Antiviral action of interferon in embryonic cells. *Lancet*, ii, 946.

ISAACS, A. & HITCHCOCK, G. N. (1960). Role of interferon in recovery from virus infections. *Lancet*, ii, 69.

ISAACS, A., KLEMPERER, H. G. & HITCHCOCK, G. (1961). Studies on the mechanism of action of interferon. *Virology*, **13**, 191.

ISAACS, A. & LINDEMANN, S. (1957). Virus interference. I. Interferon. *Proc. roy. Soc. B*, **147**, 258.

ISAACS, A. & WESTWOOD, M. A. (1959). Inhibition by interferon of the growth of vaccinia virus in rabbit skin. *Lancet*, ii, 324.

JAWETZ, E., OKUMOTO, M. & SONNE, M. (1959). Studies on Herpes simplex. X. The effect of corticosteroids on herpetic keratitis in the rabbit. *J. Immunol.* **83**, 486.

JUDAH, J. D., BJOTVEDT, G. & VAINIO, T. (1960). Protection from liver damage due to murine hepatitis virus. *Nature, Lond.* **187**, 507.

KAPLAN, A. S. (1955). Comparison of susceptible and resistant cells to infection with poliomyelitis virus. *Ann. N.Y. Acad. Sci.* **61**, 830.

KIT, S. & DUBBS, D. R. (1963). Acquisition of thymidine kinase activity by Herpes simplex infected mouse fibroblast cells. *Biochem. biophys. Res. Comm.* **11**, 55.

KLEMPERER, H. G. & PEREIRA, H. G. (1959). Study of adenovirus antigens fractionated by chromatography on DEAE-cellulose. *Virology*, **9**, 536.

KOPROWSKA, I. & KOPROWSKI, H. (1957). Enhancement of susceptibility to virus infection in the course of neoplastic process. *Ann. N.Y. Acad. Sci.* **68**, 404.

KUNIN, C. M. (1962). Virus-tissue union and the pathogenesis of enterovirus infections. *J. Immunol.* **88**, 556.

KUNIN, C. M. & JORDAN, W. S., JR. (1961). *In vitro* adsorption of poliovirus by noncultured tissues. Effect of species, age, and malignancy. *Amer. J. Hyg.* **73**, 245.

LeBouvier, G. L. (1959). The D-C change in poliovirus particles. *Brit. J. exp. Path.* **40**, 605.

Lehmann-Grube, F. & Syverton, J. T. (1961). Pathogenicity for suckling mice of Coxsackie viruses adapted to human amnion cells. *J. exp. Med.* **113**, 811.

Leymaster, G. R. & Ward, T. G. (1949). The effect of complement in the neutralization of mumps virus. *J. Immunol.* **61**, 95.

Li, C. P., Schaeffer, M. & Nelson, D. B. (1955). Experimentally produced variants of poliomyelitis virus combining *in vivo* and *in vitro* techniques. *Ann. N.Y. Acad. Sci.* **61**, 902.

Lwoff, A. (1959). Factors influencing the evolution of viral diseases at the cellular level and in the organism. *Bact. Rev.* **23**, 109.

Lwoff, A. (1962). The thermosensitive critical event of the viral cycle. *Cold Spr. Harb. Sym. quant. Biol.* **27**, 159.

Magee, W. E. (1962). DNA polymerase and deoxyribonucleotide kinase activities in cells infected with vaccinia virus. *Virology*, **17**, 503.

Mandel, B. (1958). Studies on the interactions of poliomyelitis virus, antibody, and host cells in a tissue culture system. *Virology*, **6**, 424.

Mayer, V. F., Sokol, F. & Vilcek, J. (1961). Effect of interferon on the infection with Eastern equine encephalomyelitis (EEE) virus and its ribonucleic acid RNA. *Acta Virol.* **5**, 264.

McAuslan, B. R. & Joklik, W. K. (1962). Stimulation of the thymidine phosphorylating system in HeLa cells on infection with poxvirus. *Biochem. biophys. Res. Comm.* **8**, 486.

McClain, M. E. & Hackett, A. J. (1959). Biological characteristics of two plaque variants of vesicular exanthema of swine virus, type E_{54}. *Virology*, **9**, 577.

McLaren, L. C., Holland, J. J. & Syverton, J. T. (1959). The mammalian cell-virus relationship. I. Attachment of poliovirus to cultivated cells of primate and non-primate origin. *J. exp. Med.* **109**, 475.

McLaren, L. C., Holland, J. J. & Syverton, J. T. (1960). The mammalian cell-virus relationship. V. Susceptibility and resistance of cells *in vitro* to infection by Coxsackie A9 virus. *J. exp. Med.* **112**, 581.

McLaren, A. & Sanders, F. K. (1959). The influence of age of the host on local virus multiplication and on the resistance to virus infections. *J. Hyg., Camb.* **57**, 106.

Morgan, I. M. (1945). Quantitative study of the neutralization of Western equine encephalitis virus by its antiserum and the effect of complement. *J. Immunol.* **20**, 17.

Mountain, I. M. & Alexander, H. E. (1959). Study of infectivity of ribonucleic acid (RNA) from type 1 poliovirus in the chick embryo. *Fed. Proc.* **18**, 587.

Mussgay, M. & Suárez, O. (1962). Studies with a pathogenic and an attenuated strain of Venezuelan equine encephalitis virus and *Aedes aegypti* (L) mosquitoes. *Arch. ges. Virusforsch.* **12**, 26.

Nomura, M., Okamoto, K. & Asano, K. (1962). RNA metabolism in *Escherichia coli* infected with bacteriophage T4. Inhibition of host ribosomal and soluble RNA synthesis by phage effect of chloromycetin. *J. molec. Biol.* **4**, 376.

Nungester, W. J., Gordon, J. D. & Collins, K. E. (1950). Comparison of the erythrocyte and leucocyte cell surfaces of guinea pigs by use of agglutination techniques. *J. infect. Dis.* **87**, 71.

Paucker, K. K., Cantell, K. & Henle, W. (1962). Quantitative studies on viral interference in suspended L cells. III. Effect of interfering viruses and interferon on the growth rate of cells. *Virology*, **17**, 324.

Pereira, H. G. (1958). A protein factor responsible for the early cytopathic effect of adenoviruses. *J. exp. Med.* **108**, 713.

PHILIPSON, L. & BENGTSSON, S. (1962). Interaction of enteroviruses with receptors from erythrocytes and host cells. *Virology*, **18**, 457.

PRINCE, A. N. & GINSBERG, H. S. (1957). Studies of the cytotoxic effect of New-castle disease virus (NVD) on Ehrlich ascites tumor cells. I. Characteristics of the virus-cell interaction. *J. Immunol.* **79**, 94.

QUERSIN-THIRY, L. (1961). Interaction between cellular extracts and animal viruses. I. Kinetic studies and some notes on the specificity of the interaction. *Acta Virol.* **5**, 141.

REICH, E., FRANKLIN, R. M., SHATKIN, A. J. & TATUM, E. L. (1961). Effect of actinomycin D on cellular nucleic acid synthesis and virus production. *Science*, **134**, 556.

ROIZMAN, B. (1962). Polykaryocytosis. *Cold Spr. Harb. Sym. quant. Biol.* **27**, 327.

ROIZMAN, B., MAYER, M. M. & ROANE, P. R., JR. (1959). Immunological studies of poliovirus. IV. Alteration of the immunologic specificity of purified polio-myelitis virus by heat and ultraviolet light. *J. Immunol.* **82**, 19.

ROWE, W. P. (1956). Protective effect of pre-irradiation on lymphocytic chorio-meningitis infection in mice. *Proc. Soc. exp. Biol., N.Y.* **92**, 194.

ROWE, W. P., HUEBNER, R. J., GILMORE, L. K., PARROTT, R. H. & WARD, T. G. (1953). Isolation of a cytopathogenic agent from human adenoids undergoing spontaneous degeneration in tissue culture. *Proc. Soc. exp. Biol., N.Y.* **84**, 570.

RUBIN, H. (1957). Interactions between Newcastle disease virus (NVD), antibody and cell. *Virology*, **4**, 533.

RUSTIGIAN, R. (1962). A carrier state in HeLa cells with measles virus (Edmonton strain) apparently associated with noninfectious virus. *Virology*, **16**, 101.

SABIN, A. B. (1935). The mechanisms of immunity to filterable viruses. III. Role of leukocytes in immunity to vaccinia. *Brit. J. exp. Path.* **16**, 158.

SABIN, A. B. (1950). The dengue group of viruses and its family relationships. *Bact. Rev.* **14**, 225.

SABIN, A. B. (1957). Properties of attenuated polioviruses and their behavior in human beings. In *Cellular Biology, Nucleic Acids, and Viruses*, p. 113. New York: New York Academy of Sciences.

SACHS, L. (1961). Tumor transplantation in mice inoculated with polyoma virus. *Exp. Cell Res.* **24**, 185.

SALZMAN, N. P., LOCKART, R. Z., JR. & SEBRING, E. D. (1959). Alterations in HeLa cell metabolism resulting from poliovirus infection. *Virology*, **9**, 244.

SCHLESINGER, R. W. (1950). Incomplete growth cycle of influenza virus in mouse brain. *Proc. Soc. exp. Biol., N.Y.* **74**, 541.

SCHLESINGER, R. W. (1959). Interference between animal viruses. In *Diagnostic Procedures for Virus and Rickettsial Diseases*, p. 97. New York: American Public Health Association.

SCHMIDT, J. R. & RASMUSSEN, A. F., JR. (1960). Activation of latent Herpes sim-plex encephalitis by chemical means. *J. infect. Dis.* **106**, 154.

SCHULMAN, J. L. & KILBOURNE, E. D. (1963). Induction of viral interference in mice by aerosols of inactivated influenza virus. *Proc. Soc. exp. Biol., N.Y.* **113**, 431.

SCHWERDT, C. E. & PARDEE, A. B. (1952). The intracellular distribution of Lansing poliomyelitis virus in the central nervous system of infected cotton rats. *J. exp. Med.* **96**, 121.

SHEEK, M. R. & MAGEE, W. E. (1961). An autoradiographic study on the intra-cellular development of vaccinia virus. *Virology*, **15**, 146.

SHEIN, H. M. & ENDERS, J. F. (1962). Transformation induced by Simian virus 40 in human renal cell cultures. I. Morphology and growth characteristics. *Proc. nat. Acad. Sci., Wash.* **48**, 1164.

SIGEL, M. M. (1952). Influence of age on susceptibility to virus infections with particular reference to laboratory animals. *Annu. Rev. Microbiol.* **6**, 247.

SJÖGREN, H. O., HELLSTRÖM, I. & KLEIN, G. (1961). Resistance of polyoma virus immunized mice against transplantation of established polyoma tumors. *Exp. Cell Res.* **23**, 204.

STEWART, S. E., EDDY, B. E., GOCHENOUR, A. M., BORGESE, N. G. & GRUBBS, G. E. (1957). The induction of neoplasms with a substance released from mouse tumors by tissue culture. *Virology*, **3**, 380.

STOKER, M. & ABEL, P. (1962). Conditions affecting transformation by polyoma virus. *Cold Spr. Harb. Sym. quant. Biol.* **27**, 375.

STONE, J. D. (1948). Prevention of virus infection with enzyme of *V. cholerae*. I. Studies with viruses of mumps–influenza group in chick embryos. *Aust. J. exp. Biol. med. Sci.* **26**, 49.

TEMIN, H. M. (1962). Separation of morphological conversion and virus production in Rous sarcoma virus infection. *Cold Spr. Harb. Sym. quant. Biol.* **27**, 407.

TEMIN, H. M. (1963). Further evidence for a converted, non-virus-producing state of Rous sarcoma virus-infected cells. *Virology*, **20**, 235.

TEMIN, H. M. & RUBIN, H. (1958). Characteristics of an assay for Rous sarcoma virus and Rous sarcoma cells in tissue culture. *Virology*, **6**, 669.

THOMPSON, R. L. (1938). The influence of temperature upon proliferation of infectious fibroma and infectious myxoma *in vivo. J. infect. Dis.* **62**, 307.

TODD, C. (1928). Experiments on the virus of fowl plague. I. *Brit. J. exp. Path.* **9**, 19.

VAINIO, T., GWATKIN, R. & KOPROWSKI, H. (1961). Production of interferon by brains of genetically resistant and susceptible mice infected with West Nile virus. *Virology*, **14**, 385.

VOGT, M. & DULBECCO, R. (1960). Virus cell interaction with a tumor-producing virus. *Proc. nat. Acad. Sci., Wash.* **46**, 359.

VOGT, M., DULBECCO, R. & WENNER, H. A. (1957). Mutants of poliomyelitis viruses with reduced efficiency of plating in acid medium and reduced neuropathogenicity. *Virology*, **4**, 141.

WAGNER, R. R. (1960). Viral interference. Some considerations of basic mechanisms and their potential relationship to host resistance. *Bact. Rev.* **24**, 151.

WAGNER, R. R. (1963). Cellular resistance to viral infection, with particular reference to endogenous interferon. *Bact. Rev.* **27**, 72.

WALKER, D. L. & BORING, W. D. (1958). Factors influencing host–virus interactions. III. Further studies on the alteration of Coxsackie virus infection in adult mice by environmental temperature. *J. Immunol.* **80**, 39.

WARNER, J., MADDEN, M. J. & DARNELL, J. E. (1963). The interaction of poliovirus RNA with *Escherichia coli* ribosomes. *Virology*, **19**, 393.

WENNER, H. A. (1955). Some comparative observations on the behavior of poliomyelitis viruses in animal and in tissue culture. *Ann. N.Y. Acad. Sci.* **61**, 840.

WILCOX, W. C. & GINSBERG, H. S. (1961). Purification and immunological characterization of types 4 and 5 adenovirus-soluble antigens. *Proc. nat. Acad. Sci., Wash.* **47**, 512.

WILCOX, W. C., WOOD, E. M., OH, J. O., EVERETT, N. B. & EVANS, C. A. (1958). Morphological and functional changes in corneal endothelium caused by the toxic effects of influenza and Newcastle disease viruses. *Brit. J. exp. Path.* **39**, 601.

ZEMLA, J. & SCHRAMEK, S. (1962). The action of interferon under anaerobic conditions. *Virology*, **16**, 204.

ZIMMERMAN, E. F., HEETER, M. & DARNELL, J. E. (1963). RNA synthesis in poliovirus-infected cells. *Virology*, **19**, 400.

INDEX